Structures and Mechanisms

ACS SYMPOSIUM SERIES **827**

Structures and Mechanisms

From Ashes to Enzymes

Gareth R. Eaton, Editor
University of Denver

Don C. Wiley, Editor
Harvard University

Oleg Jardetzky, Editor
Stanford University

American Chemical Society, Washington, DC

Library of Congress Cataloging-in-Publication Data

Structures and mechanisms : from Ashes to enzymes / Gareth R. Eaton, editor, Don C. Wiley, editor, Oleg Jardetzky, editor.

    p. cm.—(ACS symposium series ; 827)

  Includes bibliographical references and index.

  ISBN 0–8412–3736–0

  1. Chemistry. 2. Lipscomb, William N.

  I. Lipscomb, William N.- II. Eaton, Gareth R.- III. Wiley, Don C., 1944–2001 IV. Jardetzky, Oleg. V. Series.

QD39 S885   2002
540—dc21   2002018487

The paper used in this publication meets the minimum requirements of American National Standard for Information Sciences—Permanence of Paper for Printed Library Materials, ANSI Z39.48–1984.

Copyright © 2002 American Chemical Society

Distributed by Oxford University Press

All Rights Reserved. Reprographic copying beyond that permitted by Sections 107 or 108 of the U.S. Copyright Act is allowed for internal use only, provided that a per-chapter fee of $22.50 plus $0.75 per page is paid to the Copyright Clearance Center, Inc., 222 Rosewood Drive, Danvers, MA 01923, USA. Republication or reproduction for sale of pages in this book is permitted only under license from ACS. Direct these and other permission requests to ACS Copyright Office, Publications Division, 1155 16th St., N.W., Washington, DC 20036.

The citation of trade names and/or names of manufacturers in this publication is not to be construed as an endorsement or as approval by ACS of the commercial products or services referenced herein; nor should the mere reference herein to any drawing, specification, chemical process, or other data be regarded as a license or as a conveyance of any right or permission to the holder, reader, or any other person or corporation, to manufacture, reproduce, use, or sell any patented invention or copyrighted work that may in any way be related thereto. Registered names, trademarks, etc., used in this publication, even without specific indication thereof, are not to be considered unprotected by law.

PRINTED IN THE UNITED STATES OF AMERICA

Foreword

The ACS Symposium Series was first published in 1974 to provide a mechanism for publishing symposia quickly in book form. The purpose of the series is to publish timely, comprehensive books developed from ACS sponsored symposia based on current scientific research. Occasion-ally, books are developed from symposia sponsored by other organizations when the topic is of keen interest to the chemistry audience.

Before agreeing to publish a book, the proposed table of contents is reviewed for appropriate and comprehensive coverage and for interest to the audience. Some papers may be excluded to better focus the book; others may be added to provide comprehensiveness. When appropriate, overview or introductory chapters are added. Drafts of chapters are peer-reviewed prior to final acceptance or rejection, and manuscripts are prepared in camera-ready format.

As a rule, only original research papers and original review papers are included in the volumes. Verbatim reproductions of previously published papers are not accepted.

ACS Books Department

Dedication

Lipscomb in 1985, at the Symposium held in his honor at the University of Dover.

This book is dedicated to Professor William N. Lipscomb, creative scientist, clarinetist, Nobel Laureate, and Kentucky Colonel.

Contents

Preface..xiii

Process of Discovery (1977); An Autobiographical Sketch.......................xv
    William N. Lipscomb

Introduction

1. The Landscape and the Horizons: An Introduction
   to the Science of William N. Lipscomb..2
   Gareth R. Eaton

Inorganic Chemistry

2. Thomas Jefferson, Alice in Wonderland, Polyhedral Boranes,
   and the Lipscomb Legacy..20
   Russell N. Grimes

3. Recent Developments in the Chemistry of Main Group
   Metallacarboranes of the C_2B_4-Carborane Ligands..............................46
   Narayan S. Hosmane and John A. Maguire

4. Formation of Nanostructured Phases of Fe, Co,
   and Ni by a Freeze-Out Technique..67
   Kimloan T. Nguyen, Alfred A. Zinn, and Herbert D. Kaesz

5. Proposed New Materials: Boron Fullerenes, Nanotubes,
   and Nanotori..79
   Vladimir Dadashev, Asta Gindulyte, William N. Lipscomb,
   Lou Massa, and Richard Squire

6. Oscillations, Waves, and Patterns in Chemistry and Biology............103
   Irving R. Epstein

Theory

7. Electron Propagator Theory of Ionization Energies and Dyson Orbitals for μ-Hydrido, Bridge-Bonded Molecules: Diborane, Digallane, and Gallaborane............118
   Gustavo Seabra, V.G. Zakrzewski, and J. V. Ortiz

8. Application of Theoretical Methods to NMR Chemical Shifts and Coupling Constants............135
   Michael L. McKee

9. Does the Magnitude of NMR Coupling Constants Specify Bond Polarity?............150
   Rodney J. Bartlett, Janet E. Del Bene, and S. Ajith Perera

10. Aluminosilicate Inorganic Compounds, Minerals, and Mineral Glasses: Connections Forged by Quantum Chemistry and NMR Spectroscopy............165
    John A. Tossell

11. Bragg Diffraction and the Interference of Two Atom Lasers: An Analogy............177
    Roger A. Hegstrom

Protein, DNA, and Viruses

12. B_{12}-Dependent Methionine Synthase: A Structure That Adapts to Catalyze Multiple Methyl Transfer Reactions............186
    Martha L. Ludwig and Rowena G. Matthews

13. Metalloproteins to Membrane Proteins: Biological Energy Transduction Mechanisms............202
    Douglas C. Rees

14. Paradigms for Protein–Ligand Interactions............216
    Florante A. Quiocho

15. Structures of the Central Dogma of Molecular Biology............231
    Thomas A. Steitz

16. Analysis of Intradomain Signaling in the Multifunctional
    Protein CAD Using Novel Hybrids and Chimeric Molecules............249
    David R. Evans and Hedeel I. Guy

17. A New Engine for Cleaving Nucleic Acid..270
    Kurt L. Krause and Mitchell D. Miller

18. Activation of Hematopoiesis and Vasculogenesis in the
    Mouse Embryo: Induction and Reprogramming of
    Ectodermal Cell Fate by Signals from Primitive Endoderm..............294
    Margaret H. Baron

19. Scoring Functions Sensitive to Alignment Error Have a
    More Difficult Search: A Paradox for Threading...............................309
    Jeffrey Chang, Michelle Whirl Carrillo, Allison Waugh,
    Liping Wei, and Russ B. Altman

20. Electron Paramagnetic Resonance Techniques for
    Measuring Distances in Proteins...321
    Sandra S. Eaton and Gareth R. Eaton

21. Allostery and Induced Fit: NMR and Molecular Modeling
    Study of the trp Repressor–mtr DNA Complex..................................340
    Luciano Brocchieri, Guo-Ping Zhou, and Oleg Jardetzky

22. The 3.6 Å Structure of the Reovirus Core Particle............................367
    Karin M. Reinisch and Stephen C. Harrison

Author Index...379

Subject Index..380

Preface

A visitor to Professor William N. Lipscomb is exposed to both a great legacy of accomplishments and to proposals for new multiyear scientific explorations that are outlined with the excitement of a beginning assistant professor. This pairing is a large part of the explanation for why a group of prominent scientists gathered at Harvard on May 12–14, 2000, to somewhat belatedly celebrate the 80th birthday of "the Colonel." This book is in no way a "proceedings" of that scientific celebration, but the contributors were invited from among those who have studied with the Colonel as undergraduates, graduate students, postdoctoral research associates, and visiting faculty. The chapters in this book purposely reflect the fact that Lipscomb and his co-workers have contributed to a wide range of chemistry. We believe that this is one of the great strengths of the Lipscomb tradition, and it is intended that the breadth of the book be stimulating, and cause readers to think a little beyond the usual confines of a particular research field. We hope that the juxtaposition of, for example, boron hydrides, quantum mechanical calculations, and structure of a virus will provide a perspective on chemistry not offered by traditional texts.

An autobiographical reflection by Lipscomb, which he wrote in 1977, is included with the front matter of the book. The book begins with an introduction to Lipscomb's science by one of the editors. Following this introductory section, there are three sections of scientific reviews. The first section focuses on the "inorganic" chemistry legacy, beginning with Russell Grimes' essay on polyhedral boranes, and continuing with increasing quantum-mechanical emphasis. Because patterns have been a central theme of the Lipscomb contributions in inorganic chemistry, we end this section with Irving Epstein's chapter on oscillations, waves, and patterns in chemistry and biology. The Colonel repeatedly emphasized that theory could lead experiment, and the next section presents chemical theory of bonding and NMR parameters, and includes, appropriate to the scope of the Lipscomb legacy, interference of atom lasers by Roger Hegstrom. The final section shows the legacy of the third phase of Lipscomb's scientific career, the relation of protein structure to function. The first chapter in this section is by Martha Ludwig, who participated in the early work on the structure of carboxypeptidase. Douglas Rees discusses biological energy transduction, Florante Quiocho discusses protein–ligand interactions, and Tomas Steitz relates structures to the central dogma of biology. In addition to crystallography, structure determination by NMR and EPR is included. This section ends with comments on the relation of the retrovirus core to the function of this molecular machine by Lipscomb's last Ph.D. student, Karen Reinisch.

Thanks to the authors for the pleasure that they have given us via their chapters. If this book serves as even a mild antidote to the seemingly relentless pressure for a narrower focus in modern science, that will justify to the editors that putting this book together was not merely intellectual hedonism.

The calligraphy in the introduction was created by Jean Evans (wife of the Colonel) for this book. We thank her for this contribution.

Just as this book was nearing completion, our colleague Don Wiley tragically disappeared while attending a scientific advisory board meeting of St. Jude's Children's Research Hospital in Memphis, Tennessee. At that time, we had not yet mutually agreed upon a title. Don's latest communication to us on this topic was: "I like the title that ends Ashes to Enzymes. It is so biblical." Consequently, we decided to use that title.

Gareth R. Eaton
Department of Chemistry and Biochemistry
University of Denver
Denver, CO 80208–2436

Don C. Wiley
Department of Molecular and Cellular Biology
Harvard University
7 Divinity Avenue
Cambridge, MA 02138

Oleg Jardetzky
Stanford University School of Medicine
300 Pasteur Drive
R–320
Stanford, CA 94305–5337

Process of Discovery (1977); An Autobiographical Sketch

William N. Lipscomb

History

Early experiences

My early home environment, on the outskirts of Lexington, Kentucky from 1922 to about 1940 stressed personal responsibility and self reliance. Independence was encouraged especially in the early years when my mother taught music and when my father's medical practice occupied most of his time. In those days general practitioners still existed, and physicians even made house calls. Science hardly existed in the Victorian-like Sayre elementary school, but I studied somewhat independently, learning the distances between objects in the solar system having already become fascinated with the night sky. Also, I collected animals, insects, pets, rocks and minerals in those grade school days. Part of me will always remain close to the country.

My interest in astronomy took me to the visitor's nights at the Observatory of the University of Kentucky, where Prof. H. H. Dowing and I became lifelong friends. He gave me a copy of Baker's "Astronomy" which I read many times. I am sure that I gained many intuitive concepts of physics from this book, and from my conversations with him.

Within 100 yards, five others of my age group were to emerge two physicists (W. B. Fowler and E. C. Fowler), two physicians (W. R. Adams, J. Adams) and an engineer (R. Fish). Morse-coded messages over a wire stretched high across the street, crystal sets, Tom Swift (the boy Edison-like inventor) books, electric arcs, and lots of talk about astronomy and then chemistry occupied us in a far more interesting way than did the school work.

By the age of 12, I had acquired a small Gilbert chemistry set, but rather than abandoning it a few weeks after that Christmas day, I set about expanding it partly by ordering apparatus and chemicals from suppliers and partly by using my Father's privilege to purchase chemicals at the local drugstore at a discount. These were the years of the Financial Depression. Of course, I made my own fireworks, and entertained both willing and unwilling visitors with spectacular color changes, vile odors, and explosions with pure hydrogen and oxygen. My tolerant, but concerned mother raised a question only once, when I attempted to isolate a large amount of urea from the natural product.

Nevertheless, I had learned a considerable amount of chemistry and physics by the time I enrolled in the first chemistry course ever taught at Picadome (now Lafayette) high school. I was a sophomore in a class otherwise restricted to seniors, but ended the year by teaching qualitative analysis to this class. L. Frederick Jones, the teacher, gave me his college books on organic, analytical and general chemistry, and simply asked that I show up for the examinations. I heard most of the classes from the back of the room, where I was doing a bit of laboratory research that I thought was original: the preparation of hydrogen from sodium formate (or sodium oxalate) and sodium hydroxide. This idea was an extrapolation, from the organic text to inorganic chemistry, of the preparation of saturated hydrocarbons, e.g. the preparation of methane from sodium acetate and sodium hydroxide. This study was carefully and thoroughly done, including gas analyses and searches for probable side reactions. I later had a physics course, and took first prize in the state contest on that subject. The library was (understandably) poor in those days in a county high school, but I remember being influenced so strongly by Abbott's "Flatland" that I became very interested in special relativity. Also, I repeatedly read Alexis Carrel's "Man the Unknown," and perused unsystematically the large medical texts in my father's library. While I had no desire to become the fourth generation of physicians in the family, this influence and that of Linus Pauling years later led to my biochemical studies of recent years.

I also pursued my independent study program during my busy years (1937-41) including a music scholarship at the University of Kentucky. Dushman's "Elements of Quantum Mechanics," the Pittsburgh Staff book on "Atomic Physics" and Pauling's "Nature of the Chemical Bond" were of direct interest. Perhaps I learned more from attempting to help my fellow students than from any other source. Mathematics was a strong interest, considerably heightened by Prof. Fritz John under whom I was the only student in a summer course in Vector Analysis. We spent much of the time pursuing Maxwell's equations jointly, and he introduced me to tensors and matrices. In physics, Prof. O. T. Koppius was an especially warm friend who also taught me thermodynamics, and Prof. Bertrand Ramsey helped me with introductory quantum mechanics. Prof. F. W. Warburton asked me to give the lectures in the summer course in Electricity and Magnetism inasmuch as I was the only student, and he had three other lectures to give every day. Most influential in my early chemistry career was Prof. Robert H. Baker. When I finished the most interesting undergraduate course that I experienced, in qualitative organic analysis, he suggested that I take on a research problem: the direct preparation of derivatives of alcohols from dilute aqueous solution without first separating the alcohol and water. I suggested one method, and he suggested another. Both methods worked, leading to my first publication. In physical chemistry, I became interested in conductance of water as a solute in liquid hydrogen fluoride, and especially whether the results indicated the presence of H_4O^{++}, but I finally abandoned this work as incorrect.

Choice of graduate school was upon me in 1940-41. Prof. Maxson, Chairman of Chemistry (U of K) had decided that I would accept the standing fellowship to MIT. Northwestern offered me a research assistantship at $150/month, Caltech offered me a teaching assistantship in Physics at $20/month, and Prof. H. C. Urey then of Columbia University wrote me a nice letter of rejection. I took the Caltech offer, preferably to study theoretical quantum mechanics with Prof. W. V. Houston, but primarily because of the need for a stimulating environment, and for research problems that were more current. After one semester, I switched to the Chemistry Department under the influence of Prof. Linus Pauling to whom I was drawn by his penetrating and imaginative comments at colloquia, almost always made after everyone else had asked their own questions. At this point (December 1941) the war intervened, and for the following four years my efforts were mostly concerned with the National Defense Research Council. The first project was on analysis of smokes according to particle size, related to a project, no longer needed, to obscure the Los Angeles area optimally. However, most of my war work was concerned with nitroglycerin-nitrocellulose propellants, and involved handling pure nitroglycerin on many occasions. Rates of burning, and polarizing microscopic examination were the major areas. During these four years, I also carried out, with Verner Schomaker, several electron diffraction studies on molecular structures of gas molecules, and learned X-ray diffraction methods from Edward W. Hughes, on the C-N distance in methylamine hydrochloride. Hughes suspected that Hendrick's structure was wrong, and Pauling needed a good C-N distance from a simple molecule for his ideas on peptide bonds. I made most of this X-ray study at Caltech while Hughes was in Emeryville. Some of his replies to my research reports begin, "My God, Willie," with good reason. Actually, this is the first paper in which statistical treatment of errors was made in either the least squares or Fourier methods of X-ray crystallography.

Pauling's course in the "Nature of the Chemical Bond" was worth attending every year, because each lecture was new, and Tolman's lectures in Thermodynamics were a model of clarity and perception. Finally in January 1946, I resumed a normal program, and finished my Ph.D. that summer. Especially noteworthy were courses from Prof. J. H. Sturdivant on X-ray diffraction and one on particle physics by Prof. J. R. Oppenheimer. My research orientations of later years were strongly influenced by the structural chemistry approach, by Pauling's interest in antigen-antibody (hapten inhibition) studies, by his lectures on chemical bonding especially in boron hydrides and metals, and by the unique certainty of X-ray diffraction results coupled with the less secure interpretation of chemical behavior, from a structural point of view. I tried always to imagine what was happening to structures or reactions, in three dimensional terms. While I was able to use both the mathematical description and the more symbolic chemical symbolism, I returned continually to the structural interpretations whenever I could do so.

Minnesota. 1946-1959

By early 1946 I had applied for a National Research Council Fellowship to study crystallography at low temperatures with Bert Warren of MIT. I had perceived this area as nearly untouched, and full of promising problems such as the boron hydrides, hydrogen bonding, residual entropy studies, and a host of structures for which electron diffraction gave ambiguous or insecure results. However, I withdrew my application in order to accept an assistant professorship at the University of Minnesota at $2,880 per 9 months. My rapture at learning of the rank was modified by the salary. Harold Klug had just left Minnesota for the Mellon Institute, and his apparatus was still there. However, I tossed it all away, and set up a sealed tube unit with the aid of a power supply kindly donated by a local dentist, who demonstrated (on me) how he used hypnosis for his patients. My research started very slowly partly because of the teaching load of two courses, and also because of a lack of research funds. If my memory serves correctly, my support from the Office of Naval Research began in 1948 at a level of $9,000 per year. By then, I had initiated the series of low temperature X-ray diffraction studies first of small hydrogen bonded systems, residual entropy problems and small organic molecules. Later, after the techniques were under control, we studied the boron hydrides themselves B_5H_9, B_4H_{10}, B_5H_{11}, B_6H_{10}, B_9H_{15}, and many more related compounds in later years (50 structures of boron compounds by 1976).

John Gruner, mineralogist at the University of Minnesota, asked me to look into structure of $HMnO_2$, which he had named Groutite in honor of Prof. Frank Grout (U of Minn.). He also dug out some black crystals from a spherical iron pressure-temperature vessel in which some phosphates were also present. Having established that the X-ray powder pattern was similar to that of lazulite, he asked us (Lewis Katz and me) to do a single crystal study. It was certainly a new crystal form, tetragonal instead of monoclinic, and had iron in both II and III valence states. Although artificially prepared, Gruner named it lipscombite. It has since been found a number of locations as a real mineral.

In early 1948 I thought that there was an experimental solution of the phase problem of X-ray crystallography. The idea was to use a double reflection $\underline{h}_1$ followed by $\underline{h}_2$ which diffracts in the direction of $\underline{h}_3 = \underline{h}_1 + \underline{h}_2$. If $\underline{h}_1$ is set on the sphere of reflection so that it diffracts for any orientation of the crystal about a suitably chosen rotation axis, then $\underline{h}_1$ and $\underline{h}_2$ should show an interference effect. This idea, beautiful in principle, was defeated by the mosaic character of crystals and possibly also crystal boundary effects. Our experiment in which $\underline{h}_1$ is 040 of a glycine crystal failed, although some reflections which were forbidden as single diffractions were observed.[1] Shortly thereafter (1951) Bijvoet published his experimental solution to the phase problem by multiple isomorphous replacement methods, and I thought then that his discovery opened the way to solve protein structures. However, I did not start work in this direction until about 1958, and pursued it seriously beginning in 1961.

In the fall of 1953 the structures of B_5H_9, B_4H_{10}, B_5H_{11}, and the B_6 arrangement B_6H_{10}, were known unambiguously from our low temperature single crystal studies. Also, the B_2H_6 structure was known from thermodynamic infrared studies of Stitt, Longuet-Higgins and Bell, and Price; and the $B_{10}H_{14}$ structure had been determined by Kasper, Lucht and Harker. All of these boron arrangements were fragments of the two then-known polyhedra, the B_6 octahedron, and the B_{12} icosahedron. Our X-ray studies were difficult because of the instability of the preparations and the tendency for explosion if a vacuum line cracked. William Dulmage had such an explosion, which just missed him as he bent over in handling B_5H_9 on a vacuum line. The folklore regarding $H_2B_2O_3$ was not then known: use dry ice for traps, not liquid nitrogen, so that the $H_2B_2O_3$ does not accumulate. Maintenance of the liquid nitrogen supply led to emergency trips in cars or streetcars with Dewars to be filled at the local plant, and on occasion led to a watchman's question about whether we were stealing apparatus from the laboratory. The first B_6H_{10} study was terminated when an electrician dropped a wrench on a transformer, shorting it out, and years later (1957) the second B_6H_{10} investigation was ended prematurely when an undergraduate assistant tripped on the stairs in the very early morning, on his way to fill the Dewar, and lay unconscious for about an hour.

In puzzling over these structures, I had already noticed the similarity of bonding in B_2H_6 to that in the bridge regions of B_4H_{10}, B_5H_9, B_5H_{11} and $B_{10}H_{14}$. These similarities led me to write in the 1954 paper, "These ideas suggest that to a considerable degree the hybridization about boron in many of these higher hydrides is not greatly different from the hybridization in diborane. In addition the probable reason for the predominance of boron triangles is the concentration of bonding electron density more or less towards the center of the triangle so that the bridge orbitals (π orbitals in B_2H_6) of the three boron atoms of the triangle overlap. It does seem very likely from these structural comparisons that the outer orbitals of an atom are not always directed toward the atom to which it is bonded. This property is to be expected for atoms which are just starting to fill new levels and therefore may be a general property of metals and intermetallic compounds." This paper was a companion to "The Valence Structures of the Boron Hydrides," by W. H. Eberhardt, B. Crawford, Jr. and myself, in which the generality of three center bonds in the boron hydrides was first recognized. We also formulated a rather simple set of rules which gave three relations between the chemical formula B_pH_{p+q} and the constituent BHB bridges (s), three-center BBB bonds (t), two-center BB bonds (y) and BH_2 groups (x) : s+x = q, s+t = p, and p = t+y+q/2. There are three equations and four unknowns (styx) for a given formula so that ambiguities remain even when one requires that s,t,y and x shall not be negative. After using these equations and structural elements among these and related compounds, I finally decided that in many compounds the ambiguities are quite real, representing different isomers or pathways of transformations.

In those days X-ray intensities were measured visually with the use of a standard scale. A major problem was to locate the number and position of hydrogen atoms in these compounds. High quality data were required. For B_9H_{15} even the number of boron atoms was not known until the structure was solved by the method of Holmes (". . . when all other contingencies fail, whatever remains, however improbable, must be the truth." From *The Complete Sherlock Holmes*, Garden City Publishing Company, Inc. 1089; see also pp. 118, 360, 1192). The number of hydrogen atoms was also established in further X-ray analysis of this material. Another example is the complete formula and structure of B_8Cl_8 both of which were established by the X-ray data alone. At least our speculations of bonding and reactivity were based on secure structures. "We have even ventured a few predictions, knowing that if we must join the ranks of boron-hydride predictors later proved wrong, we shall be in the best of company" (Eberhardt, Crawford, Lipscomb, 1954). The spirit in which these and related later extrapolations, guesses and predictions were made was the realization that a large body of new chemistry existed, if we could only persuade ourselves and other chemists to look for it. Anyway, it has been fun to guess and to be proved right, as well as distressing to be proved wrong.

In 1954-5 on a Guggeheim Fellowship to Oxford, I spent most of my efforts on theoretical chemistry, half-intending to leave both X-ray diffraction and boron chemistry. However, the impressive and highly intuitive, solution of the vitamin B_{12} structure was being obtained by Dorothy Hodgkin and her associates there, and I resolved to go to larger structures with the aid of high speed computers when I returned. In boron chemistry, Riley Schaeffer, Herman Schlesinger and, later, M. Frederick Hawthorne sent us new boranes, boron halides and derivatives of boranes for structural studies. As an aid in solving the crystal structures and for predicting new boranes, I developed a very intuitive approach, which Richard Dickerson and I formalized into a topological set of rules.

One misses a lot of important discoveries on the way. I remember one day in Oxford in the fall of 1954 when Coulson mentioned the single nuclear magnetic resonance in PF_5. Donald Hornig, with whom I shared the office, suggested intermolecular exchange in order to produce this unexpected equivalence of the equatorial and axial fluorine atoms, whereupon I suggested an intramolecular rearrangement process. This is, of course, the pseudorotation proposed by Steven Berry a few years later. I did not publish this result, or even think of this area as a new and promising one at that time. In a sense, however, this discussion made it much easier for me to understand the equivalence of all boron and all hydrogen atoms in the ^{11}B nuclear magnetic resonance spectrum of $B_3H_8^-$ by use of a pseudorotation process. This occurred in December of 1958 while refereeing a communication by Earl Muetterties and W. D. Phillips, when I found that their incorrect static structure would not give the relative intensities which they had found experimentally. Actually this spectrum was not well above background so that I had to assume that the observed septet was really a nonet in which the two outer lines were weak. Another preparation for this interpretation was the difference in hydrogen positions in $B_{10}H_{14}$ (four bridges) and $B_{10}H_{14}^{2-}$ then extrapolated from the $B_{10}H_{12}$ $(NCCH_3)_2$ structure as a two-bridge, two-BH_2 structure. The detailed mechanism of the $B_3H_8^-$ rearrangement had to preserve the coupling constant between B and H averaged on the nmr time scale. Later I developed a general theory of pseudorotation in polyhedral species (1966). Internal rearrangement in B_6H_{10} was foreshadowed by the ambiguity in the hydrogen arrangement (*J. Inorg. Nucl. Chem.* **11**, 1 (1959)), in an article in which I attempted to systematize the chemical transformations then known for the boranes. One rather successful prediction was the probable existence of $B_4H_7^-$ and $B_6H_{11}^+$, which have framework orbitals like those in B_5H_9 and indeed have a σ, π bonding scheme rather like that in the aromatics $C_5H_5^-$, C_6H_6 and $C_7H_7^+$. Both B_4H_7 and $B_6H_{11}^+$ were eventually prepared in other laboratories.

Many times, following the 1954 ECL paper, I decided to leave boron chemistry, only to find myself thinking of a new way to look at the structures, or hearing of a new interesting preparative success. It was and still is, an area of chemistry to which I returned almost subconsciously within half a year of each decision to move on to other research areas. The most important other direction was the determination of large organic structures, preferably natural products. In 1953 I wrote, "The elucidation of complete molecular structures of relatively complex organic compounds, often by methods which require a minimum of chemical information, by X-ray diffraction techniques has been one of the most significant advances in recent years. The total effort in man hours required for such a complete structure determination is sometimes less that expended in the determination of the molecular structures by the more classical organic methods. Lack of realization of this fact by most organic chemists is matched only by the reluctance of most crystallographers to under-take such investigations" (*Ann. Rev. Phys. Chem.* 1953). This optimism was only later to be fully realized with the advent of the high speed digital computer and the automated X-ray diffractometer. Our studies of natural products in my later years at Minnesota were a prelude to later work at Harvard in the three dimensional structure and function of enzymes.

Of the many other studies done at Minnesota, I may mention the study of Roussin's black salt ($Fe_4S_3(NO)_7^-, Cs^+$) from which I failed to realize that the Fe_4S_4 cluster later found in oxidation-reduction proteins, might be a reasonable extrapolation. On the more humorous side, I comment on the SiF_4 structure, obtained when I tried to do the HF structure in pyrex capillaries (later we did solve the HF crystal structure), and the cyclic silane $(SiO_3)_4$ obtained from a vacuum line preparation of boron hydrides in which silicone stopcock grease was used. Also, countless single crystals of ice were photographed from condensate on the outside of our capillaries, which were cooled to low temperatures for X-ray diffraction work. However, we took full advantage of the low humidity of the Minnesota winter by opening the windows of the X-ray laboratory.

Harvard 1959

While not at all unhappy at Minnesota, I was aware that the Harvard Chemistry Department was one of excellence and certainly that the graduate students there were the best in the country. The circumstances under which I was "called to Harvard" (an expression which, I hope, is in its sunset) were somewhat embarrassing: all of the members of the Physical Chemistry Division of the U of M were in my office.

Several graduate students and staff came to Harvard with me, along with one infamous piece of apparatus: a Rigagu-Denki rotating anode X-ray unit. That very early model, purchased for about $4,000, was rebuilt by William Streib and Sandy Mathieson many times. When it finally worked I could not persuade anyone to use it because of its history. So I gave it away to the Laboratory of Applied Physics if only they would pay the cost of moving it. Shaughnessy, the movers, had to take it out of the window but they used a long beam to which they attached a rope, which broke. Just before that one of the students said, "I wonder what would happen it the rope broke?" It fell three floors, ended one foot into the solid ground, and was sold for junk.

While visiting Harvard in the spring of 1959, I talked with Roald Hoffmann about programming extended Hückel theory for the new IBM computer which I influenced Harvard to obtain. It was only after he spent a year in Moscow that he returned to start the molecular orbital work which was first successfully used on the polyhedral borane anions, and now has been applied to many parts of the periodical table, including (years later) tests during the development of the

Woodward-Hoffmann rules. It was in 1959 that I first visited M. Frederick Hawthorne, then at Huntsville, Alabama, where we began a collaborative effort which has lasted for years. I saw the $B_{10}H_{10}^{-2}$ ^{11}B nmr spectrum, and realized a few days later what the correct polyhedral structure must be, as well as its mechanism of formation from collapse of the more open (nido) framework of $B_{10}H_{12}L_2$ where L is a ligand such as trimethylamine (*J.A.C.S.* **81**, 5833 (1959)). Shortly thereafter Hawthorne gave us a salt of $B_{12}H_{12}^{-2}$ which we showed had an icosahedral structure, as had been predicted by M. deV. Roberts and H. C. Longuet-Higgins, and by us. Hawthorne supplied us with many compounds in those days, and Russell Grimes made some more new ones here. It was satisfying to see the enormous development of this area of chemistry, and to feel that the theoretical and X-ray structural approaches were, for once, leading the way. I continued to look for new structural principles and to try to formulate principles of chemical reactions, partly summarized in my 1963 book "Boron Hydrides" which is long since out of print, having sold only 4000 copies. Even the hypho-boranes are foreshadowed in this book.

Prepared by the earlier work on pseudorotation noted above, and by the independent interpretation by Robert E. Williams and me of pseudorotation as the mechanism for the 1 to 2 shift of substituents R in B_5H_8R, I studied this process further. I remember recalling the juxtaposition of the icosahedral B_{12} unit and the cubeoctahedral B_{12} unit in a figure in the paper published in 1960 with Doyle Britton on Valence Structure of the Higher Borides.

I wondered if the cubeoctahedron could close its square faces in one or the other of two different ways, and thus give a pathway from one icosahedral to another icosahedral structure. Not long after a thought occurred to me of this same process, more locally, in $B_{10}H_{10}^{-2}$ in which a B_4 diamond-shaped unit could open to square, and then close the other way, to produce an isomerization. I remember Roald Hoffmann asking, "How do you think of ideas like that?" This was to become a general paper on pseudorotation mechanisms in polyhedral species (*Science* **153**, 373 (1956)), an idea which was worked out at home with models when I was recovering from influenza.

Actually the programming of extended Hückel theory was begun in my laboratory by Lawrence L. Lohr, Jr., whose interests in transition metal chemistry required d orbitals. When Hoffmann returned from Russia, he also started, but needed only s and p orbitals, so he finished his program first. We applied the new program to many compounds, but I received so much criticism about the basis of the theory, mostly from E. B. Wilson, Jr., that I suggested to Marshall Newton and F.P. Boer that they look into improving the theory. I had previously started C. William Kern, Richard M. Stevens and Russell M. Pitzer on self-consistent field theories of molecules, and of certain second order properties including the chemical shift. Although even now (1977) we have failed to calculate ^{11}B chemical shifts reliably in a complex borane, we did succeed for many diatomics, finding paramagnetism in diatomic BH even though all electrons are paired. The SCF theory of the ground state was, however, applicable to large molecules. I took particular delight in asking Russell Pitzer to make the first correct calculation of the barrier to internal rotation in ethane: it was his father, Kenneth S. Pitzer, who found this barrier experimentally, before World War II. I am told that I confused theoretical chemists of that time by using relatively exact SCF theory and relatively crude extended Hückel theory. The purpose was eventually to produce a relationship which would then allow the calculation of much more reliable wave functions for large molecules and for reaction pathways. The results of the SCF theory became an objective, and we proceeded from a non-empirical theory to our latest one (1977), partial retention of diatomic differential overlap (PRDDO), programmed by Thomas A. Halgren. He also programmed a "synchronous transit" method for reaction pathways of internal rearrangement ring opening, or of chemical reactions between different molecules.

The presence of these new theoretical methods, made possible by high speed computers, allowed us to reinvestigate the bonding in boranes, carboranes and related compounds in a totally objective way. Many of the simplest ideas of three-center bond theory were confirmed, and the limits of validity (about 0.1 to 0.2 electrons/bond) were established. Also, preferred single valence structures, sometimes possessing fractional bonds, emerged where previously we had to use many resonance structures in a hybrid description. These methods of localizing molecular orbitals (Lennard-Jones, Pople, Boys, Edmiston, Ruedenberg, all of other laboratories) greatly simplified bonding descriptions of these boron compounds. Also, they produced a vivid connection between the highly delocalized symmetry molecular orbitals and the localized bonds in which chemists believe so strongly. When I reviewed this work at the American Chemical Society on the occasion of the Debye Award, H. F. Schaefer commented to me, "Other people had looked at localized orbitals, but you are the only one to do so much chemistry with them." The work has only begun. These localized orbitals are more complex than the usual chemical bonds, and we have hardly begun to look at the delocalized "tails" which are mostly neglected, but I am convinced that these orbitals are a more accurate description of electron pair bonds than are more conventional bonds. Even so, they need drastic modification when electrons are unpaired in chemical reactions. Our studies of these localized orbitals since 1969 were mentioned by others, along with the many earlier X-ray, bonding, and chemical studies in my conversations in Sweden in December 1976.

In probing for new areas, during those intervals in which boron chemistry would allow me to do so, my students and I explored several other areas related to structure and function. William E. Streib built a helium cryostat which allowed us to solve the crystal structures of β-N_2, α-N_2, β-F_2, γ-O_2, CH_4 and B_2H_6. A study of $(OC)_3FeC_8H_8$ gave a crystal structure in which the $(OC)_3Fe$ was bonded to only four C's of the cyclooctatetraene, while all H atoms were equivalent in the nmr spectrum. From these facts I guessed that the ring was rotating through the bonding region on the nmr time scale. This was one of the first proposals of dynamical effects in transition metal complexes (1962), but not as early as one of Geoffrey Wilkinson's proposals. Of the organic natural products, perhaps the structure of vincristine (with J. William Moncrief, *J. Am. Chem. Soc.* **87**, 4963 (1965)) was most important because of the previously incorrect stereochemistry, the use of anomalous scattering in the solution, and the success of this compound in the treatment of childhood leukemia. From the molecular orbitals of hypothetical B_4H_4, I once concluded that cyclobutadiene might be a tetrahedral molecule, which it is certainly not, but where else could one publish such a paper but Tetrahedron Letters? One disappointment was that the National Science Foundation refused to support the work started by J. Gerratt and me on spin-coupled wave functions (*Proc. Natl. Acad. Sci. U.S.* **59**, 332 (1968)). Unfortunately, they also abruptly discontinued my only NSF Grant in 1964 after only four years of support of the research on boranes. Support from NSF was, however, resumed in 1977 subsequent to an event in December 1976. More recently, Daniel Jones and I studied the effect of chemical bonding on the distances in B_2H_6 as determined by X-ray methods. This is a general area of direct determination of electron densities throughout a molecule, but I decided not to make a major effort in this direction because it would require complete and accurate X-ray data at the temperatures of liquid N_2 as well neutron diffraction results on the nuclear positions. Also, for a few years, we tried to attach borane and carborane cages to organic molecules which were then to be attached to specific antibodies to tumors for treatment by neutron irradiation (*J. Medicinal Chem.* **17**, 785 and 792 (1974)).

My interest in biochemistry goes back to my perusal of medical books in my father's library and to the influence of Linus Pauling from 1942 on. After trying unsuccessfully (with Scott Matthews) to isolate a small enzyme from tetrahymena pyroformis in Minnesota days, I started the carboxypeptidase A work in 1961. However, the project did not really get under way until Martha Ludwig arrived in May of 1962. In August of 1963 we published the solution of the lead derivative, but progress slowed after that for a time. Morale problems were due to the pessimistic outlook of two members of the rather large group that assembled. In retrospect, optimism was well justified- the derivatives were good; the intensity changes were not primarily due, after all, to conformational changes. Those who stayed with the project (Martha Ludwig, Jean Hartsuck, Thomas Steitz, James Coppola, Florante Quiocho, Hilary Muirhead, George N. Reeke, Jr. and Paul Bethge) were most cooperative, and have distinguished themselves since in independent studies. Is it the nature of the cooperative research, the field of biochemistry, the dilution of individual identity, or some other factor that makes biochemists generally, but not always, more sensitive than chemists to their ambitions? I am not referring to those members of my group as listed above.

The accomplishments on the carboxypeptidase A study were impressive. First, I located the disulfide bridge in spite of the overwhelming evidence of Vallee that there could not be one because one of the two sulfhydryl groups was bound to zinc. It was not. We identified the three ligands to zinc as 69,72 and 196 in the polypeptide chain without a sequence, and properly identified 69 as His and 72 as Glu, but misidentified His 196 as Lys. In our proposal of a mechanism, Glu 270 either attacks the substrate or promotes the attack of water, and Tyr 248 (or water) is the proton donor[2] when peptides are cleaved. At the binding stage Arg 145 forms a doubly hydrogen bonded salt link to the C-terminal carboxylate group of the substrate. High resolution was achieved in August 1966, as reported at a Gordon Conference. Only myoglobin and lysozyme had reached high resolution at that time.

It is sometimes extremely difficult to repeat a biochemical experiment, and the preparation of the crystalline form which we studied (a= 51.41 Å, c = 47.19 Å and β = 97°35') was just such a problem. Only about two of some thirty preparations crystallized with these cell dimensions. On the other hand most preparations, including the commercial preparation, give cell dimensions of a= 50.9 Å, b = 57.9 Å, c = 45.0 A and β = 94°40'. The activities of these two forms are very different in the solid state, 1/3 and 1/300 of that of the material in solution (*Proc. Nat. Acad. Sci.* **70**, 3797 (1973)). The arsanilzao Tyr 248 derivative of the former crystals behaves like the material in solution, but this derivative of the latter crystals is quite different from the material in solution. Vallee continues to discount the results of our crystallography, done on the former unit cell, on the basis of his results on the commercial form, which most likely has the latter unit cell. Probably the main difference is in the intermolecular contacts in the two crystalline forms. Since the crystals of the X-ray study are active, the conclusion of this work should be tested on that basis. Our present plan (1977) is to elucidate the structures of the protein inhibitor of carboxypeptidase A isolated from potatoes, and of its complex with the enzyme in the hope that atomic displacements toward the transition state can be recognized.

The glucagon structure was a different matter. William Haugen, not a careful experimentalist, worked only at high pH where the intensity changes upon addition of heavy atom salts were mostly conformational changes. Jean Hartsuck made that observation. The low resolution structure was noted in 1969, but it was Thomas Blundell, one of Dorothy Hodgkin's former students, who later solved the structure at a lower pH. We sent him all of our results, and I have the highest praise of him as a scientist and a gentleman.

Concanavalin A, was solved here to 4 Å resolution, by Florante Quiocho. The refinements are detailed in the Ph.D. thesis of Brian Edwards, who did most of the calculations on Quiocho's data. We entered a collaboration with Gerald Edelman at Rockefeller University, where the sequence was being done and where George Reeke, Jr. was just then setting up a new laboratory for protein structure research. A joint publication (*Proc. Nat. Acad. Sci.* **68**, 1853 (1971)) is based entirely on Quiocho's data.

Since 1967 our major effort has gone into the determination of the various structures of an allosteric enzyme, aspartate transcarbamylase from E. coli. Our first publication has an error, due to our belief in Schachman's result that the enzyme was a tetramer (C_4R_4), and also due to our failure to compute the water content correctly: We found our error, and discovered both a two-fold and a three-fold axis in the molecule, and therefore the hexameric nature of this molecule (Wiley and Lipscomb, *Nature* **218**, 1119 (1968)). Simultaneously, Klaus Weber, in his own laboratory here at Harvard, found the six-fold symmetry from analysis of the N-terminal groups, and from the sequence of the regulatory chain (R). The catalytic chain (C) has now (1977) been almost completely sequenced by William Konigsberg of Yale, who has kindly sent us his results from time to time. Thus the enzyme has the composition C_6R_6, where C is about 34,000 and R is about 17,000 in molecular weight. The work is at about 3 Å resolution for the native enzyme and for its complex with CTP (Cytidine triphosphate is the allosteric inhibitor); and crystals have been obtained of the complex of this enzyme with a substrate analogue. Finally, this problem which may go on for more years, is moving well.

Why was it so difficult earlier? The major problem was to obtain at least one derivative in which the heavy atom could be located. The problems were (1) the high molecular weight (310,000) and hexameric nature which made it difficult to locate heavy atoms in multiple site derivatives, (2) the sensitivity of the enzyme to conformational changes which destroyed the single crystal, and (3) the greater reactivity (destroying the crystal) of the sulfhydryls of the regulatory chain as compared with that of the single sulfhydryl of each catalytic chain. It was Cecil McMurray who solved this problem by making a large number of closely related mercurials, of which 2-chloromercuri-4-nitrophenol was most successful in its preference for the sulfhydryl of the catalytic chain. With one heavy atom per C chain, we were then able to solve for the heavy atom positions in the uranyl, platinum and gold derivatives. Even now there are problems because the X-ray data fade out at about 3 Å and because our experimental intensities have a background which is too high. Also, of these four derivatives, only the Hg and U are good to 3 Å resolution, so that we shall have to use averaging of non-crystallographic symmetry in order to improve the phasing of the diffraction maxima.

However, the research problem is well worth the effort. The enzyme shows sigmoidal kinetics indicating positive cooperativity, inhibition by cytidine triphosphate, stimulation by adenosine triphosphate, and negative cooperativity in binding of CTP. Also, Elita Pastra-Landis and David Evans have shown here that the fragment C_6R_4 has reduced, but very definite cooperativity. The distance between the allosteric (regulatory) site and the catalytic site is 43 Å, an amazingly long distance for propagation of an allosteric signal. Also, the mechanisms of catalysis of the reaction of carbamyl phosphate with aspartate to give carbamyl aspartate, the first step of the pyrimidine pathway, is still a mystery.[3]

Although the theme of X-ray crystallographic studies of structures runs through much of this account, it has been of great interest to us to carry on parallel experimental, computer or theoretical studies in order to illuminate the relationships of the structures to the rest of the chemistry. On occasion this diversity has even helped the research. The excellent crystals of unliganded aspartate transcarbamylase were obtained by accident from an nmr study during a kinetics experiment. Certainly, the interests of the research group have been broadened, and I have normally been able to ask for a Ph.D. thesis in more than one technique on a particular subject. In a real sense, we have thus generated our own research problems, rather than starting from the open literature. Seemingly unrelated areas come together, for example our recent attempts to use molecular orbital theory to elucidate the sequential steps in a model of the active site of an enzyme.

Acknowledgments

I wish to thank the author for permission to add three footnotes to the original 1977 manuscript.

Footnotes

[1] However, see B. Post, *Phys. Rev. Lett.* **39**, 760 (1977).

[2] In 1982 D. C. Rees and I added Glu 270 as a possible proton donor (*J. Mol. Biol.* **160**, 475).

[3] See J. E. Gouaux, K. L. Krause and W. N. Lipscomb, *Biochem. Biophys. Research Commun.* **142**, 893 (1987).

Introduction

Chapter 1

The Landscape and the Horizons: An Introduction to the Science of William N. Lipscomb

Gareth R. Eaton

Department of Chemistry and Biochemistry, University of Denver, Denver, CO 80208

William Nunn Lipscomb, Jr

The Lipscomb genealogy

Claude Louis Berthollet	1748-1822	
Joseph Gay-Lussac	1778-1850	
Justus von Liebig	1803-1873	
Carl Schmidt	1822-1894	
Wilhelm Friedrich Ostwald	1853-1932	(Nobel Prize 1909)
Arthur Amos Noyes	1866-1936	
Roscoe Gilkey Dickinson	1894-1945	
Linus Carl Pauling	1901-1994	(Nobel prizes chemistry 1954, peace 1962)
William Nunn Lipscomb	1919-	(Nobel Prize 1976)
Roald Hoffmann	1937-	(Nobel Prize 1981)

© 2002 American Chemical Society

From an interview, as published in a newspaper, after receiving the Nobel prize:
"Q.: Did you expect this honor – today or ever?
A.: Oh, no. You never think about it. You just try to keep from falling flat on your face.
Q.: What does winning the Nobel Prize in Chemistry mean to you, Professor?
A.: Winning the Nobel Prize means that someone was reading your articles all those years, after all."

Professor William N. Lipscomb received the Nobel Prize in 1976. He is also an official Kentucky Colonel. Indeed, most of us who have worked with him know him as the Colonel.

The Colonel

When the Colonel received the Nobel Prize, President Bok of Harvard wrote to him: "In a period of such high specialization, it is very heartening to know that it is still possible to be a good tennis player, an excellent clarinetist, and an outstanding scientist." (The Colonel still pays the clarinet, and played classical music at the celebration held in his honor at Harvard in May 2000.)

Considering the vast number of papers published by the Colonel, it is a challenge to review their significance. This introduction attempts to give the reader a feel for the scope of his work, rather than pick out a few selected papers as the "most important." It might be argued that the most important paper any scientist writes is his first one. Most people would probably not guess that the Colonel's first paper was on the identification of alcohols (*1*).

The constant thread through the Colonel's research has been the elucidation of the relationship between the geometrical and electronic structures of molecules and their chemical and physical properties. Since most of his work has focused at the molecular level, it is appropriate to label the Colonel as a chemist, although he started graduate school at Cal Tech in physics. When I was an undergraduate, and a little confused by some of the labels, I asked him "are you an inorganic chemist or a physical chemist?" He answered somewhat mischievously "some of my papers are on a small organic compounds - so I am an organic chemist too." Since he has a mineral named for him (Lipscomb – there is a sample on display in the science museum in London), he has at least a reasonable claim to be an inorganic chemist. In fact, the Colonel has been very productive in several mainstream areas of chemistry. At the same time, he has been a real pioneer at the boundaries, interfaces, or horizons. In many cases, his productivity has converted a horizon into a mainstream area. Certainly, this is true both of boron hydride chemistry and protein crystallography.

The Colonel taught chemists to think about (a) three-center bonds and fractional bonds, (b) intra-molecular dynamics, (c) boron hydrides, and (d) the

role of metals coordinated to proteins. If we try to force the Colonel's work into a simplified outline, we get something like the following:
> instrumentation and interpretation of X-ray diffraction
>> low temperature techniques
>> data acquisition and analysis
>
> magnetic resonance, especially of boron-11
> crystallography of selected small molecules
> enzyme structure and function

One might wonder how ideas germinate. The Colonel tells of hearing Pauling lecture about the boron hydrides at Caltech (1941-46), and explain wrong structures with valence bond theory. He says "I believed none of it." In 1946 he applied for a fellowship, and proposed to develop low-temperature X-ray diffraction techniques with boron hydride structures as a goal. "Pauling did not think that this was interesting." At Minnesota, he began these experiments. The Colonel has been a leader in experimental crystallography for several decades. He especially has guided the development of techniques for low temperature crystallography (*2-4*). This resulted in, for example, the crystal structures of N_2 and F_2 (*5, 6*). Determination of the structure of diborane, which is also a gas at STP, was accomplished with these low temperature techniques (*7*).

He has also contributed to the theoretical aspects of crystallography over several decades. Both in crystallography and in calculations of molecular electronic structure, Lipscomb was at the forefront of the use of computers in chemistry. He summarized the state of the art in the late 50's in a paper titled "an Account of Some Computing Experiences," which is notable for its emphasis on the utility of color in computer graphics (*8*).

The Colonel and his coworkers have done the crystal structures of a large number of molecules such as hydrazine, nitric oxide, metal-dithiolene complexes, and methyl ethylene phosphate (*9*). Recall that many of these were done at a time when a crystal structure involved a much larger effort than at present. It is striking what a large fraction of these structures provided the crucial evidence in the respective fields of chemistry. Some of them have become the standard "text-book" examples. He has had a knack for selecting a compound whose structure would turn out to be important. One small molecule studied was 1,4-dithiadiene, which was an important structure for sorting out the bonding of sulfur (*10*). The paper was rejected by the Journal of Organic Chemistry with the comment "I fail to see any excuse at all for publishing Lipscomb's paper in our journal. It is already known to be a cyclic compound, which is all that is needed by the average organic chemist."

The Colonel did not discover boron hydrides, nor was he even the first to determine the structure of a boron hydride. However, he was the first to see the patterns. He taught the world the structural systematics and showed how the bonding in the boron hydrides could be understood in ways that had predictive utility. Two papers by Lipscomb (*11, 12*), one coauthored with Eberhart and Crawford, in 1954 systematized the known structures of boron hydrides and "ventured a few predictions, knowing that if we must join the ranks of boron-hydride predictors later proved wrong, we shall be in the best of company." Of course, it is now well known that the semi-topological valence postulates, including the concept of the localized three-center bond, had great predictive utility. When Lipscomb went on to predict possible boron hydride ions such as $B_{12}H_{12}^{2-}$, $B_{10}H_{14}^{2-}$, $B_6H_6^{2-}$, and $B_3H_6^+$, the reviewers were not very appreciative. One stated that it should not be "launched as naked prophecy," and the other called it "more an exercise in the use of topological theory than an important contribution." The paper was rejected by JACS and eventually published as a "Note" in Journal of Chemical Physics (*13*). Often he confirmed his own predictions. Everyone who has worked in the Lipscomb group has heard the proud "I predicted that!" more than once. In 1959, the Colonel predicted the D_{4d} structure of $B_{10}H_{10}^{2-}$ (*14*). My undergraduate research included showing that the ^{11}B and ^{1}H NMR spectra were consistent with the predicted structure (*15*), which was also confirmed by x-ray crystallography (*16*).

In one case in which the theory was developed after experimental observations, it was stated "We admittedly make this observation with the benefit of hindsight. This science is known as retrospectroscopy" (*17*). The Colonel has enjoyed (in retrospect at least) these and other amusements during his career. He liked to imbed in his papers little quips, such as thanking the computer for valuable suggestions (*18*), thanking Al Powder for assistance in the calibration (*19*), or citing Sherlock Holmes (*20*) or the "Hunting of the Snark" (*21*). Lipscomb's appreciation of the value of such quotations is revealed in his review of the book "Spectacular Experiments and Inspired Quotes" (*22*).

X-Ray crystallography was a indispensable tool for finding new structural principles in the boron hydrides. The $B_{10}H_{12}(NCCH_3)_2$ structure corrected previous ideas about the acetonitrile groups (*18*), and the $B_{10}H_{16}$ structure found a new type of bond - the direct B-B bond between B atoms without terminal H's (*23*). A sample believed to be a B_8 hydride was found to be a B_9 species by X-ray diffraction, and identified as B_9H_{15}. The hydrogens could not be located in the diffraction pattern. They were determined by application of the styx principles, and by arguments about the volume of the molecule (*20*).

Experimental technique had to be developed for some of these studies. In view of the rapidity with which small molecule crystallography studies can be accomplished today, it will astonish some to read that determination of the structure of $(CH_3)_2B_5H_7$ required sealing the air-sensitive liquid in a capillary, growing a crystal in a stream of cold nitrogen, and keeping it at $-50°$ C for two months while recording photographic x-ray diffraction data (*24*).

In 1957 the Colonel showed that Roussin's black salt contained a Fe_4S_3 cluster (25). In retrospect, he has said that he should have seen the extension of this to the FeS clusters in redox proteins. If he had, he would have been decades ahead of the pack.

Throughout the studies discussed above, there has been a close interplay between structural measurements and development of theoretical models of molecular electronic structure. Lipscomb's contributions have included topological description of the boron hydrides and fundamental theory. Several coworkers contributed in this area, but most notable is the work with Roald Hoffmann, which changed the way chemists approach the theory of molecules of interesting complexity (see, e.g., the comments in (26)). The "extended Huckel method" was developed in the Lipscomb's group by several people, including especially L. L. Lohr, Jr., and Roald Hoffman (see Roald's recollections in Current Contents Citation Classic, May 8, 1989). Although the method probably contributed more to chemistry than any other method until very recently, Lipscomb recalled decades after he introduced it that "this method received intense criticism, even denouncement." More exact theory led to the first correct calculation of the rotational barrier in ethane (with R. M. Pitzer in 1963).

The range of stimulating theoretical work the Colonel did outside the boron hydrides encompassed sodium in liquid ammonia (27), a suggestion that cyclobutadiene could have a stable tetrahedral structure similar to that of B_4Cl_4 (28), and the molecular symmetry of XeF_2 and XeF_4 (29). His paper on sodium in liquid ammonia described the polaron, but he did not name the phenomenon, so someone else later received credit for it.

A structural perspective gave rise to predictions about reasonable rearrangement pathways for polyhedral molecules. He also measured isomerizations that confirmed the predictions. The calculations that describe the boron hydride structures also gave basis for predicting which atoms would be most reactive. These predictions stimulated much synthetic chemistry.

Many people make the mistake of equating the study of three-center bonds to the study of the boron hydrides. Certainly, the Colonel was very successful in applying 3-center bond concepts to boron hydrides. However, we should see the larger applicability of the general concept to noble gas compounds, localized orbitals for species such as CO_3^{2-}, metals, metal- cluster compounds, etc. Someplace between the localized 2e-2c bonds beginning students (and some elderly chemists) love and cherish, and the fully-delocalized molecular orbitals we so often find useful, lies a localized multicenter bond. This approach to localized bonds looks like it will be very useful to chemists when it is more fully developed. They have some resemblance to the intuitive "best" valance structure selected from resonance hybrids. One should be able to do chemistry with these orbitals. Fractional bonds may be uncomfortable at first, but note that the octet rule applies to electrons, not orbitals. One of the most important contributions the Colonel and his coworkers have made to inorganic chemistry is to show the way to selecting the appropriate level of calculation for bonding in molecules.

Anyone who tries truly innovative ideas will fail once in a while. E. B. Wilson once commented after a seminar that was presented as if everything was obvious and flowed so smoothly, "I wonder how full his wastebasket is"! The Colonel has missed a few. An attempt to study the structure of HF led to a paper on SiF_4! One "new borane" turned out to be $(SiO_3)_4$. The Colonel and his coworkers tried unsuccessfully to react boron hydrides with organometallic compounds - for example, B_5H_9 and $Fe(CO)_5$. They did make metal "salts" - actually rather covalent species - with anionic boranes, e.g., $Cu_2B_{10}H_{10}$. The real breakthrough in this area, though, was when Fred Hawthorne recognized the similarity to $C_5H_5^-$ of the orbitals the Colonel presented for the open face of $C_2B_9H_{11}^{-2}$. Many days were spent in fruitless chase of the brilliant yellow color that always accompanied formation of boranes in the electrical discharge apparatus. Every "purification" gave white product and a small amount of more intense yellow stuff. We still wonder where the yellow went.

Sometimes the Colonel had the right idea, but didn't recognize the importance that others would attach to it. A dramatic case involves intramolecular dynamics. Dynamic intramolecular rearrangements, recognized in $B_3H_8^-$ and in $Fe(CO)_3(COT)$, pointed to extensive new insights about the behavior of molecules. The Colonel saw the power of NMR to elucidate the structure of molecules in solution, and early showed how solution and crystal structures could be interpreted in terms of the concept of dynamic rearrangement of atoms on the NMR time scale. While the Colonel was on sabbatical leave at Oxford in the fall of 1954, Coulson mentioned the NMR equivalence of all 5 F in the NMR spectrum of PF_5 (which had been observed in 1953 by Gutowsky, McCall, and Slichter). The Colonel suggested an intramolecular rearrangement, but recalls that he did not think of it as worth publishing. There is a hint in some of the Colonel's light-hearted comments that he feels ignored. One of his famous footnotes thanked a person for reading his paper. There is also the famous comment he made to a reporter the day he received the Nobel Prize ("Winning the Nobel Prize means that someone was reading your articles all those years after all."). No one can read the papers if you don't publish them.

In 1956 Wilkinson and Piper published NMR spectra of metal-cyclopentadienyl complexes, which exhibited only a single line for the cp ring, rather than the complex spectrum they expected. They speculated that the metal was executing a 1,2 rearrangement, and that "the cyclopentadienyl group may be regarded as rotating." They proposed the 1,2 shift without considering alternatives. When Al Cotton with John Waugh's assistance obtained the first C-13 NMR spectrum of a metal carbonyl compound, and found only a single line for the C's in $Fe(CO)_5$, R. Steven Berry wrote to Cotton, proposing what is now called the Berry pseudorotation mechanism. Berry published this in 1960.

Before one gets too wound up in the issue of priority here, we should recall that the concept of intramolecular rearrangements to degenerate structures had been in the literature for a very long time. Maybe the first expression of pseudorotation was Pitzer's description of cyclopentane in about 1930. Another example is that Ray and Dutt (*30*) had described what we now call the Ray-Dutt twist for trischelate complexes in 1943. The well-known Bailar twist followed 15 years later (*31*). Furthermore, of course, rotations of methyl groups are in the same category of reactions. Although the Colonel did not publish anything about the interpretation of the NMR spectrum of PF_5, the discussions prepared him to think in terms of intramolecular motion when he confronted the experimental data on $B_3H_8^-$. He saw the data in 1957 in advance of publication, and communicated his ideas to W. D. Phillips, but they were not accepted, and Phillips published the wrong interpretation. Then, when the Colonel tried to publish the correct interpretation, the paper was rejected because it didn't present new experimental evidence, and was "at best an interesting suggestion." Eventually his ideas were published in 1959 (*21, 32*).

Even greater controversy surrounded the interpretation of cyclooctatetraene (COT) complexes. Once more, an isolated crystal structure by the Colonel turned out to be the right compound at the right time. Please recall that this was a more significant achievement 30 years ago than today. Today one can pay a commercial service to do a structure, and large drug companies find X-ray cheaper than more sporting methods of identifying compounds. However, the early crystal structure papers each resulted from a year or more of effort. One had to choose problems more carefully then than now. So the decision to have Brian Dickens work on $Fe(CO)_3(COT)$ was a major one. The complex was prepared in several laboratories almost simultaneously. The laboratories of Wilkinson and of Stone observed only one line in the NMR spectrum. A planar octagonal ring was favored by many workers as an explanation for this, and many rushed into print with proposals while the crystal structure was being determined. All were proven wrong when the structure turned out to be analogous to $Fe(CO)_3(1,3$-butadiene) (*33, 34*). In this paper it was proposed that the NMR spectrum could be interpreted in terms of "a dynamical effect amounting to permutation of the C atoms of the ring relative to the $Fe(CO)_3$ group..." and it was suggested that VT NMR might verify this proposal. It turned out that the activation energy was so low that the limiting spectrum was inaccessible even at $-155°$ (at the 100 MHz proton frequency then available). Eventually, it was found that the reaction occurred more slowly in the Ru analog, and that spectrum was interpreted in terms of 1,2 shifts with the help of computer simulation. Subsequently, vast numbers of intramolecular rearrangements have been studied. The idea of dynamic intramolecular rearrangements of atoms in boranes and in organometallic compounds is now fundamental to the field, and any scientist, working with any class of compounds needs to consider intramolecular dynamics as one aspect of interpretation of spectroscopic and structural information.

In 1968, although enrolled at MIT, I attended the Colonel's class at Harvard on symmetry and MOs. I recall that after a couple lectures he came to where I was sitting and said "aren't you glad you had Al Cotton's class first?" I was, because the Colonel had covered all of the one-semester MIT course in a few lectures. One day the Colonel briefly mentioned the Longuet-Higgins treatment of non-rigid groups in his class. Years later, while puzzling about Sandy's results on NMR of trischelate complexes, I remembered the Colonel's lecture. This resulted in our group-theoretical treatment of rearrangements of transition metal trischelate complexes, sorting out a field in which there had already been hundreds of papers (*35, 36*). The Colonel's formal teaching in this way directly contributed to our work on non-rigid molecules. I think it is safe to say that any compound that can be seen to have multiple isomers or conformers of closely similar energy should be expected to be undergoing rapid intramolecular rearrangements. (Unless there is a high barrier to them, or more facile intermolecular or dissociative rearrangements dominate.)

In the early 1960s, the Colonel began working on x-ray crystallography of enzymes. Studies of the large metal chelate complex carboxypeptidase (CPA) served as a predictor of the atomic detail in which one could understand the mechanism of metalloenzymes. The Colonel was not the first to apply the methods of x-ray crystallography to metalloproteins. But, he was the first to combine the structural insights from x-ray with the mechanistic insights that come from quantum-mechanical interpretations of reactivity (an outgrowth of the boron hydride work) to discuss in detail the way metals effect biological transformations. The first X-ray study of CPA published showed the location of the Pb atoms in the heavy atom derivative. It took several years of work with many coworkers to get down to 2 A resolution. After about two decades of work in the Colonel's lab, combining theory and crystallography, as he had done with boron hydrides, more was known about this enzyme than any other. The purpose of the study was to understand mechanism, not just to see the structure. The importance of the CPA work, from an inorganic perspective is not merely the elucidation of the mechanism of CPA per se. The real importance is that some metalloprotein has been studied with the same tools, and at almost the same level of detail, as had previously been characteristic of small molecules. Indeed, we now can see that CPA is just a big special-purpose ligand to $Zn(II)$, $Co(II)$, etc. Nature designed this ligand for a special function, just as we design small ligands to have special properties. The refinement of the structure-function of CPA poses a challenge to inorganic chemists to understand the ligand environment of various metals. Once bioinorganic chemistry was mimicry, the model, or analog, of the "real" system. Now we can see there is a true two-way street, with insights coming from the study of small molecules and of bio-molecules each enriching the other.

The structure determination of carboxypeptidase will go down in history as a milestone achievement for still another reason. All other protein models of that era – myoglobin, lysozyme, ribonuclease, papain, chymotrypsin, staphyl-ococcal nuclease – were constructed by fitting the independently determined amino acid sequence to the electron density. An excellent 2.8 Å map became available in mid-1966, but the complete sequence did not become available until 1969. In the interim, the structure was refined to 2.0 Å and the sequence became apparent in the electron density. Attention was then focused on understanding the mechanism by a combination of structural studies of enzyme-inhibitor complexes and theoretical analysis. A rather complete understanding of the mechanism was presented in 1989 (*37*). Of the moieties originally considered as probable proton donors – H_2O, Tyr248, and Glu270 (*38*), Tyr248 was eliminated by the mutation of Tyr248 to Phe in Rutter's lab in 1985.

In the meantime, a series of other protein structures were being determined – concanavalin A, aspartate transcarbamylase, fructose 1,6-biphosphatase, leucine aminopeptidase, and two different chorismate mutases. In each case what emerged was not a single structure from a single source, but multiple structures, including mutants, with and without ligands (inhibitors and substrate analogs, activators and inhibitors of allosteric control) from multiple sources. These structures were accompanied by molecular orbital and other theoretical calculations aimed at understanding the mechanisms involved from first principles, and in detail. Aspartate transcarbamylase became the foremost example of such studies. In 1968, Wiley and Lipscomb discovered (simultaneously with Klaus Weber) the hexameric character of the molecule, consisting of 6 catalytic and 6 regulatory units. The experts previously had declared it to be a tetramer. I happened to be in the lab at the time, and vividly recall Don's excitement. I also recall teasing the Colonel that if he published in Nature (*39*), it would be published so fast that there would not be time for second thoughts. Over a twenty-year period, the structures of the enzyme from E. coli, several mutants and several complexes were solved at 2.0 to 2.5 Å resolution. Using site-specific mutagenesis, it was shown that Arg54 in the active site of ATCase from E. coli was essential for catalysis, and molecular modeling studies by Gouaux, Krause and Lipscomb led to a clear concept of the catalytic mechanism of the enzyme by 1987 (*40*). The allosteric regulatory mechanism presented a more formidable challenge.

As the mechanism of allostery became the main focus of Lipscomb's interests in protein structure and function in the 1990s, again the comparative approach was brought to bear on the problem. The remarkable fact, first demonstrated in the case of hemoglobin (initially by Felix Haurowitz and ultimately in detail by Max Perutz), that allosteric enzymes underwent a dramatic change in quaternary structure in the transition from an active to an inactive state, was shown to be general. In all cases studied, including fructose 1,6-biphosphatase, aspartate transcarbamylase, and chorismate mutase, one finds a rotation of the subunits relative to each other. An explanation, or rather a parallel, is in the change in intersubunit contacts. An insightful summary of known cases, including hemoglobin (Perutz), glycogen phosphorylase (Johnson), phosphofructokinase (Evans), lactose dehydrogenase (Iwata), and others, was published by Lipscomb in 1998 (41), along with a revised version of his 13 "principles for allosteric behavior." The first of these principles is that "enzymes rely on oligomerization in which the subunits undergo relative rotational and translational shifts in order to achieve either homotropic (substrate only) or heterotropic regulation." Rotations ranging from $3.8°$ to $17°$ have been observed. A particularly striking structural change is seen in aspartate transcarbamylase, involving a translocation as well as a rotation of both the catalytic and regulatory subunits. Three of the catalytic subunits rotate by $12°$ relative to the other three, and all three of the R2 regulatory units rotate by $15°$. In addition, there is an unusually large increase of 11 Å in the distance between the catalytic units (42). Most distance changes observed in other proteins were of the order of 1 Å. Other key principles of allosteric transitions are "The symmetry before and after the allosteric transition is largely preserved (except for partial occupancy by ligands)" and "The quaternary transition changes the affinity to the substrate."

The ultimate detailed mechanism by which information is transmitted through the tertiary structure from the allosteric site to the intersubunit contacts and effect the active site has, of course, thus far proven to be elusive. This is explicitly acknowledged in one of Lipscomb's principles: "Information transfer over long distances requires organized secondary and tertiary structure." Many more studies are required to provide structural mechanisms and relate them to the biochemistry and molecular biology of these systems. In light of this principle, some of the remaining principles are likely to undergo modification. The simple push-pull mechanism originally proposed by Perutz for hemoglobin is ruled out by the fact that the quaternary structure changes are not accompanied by any significant detectable changes in the tertiary structure. In the latest discussion of this problem, Lipscomb suggested that there is no unique mechanism, or pathway, of transmission. Instead there may be multiple pathways, mobilized depending in part on the dynamic state of the protein at the time of ligand binding. There is no evidence for this hypothesis, but if the evidence were to be found over the next decades, it would not be the first time that an unprovable Lipscomb idea were to be proven right in the long run.

Interspersed with the massive systematic attack on the major problems of structural biochemistry (the relation of structure and function, the mechanisms of enzyme catalysis and allosteric control) one finds truly pioneering theoretical investigations of mechanisms of catalysis using molecular orbital theory. Examples include the charge relay system and tetrahedral intermediate in serine proteases (1975) and substrate and inhibitor binding to carbonic anhydrase. Very recently, interesting contributions to active site mechanisms have come from his use of molecular mechanics (MM) and quantum mechanical molecular mechanics (QM/MM), especially in the case of chorismate mutase.

Examples were given above of Lipscomb's enjoyment of aesthetics, and especially humor, in science. In this spirit is Lipscomb's 1971 report (*43*) of the discovery of a new enzyme, ultimately named autoinvertase, in which the reader is led from an initial, seemingly plausible finding, through a series of increasingly implausible arguments, supported by nonevidence in the form of a blank figure to the, by then obvious, conclusion that exposure to the enzyme, which inverts l- and d-amino acids in peptide chains, leads the investigators to drive on the left, rather than the right, side of the road. This was apparently observed in Central Europe, and the study remained incomplete in failing to state what was happening to British drivers.

Lipscomb was among the first chemists to recognize the importance of NMR, and probably the first to anticipate its importance in biological research. When a Minnesota graduate student (Oleg Jardetzky) came to him in 1955 with a thesis proposal to investigate active transport in biological membranes using statistical mechanics, he wrinkled his nose and said "there is this new technique NMR that might bring much more information on the subject than statistical mechanics. Why don't you look into it and talk to John Wertz, who has just written a review on it?" This proved to be the beginning of biological applications of NMR. The first papers acknowledged Lipscomb's contribution, but it was typical of him not to claim credit even for important suggestions if he did not work on it himself. Linus Pauling one said that the secret of success is to have many ideas and throw away the bad ones. Lipscomb could be said to have gone one step further, tossing away or giving away even some of the good ones. The contribution that this kind of natural generosity makes to the progress of science, and perhaps even more important to the spirit in which science is done, should never be underestimated.

When the Colonel was asked about teaching and research some time ago he replied "They are one and the same. When I do research, I get materials for my lectures, and when I teach I get ideas for my research." Many of us have also learned from the Colonel how much one can teach while seeming to merely inquire about the day's research progress. Certainly, his research provided the rest of us with material for our teaching, and in many areas. His work on boranes and on carboxypeptidase are the standard textbook treatments already. He used group theory and symmetry extensively in his research., and his Harvard course on Molecular Symmetry and Molecular Orbitals provided a perspective not available from any of the standard texts. Writing - of papers, of books, of letters of opinion in scientific journals - is part of research and an extension of teaching beyond the local classroom. The quantity the Colonel has written is remarkable. Also notable is the quality, not only of the work reported but of the writing per se. His papers convey the nature of the conduct of the research, including dramatically spilled samples. His papers convey excitement about ideas. His papers dare to predict - and he usually turns out to be right. The book "Boron Hydrides" is the classic in the field. Though published many years ago, it remains a fountain of ideas for a field continually enriched by his contributions.

Less obvious to many people - and notable because his efforts were not obvious - is the Colonel's role as an administrator. A research effort of the size and vitality of his is a major administrative task by itself. Even as a relatively junior faculty member at Minnesota, he was chief of the Physical Chemistry Division. At Harvard, he took his turn as Chairman of the Chemistry Department. His coworkers have been impressed with the facility with which he handled these tasks while keeping his research going. He did the jobs well, though not with relish. He certainly did not become a fatality of his administrative responsibilities.

The foremost judgment to express about the Colonel is the vitality of his work. Equally important as an overall lesson is the fact that his work defies conventional labels. We live in a time when funding patterns force many researchers to try to package their work so that it fits a heart and lung program, or a chemical dynamics program, etc. We live in a time when some famous scientists build their careers on adding a hyphen between the names of two techniques, or an adjective to a branch of chemistry. It is a time when we label people as organic chemists, or analytical chemists, etc. A research proposal which is analytical in nature can be rejected because you are not a "card-carrying" analytical chemist. Maybe the title of inorganic chemist is best because it encompasses everything in the periodic table. In this social environment of modern science it is well to reflect that in this book we celebrate a career that cannot be compartmentalized neatly. Indeed, when the first reports of the Nobel Prize came without any details, my colleagues immediately asked "which area of his work did they award it for?"

I am not a sociologist of science, but I will venture a bit into that territory to provide a perspective on this discussion of the Colonel's work. There are three types of leadership contributions a scientist can make: first discovery, correct interpretation, and building a field. Some scientists are well-known for taking a discovery (maybe not their own) and building a field. Almost every subdiscipline of science that is known by a label or a set of initials has a leader who named and built the field. This type of leadership is important. It creates new paradigms, it focuses federal funding, it trains a set of clones of the leader, for good or ill. The Colonel has not been a leader in this sense. We should also note that most awards go to leaders of this type. The Colonel was probably the most under-rewarded person in academic science until the Nobel Prize made the other un-received awards irrelevant. The Colonel has made a lot of discoveries, and has given a lot of correct interpretations. He has emphasized that it is possible to be both first and right. Many times, though, he has failed to foresee the importance others would attach to discoveries and has not followed up with the papers that would establish the paradigm. Sometimes he has not published at all, thinking the result was trivial. More importantly, he has often predicted results long before their discovery in the laboratory.

The diversity of topics to which the Colonel applied his mind are reflected in the diversity of topics that currently engage the minds of his former coworkers, as reflected in this book. From another perspective, the wide diversity of papers the Colonel has published has another common thread - the lesson that theory can guide experiment that will enrich theory. Also, I recall him teaching emphatically that you can be both first and right.

COLONEL LIPSCOMB

The blackboard in the Colonel's office was always cluttered; there was always room on this blackboard for at least one more idea. (This photograph, provided by Lipscomb, is reproduced by permission of the University of Minnesota Archives.)

This may be the longest introduction ever given for a person who needs no introduction (and I left out a lot of important topics!). Nevertheless, it may have been of some value to those who know Lipscomb the boron hydride chemist, to introduce Lipscomb the protein crystallographer, and to the protein chemists to introduce Lipscomb the chemical bonding theoretician. But most importantly, it was the Colonel who introduced many of us to the excitement of science. His work has ranged over much of the landscape of chemistry, and he has seen horizons others have not seen.

Acknowledgments

I thank the Colonel for providing copies of the referee comments cited in this introduction, and guiding me to some of the quotations in papers cited. The section on proteins is due to the generous assistance of Oleg Jardetzky, and much of the wording is his. Russell Grimes caught an error in one of the boron hydride formulae. The "Genealogy Database" maintained by Vera V. Mainz and Gregory S. Girolami was the reference for the full names and dates in the genealogy at the beginning of this introduction. Professor Sandra Eaton helped bring this to publishable form, just as she did the book on NMR of boron hydrides long ago.

References

1. Lipscomb, W. N.; Baker, R. H., The Identification of Alcohols in Aqueous Solution. *J. Amer. Chem. Soc.* **1942**, *64*, 179-180.
2. Lipscomb, W. N. Experimental Crystallography. *Ann. Rev. Phys. Chem.* **1953**, *4*, 253-266.
3. Lipscomb, W. N., Low Temperature Crystallography. *Norelco Reporter* **1957**, *IV*, No. 3, p.54ff.
4. Streib, W. E.; Lipscomb, W. N., Growth, Orientation, and X-ray Diffraction of Single Crystals Near Liquid Helium Temperatures. *Proc. Natl. Acad. Sci. USA* **1962**, *48*, 911-913.
5. Jordan, T. H.; Smith, H. W.; Streib, W. E.; Lipscomb, W. N., Single-Crystal X-Ray Diffraction Studies of α-N_2 and β-N_2. *J. Chem. Phys.* **1964**, *41*, 756-759.
6. Jordan, T. H.; Streib, W. E.; Lipscomb, W. N., Single-Crystal X-Ray Diffraction Study of β-Fluorine. *J. Chem. Phys.* **1964**, *41*, 760-764.
7. Smith, H. W.; Lipscomb, W. N., Single-Crystal X-Ray Diffraction Study of β-Diborane. *J. Chem. Phys.* **1965**, *43*, 1060-1064.
8. Rossmann, M. G.; Jacobson, R. A.; Hirshfeld, F. L.;. Lipscomb, W. N, An Account of Some Computing Experiences. *Acta Cryst.* **1959**, *12*, 530-535.
9. Steitz, T. A.; Lipscomb, W. N., Molecular Structure of Methyl Ethylene Phosphate. *J. Am. Chem. Soc.* **1965**, *87*, 2488-2489

10. Howell, P. A.; Curtis, R. M.; Lipscomb, W. N. The Crystal and Molecular Structure of 1,4-Dithiadiene. *Acta Cryst.* **1954**, *7*, 498-503.
11. Lipscomb, W. N., Structures of the Boron Hydrides. *J. Chem. Phys.* **1954**, *22*, 985-988.
12. Eberhardt, W. H.; Crawford, B.; Lipscomb, W. N., The Valence Structure of the Boron Hydrides. *J. Chem. Phys.* **1954**, *22*, 989-1001
13. Lipscomb, W. N., Possible Boron Hydride Ions. *J. Phys. Chem.* **1958**, *62*, 381-382.
14. Lipscomb, W. N.; Pitochelli, A. R.; Hawthorne, M. F., The Probable Structure of the $B_{10}H_{10}^{2-}$ Ion. *J. Am. Chem. Soc.* **1959**, *81*, 5833-5834.
15. Eaton, G. R.; Lipscomb, W. N., *NMR Studies of Boron Hydrides and Related Compounds*, W. A. Benjamin, Inc., New York, NY, 1969, pp. 216-225.
16. Dobrott, R. D.; Lipscomb, W. N., Structure of $Cu_2B_{10}H_{10}$. *J. Chem. Phys.* **1962**, *37*, 1779-1784.
17. Laws, E. A.; Stevens, R. M.; Lipscomb, W. N., A Self-Consistent Field Study of Decaborane(14). *J. Am. Chem. Soc.* **1972**, *94*, 4467-4474.
18. van der Maas Reddy, J.; Lipscomb, W. N., Molecular Structure of $B_{10}H_{12}(CH_3CN)_2$. *J. Chem. Phys.* **1959**, *31*, 610-616.
19. Schwalbe, C. H.; Lipscomb, W. N., Crystal Structure of Bis(triethylammonium) Octadecahydroicosaborate. The Structure of Octadecahydroicosaborate(2-). *Inorg. Chem.* **1971**, *10*, 151-160.
20. Dickerson, R. E.; Wheatley, P. J.; Howell, P. A.; Lipscomb, W. N.; Schaeffer, R., Boron Arrangement in a B_9 Hydride. *J. Chem. Phys.* **1956**, *25*, 606-607.
21. Lipscomb, W. N., Recent Studies of the Boron Hydrides. *Adv. Inorg. Radiochem.* **1959**, *1*, 117-156.
22. Lipscomb, W. N., review of "Spectacular Experiments and Inspired Quotes. Chemical Curiosities," by H. W. Roesky and K. Mockel. *Angew. Chem. Intl. Ed. Engl.* **1997**, *36*, 169.
23. Grimes, R.; Wang, F. E.; Lewin, R.; Lipscomb, W. N., A New Type of Boron Hydride, $B_{10}H_{16}$. *Proc. Natl. Acad. Sci. USA* **1961**, *47*, 996-999.
24. Friedman, L. B.; Lipscomb, W. N., Crystal and Molecular Structure of $(CH_3)_2B_5H_7$. *Inorg. Chem.* **1966**, *5*, 1752-1757.
25. Johansson, G.; Lipscomb, W. N., Structure of Roussin's Black Salt. *J. Chem. Phys.* **1957**, *27*, 1417.
26. Jorgensen, W. L., Free Energy Calculations: A Breakthrough for Modeling Organic Chemistry in Solution. *Acc. Chem. Res.* **1989**, *22*, 184-189.
27. Lipscomb, W. N., Solutions of Sodium in Liquid Ammonia. *J. Chem. Phys.* **1953**, *21*, 52-54.
28. Lipscomb, W. N., Tetrahedral "Cyclobutadiene"? *Tetrahedron Lett.* **1959**, No. 18, 20-23.

29. Lohr, L. L., Jr.; Lipscomb, W. N., Molecular Symmetry of XeF$_2$ and XeF$_4$. *J. Amer. Chem. Soc.* **1963**, *85*, 240-241.
30. Rây, P.; Dutt, N. K., Kinetics and Mechanism of racemization of Optically Active Cobaltic Trisbiguanide complex. *J. Indian Chem. Soc.* **1943**, *20*, 81-92.
31. Bailar, J. C., Jr., Problems in the stereochemistry of coordination compounds. *J. Inorg. Nucl. Chem.* **1958**, *8*, 165-175.
32. Lipscomb, W. N., Structure and Reactions of the Boron Hydrides, *J. Inorg. Nucl. Chem.* **1959**, *11*, 1-8
33. Dickens, B.; Lipscomb, W. N., Structure of (OC)$_3$Fe(C$_8$H$_8$)Fe(CO)$_3$. *J. Amer. Chem. Soc.* **1961**, *83*, 489-490.
34. Dickens, B.; Lipscomb, W. N., Molecular Structure of C$_8$H$_8$Fe(CO)$_3$. *J. Amer. Chem. Soc.* **1961**, *83*, 4862-4863.
35. Eaton, S. S.; Eaton, G. R.; Holm, R. H.; Muetterties, E. L., Intramolecular Rearrangement Reactions of Tris-Chelate Complexes. IV. Further Investigation of the Rearrangements of Tris (α-isopropenyl- and α-isopropyl-tropolonato)metal(III,IV) Complexes *J. Amer. Chem. Soc.* **1973**, *95*, 1116-1124.
36. Eaton, S. S.; Eaton, G. R., Symmetry Groups of Nonrigid Tris-Chelate Complexes. *J. Amer. Chem. Soc.* **1973**, *95*, 1825-1829.
37. D. W. Christianson and W. N. Lipscomb, Carboxypeptidase A, *Accounts Chem. Res.* **1989**, *22*, 62-69.
38. D. C. Rees and W. N. Lipscomb, Refined crystal structure of the potato inhibitor complex of carboxypeptidase A at 2.5 Å resolution. *J. Mol. Biol.* **1982**, *160*, 475-498.
39. D. C. Wiley and W. N. Lipscomb, Crystallographic determination of the symmetry of aspartate transcarbamylase: Studies of trigonal and tetragonal crystalline forms of aspartate transcarbamylase show that the molecule has a three-fold and a two-fold symmetry axis. *Nature* **1968**, *218*, 1119-1121.
40. J. E. Gouaux, K. L. Krause, and W. N. Lipscomb, The catalytic mechanism of Esherichia coli aspartate transcarbamylase: A molecular modeling study. *Biochem. Biophys. Res. Commun.* **1987**, *142*, 893-897.
41. W. N. Lipscomb, Multisubunit allosteric proteins. *NATO ASI Ser., Ser. A*, **1998**, *301* (Protein Dynamics, Function, and Design, ed. by O. Jartetzky, J.-F. Lefèvre and R. E. Holbrook, Plenum Press, New York), 27-35.
42. E. R. Kantrowitz and W. N. Lipscomb, Escherichia coli Aspartate transcarbamylase: The molecular basis for a concerted allosteric transition. *Trends Biochem. Sci.* **1990**, *15*, 53-59.
43. W. N. Lipscomb, A New Enzyme (Spoof). *Nachrichten aus Chemie und Technik* **1971**, *19*, 122

Inorganic Chemistry

Chapter 2

Thomas Jefferson, Alice in Wonderland, Polyhedral Boranes, and the Lipscomb Legacy

Russell N. Grimes

Department of Chemistry, University of Virginia, Charlottesville, VA 22901

The impact of the studies of William N. Lipscomb and his collaborators on the development of polyhedral borane chemistry, and more broadly on the general field of polyhedral clusters, is reviewed. After a brief recapitulation of the state of knowledge of borane structures and bonding at the start of Lipscomb's career in the 1940s, the influence – direct and indirect – of his work on the shaping of this area, and on the development of cluster science into its current multifaceted form, is explored with the aid of selected examples. The role of scientific imagination in chemistry, as exemplified by minds as diverse as those of Jefferson, Lewis Carroll, and Lipscomb, is highlighted.

"The time has come, the Walrus said,
To talk of many things:
Of shoes, and ships, and sealing wax,
Of cabbages, and kings,
And why the sea is boiling hot,
And whether pigs have wings"

-- Lewis Carroll, *The Walrus and the Carpenter* (1)

Introduction

The remarkable scientific career of William N. Lipscomb (named a Kentucky Colonel by Governor Happy Chandler years ago) has spanned more than half a century, during which most areas of chemistry have experienced revolutionary change, but none more so than the study of polyhedral boranes, metal clusters, and related cagelike molecules. One need only look at the general state of knowledge of this field of study in the late 1940s when the Colonel as an Assistant Professor began his investigations of the boron hydrides, to realize the astonishing advances that have been recorded, including the opening up of entire subfields and the development of borane-based applications that were then scarcely imaginable. This has been an achievement of many groups and individuals, including several of the most prominent synthetic and theoretical chemists of the 20th century, but no one has had greater impact from the beginning than Lipscomb. His contributions to this revolution have been documented many times, especially in the wake of his Nobel award in 1976 (2). The theme I wish to develop in this paper is somewhat broader and focuses on the impact of his work in the general area of polyhedral cluster chemistry, and especially on his role in expanding general perceptions of covalent bonding beyond the traditional Lewis-type model.

The title of this paper underlines the vital role of *imagination* in scientific discovery. Anyone who has interacted with the Colonel, who has read the works of Lewis Carroll (Charles Lutwidge Dodgson, a mathematics professor at Oxford and the author of Alice in Wonderland), and who has some familiarity with the writings of Thomas Jefferson, knows that two attributes shared by these men are a towering intelligence and an active imagination – precisely the qualities that are most often cited as the main requirements for success in math and science. But there are more connections than that. The Colonel has respected Carroll enough to quote him in papers on bonding theory (3). And his week in residence in Mr. Jefferson's University in Charlottesville some 20 years ago, complete with a rare visit to the Dome Room in Monticello -- an area normally inaccessible to tourists -- still reverberates in my memory.

In this short paper I will attempt to illustrate how fundamentally the knowledge, understanding, and application of borane structures and covalent bonding has changed in the course of the Colonel's career over the last half-century, and to suggest how much of modern cluster science has his fingerprints on it.

Borane Chemistry in the 1930s and '40s

Alfred Stock, a great experimental genius, pioneered the synthesis, isolation, and characterization of the lower boron hydrides between 1912 and 1936. His research began under the most difficult circumstances imaginable: almost everything that had been reported in the scientific literature on these compounds, *including their elemental formulas*, was wrong! Using evacuated glass vacuum apparatus of his own design, equipped with mercury float valves (another Stock invention) in lieu of greased stopcocks, and working with dangerously pyrophoric volatile materials, Stock and his colleagues proved that the substances previously reported as BH_3 and B_3H_3 had been incorrectly formulated (the former was shown to be B_2H_6 and the latter apparently a mixture). They prepared and characterized, via elemental analysis and non-spectroscopic physical properties, the hydrides B_2H_6, B_4H_{10}, B_5H_9, B_5H_{11}, B_6H_{10}, and $B_{10}H_{14}$, of which all except the last were gases or volatile liquids (4). Overcoming severe bouts of sickness that he himself eventually determined to be caused by exposure to mercury in his laboratory, Stock meticulously analyzed, weighed, and studied the reactivity of all of these then-novel species and set a standard for synthetic achievement that has rarely been matched in any area of chemistry.

Despite these spectacular successes, the molecular structures of the boranes were entirely unknown and basically unknowable at that time. Lacking modern tools such as NMR and single-crystal X-ray diffraction techniques (5), and with ideas of covalence in that era dominated by Lewis-style electron pair bonding as found in organic compounds, Stock and his contemporaries could only speculate about structure. For lack of a better idea, they assumed that the boron hydrides, notwithstanding an apparent deficiency of electrons, must adopt hydrocarbon-like chain structures such as the examples shown in Chart 1, for which both non-ionic and ionic models were suggested (4).

Chart 1. Borane structures proposed by Alfred Stock, 1930s

B_2H_6

$$H\cdots\underset{\underset{H}{|}}{\overset{\overset{H}{|}}{B}}=\underset{\underset{H}{|}}{\overset{\overset{H}{|}}{B}}\cdots H \quad \text{or} \quad \left[\underset{\underset{H}{|}}{\overset{\overset{H}{|}}{B}}=\underset{\underset{H}{|}}{\overset{\overset{H}{|}}{B}}\right]^{2-} 2H^+$$

B_4H_{10}

$$H-\underset{\underset{H}{|}}{\overset{\overset{H}{|}}{B}}=\underset{\underset{H}{|}}{\overset{\overset{H}{|}}{B}}-\underset{\underset{H}{|}}{\overset{\overset{H}{|}}{B}}=\underset{\underset{H}{|}}{\overset{\overset{H}{|}}{B}}-H \quad \text{or} \quad \left[H-\underset{}{\overset{\overset{H}{|}}{B}}=\underset{}{\overset{\overset{H}{|}}{B}}-\underset{}{\overset{\overset{H}{|}}{B}}=\underset{}{\overset{\overset{H}{|}}{B}}-H\right]^{4-} 4H^+$$

$B_{10}H_{14}$

$$\left[\overset{..}{\underset{..}{B}}=\overset{..}{\underset{..}{B}}=\overset{..}{\underset{..}{B}}=\overset{..}{\underset{..}{B}}=\overset{..}{\underset{..}{B}}=\overset{..}{\underset{..}{B}}=\overset{..}{\underset{..}{B}}=\overset{..}{\underset{..}{B}}=\overset{..}{\underset{..}{B}}=\overset{..}{\underset{..}{B}}\right]^{14-} 14H^+$$

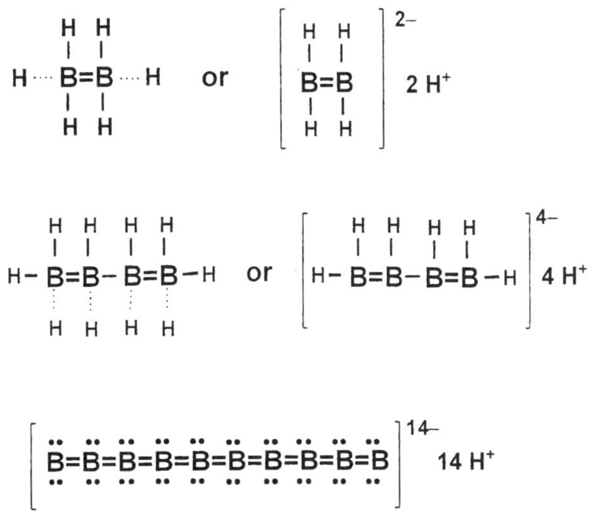

Textbooks of that period also convey the aura of mystery that surrounded these compounds, whose formulas made little sense when compared to the well-ordered world of hydrocarbons and the tetrahedral carbon atom. Thus the *entire section* on boron hydrides in my freshman college text, published in 1952 (6), consisted of four sentences: "Boron forms a number of hydrides, called boranes; this illustrates the diagonal relationship of that element to silicon ...The simplest borane, BH_3, has never been prepared. The more important boranes are diborane, B_2H_6; tetraborane, B_4H_{10}; pentaborane, B_5H_9; and hexaborane, B_6H_{12}. The molecular structures of the boranes are not definitely known." Terse as this is, it exceeds the treatment of boranes in another general chemistry text (7), which contains no mention whatever of any hydrogen compounds of boron! Somewhat better is the coverage in Moeller's advanced inorganic textbook of 1952 (8), which devotes several pages to the chemistry of the boranes but, oddly, highlights older structural proposals such as those of Pitzer (9), shown in Chart 2, that had in fact been superseded by experimental findings. The book briefly mentions X-ray and electron diffraction studies of B_5H_9 (10, 11), but makes no reference to the earlier 1948/1950 landmark X-ray crystal structure analysis of $B_{10}H_{14}$ by Kasper, Lucht, and Harker (12) that revealed its cagelike geometry. Some of the confusion about structures in that period can be attributed to electron diffraction studies of the boranes that that mistakenly indicated open chain or ringlike structures.

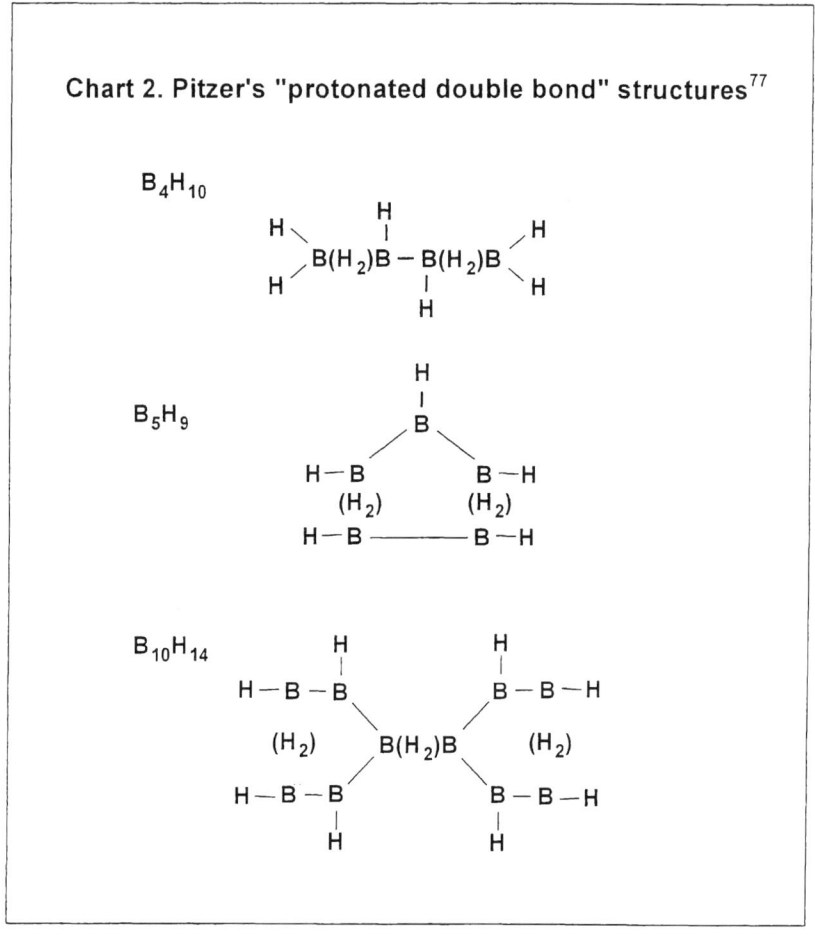

Chart 2. Pitzer's "protonated double bond" structures[77]

Such was the state of general awareness of borane structural chemistry in the early years of Lipscomb's investigations. Incorrect notions about structures and bonding seemed to persist despite experimental evidence to the contrary (e.g. the $B_{10}H_{14}$ X-ray study already mentioned) and the prediction of the icosahedral geometry of $B_{12}H_{12}^{2-}$ by Longuet-Higgins (13a), who also pioneered the concept of the 3-center B-H-B bond (13b). In fact, most chemists of the time were perplexed by the fundamental bonding problems posed by the boranes and simply ignored the issue. It must have been hard to recruit graduate students to work on the synthesis and chemistry (as opposed to crystallography and theoretical studies) of the polyboranes; in Lipscomb's laboratory, none were until 1958. Nevertheless, the Colonel and his crystallographic coworkers, working with samples of B_4H_{10}, B_5H_9, B_6H_{10}, B_5H_{11}, B_4Cl_4, and B_9H_{15} acquired from other laboratories, were able to collect X-ray diffraction data on these compounds from crystals grown in Pyrex capillary tubes that were cooled in a stream of cold nitrogen gas (3,14). This experimental work, coupled with theoretical studies that were conducted in parallel, led to the development of a general topological theory of bonding that (as all successful theories must) not only rationalized known structures but allowed predictions of new ones. The essence of this theory was the assumption that all borane frameworks are comprised of combinations of two-center B-B bonds (i.e. ordinary covalent interactions containing an electron pair shared by two atoms) and three-center B-B-B or B-H-B bonds in which three atoms are bound by a single electron pair. For any given molecule, the number of valence electrons and valence orbitals, together with restrictions on valence bond angle geometry and other considerations, expressed in the "styx rules" (15) lead to a unique structure or to a very small number of possible geometries. Refinements of the original approach (14b), as in the introduction of partial 3-center bonds and especially the application of molecular orbital methods, have added to its utility; however, the most important consequence of this theory is its utilization, direct and indirect, in a remarkably broad spectrum of applications. In the following section I outline a few selected examples.

"Error of opinion may be tolerated where reason is left free to combat it"

-- Thomas Jefferson, First Inaugural Address, 1801

"I would rather believe that Yankee professors would lie, than to believe that rocks can fall from the sky"

-- Thomas Jefferson, on reports of the finding of a meteorite in New England by two Yale professors

The Topological Theory of Boranes: Widening Impact

The influence of Lipscomb and his collaborators, including R. E. Dickerson, W. H. Eberhardt, and Bryce Crawford, and later Roald Hoffmann, on the development of borane structural theory extends well beyond their own publications, and is clearly apparent in the efforts (and the thinking) of others. A prime example is afforded by the work of Kenneth Wade (16), whose skeletal electron-counting rules, proposed thirty years ago and subsequently elaborated by Rudolph (17), Mingos (18), Williams (19) and others (20,21), are now solidly entrenched as an important tool for dealing with polyhedral boranes, metallaboranes, carboranes, and other types of metal and nonmetal clusters. In his seminal paper on this topic in 1971 (16a), Wade adopted the nido/closo/arachno classification of Williams and the Lipscomb styx rules to provide a simple correlation between cluster geometry and the number of skeletal electrons available for binding the cluster together. The number of skeletal electron pairs (sep) is equal to the total number of valence electron pairs less the pairs stored in nonbonding orbitals and those used for binding skeletal atoms to peripheral (ligand) groups. Thus, n-vertex closo polyhedra (cages having only triangular faces) normally have sep = n + 1, while higher numbers of

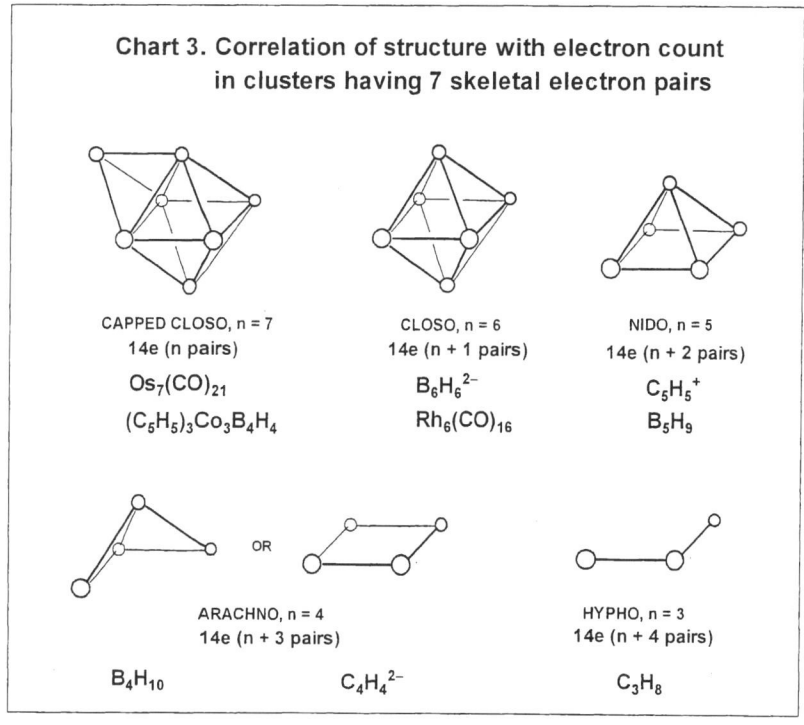

Chart 3. Correlation of structure with electron count in clusters having 7 skeletal electron pairs

CAPPED CLOSO, n = 7
14e (n pairs)
$Os_7(CO)_{21}$
$(C_5H_5)_3Co_3B_4H_4$

CLOSO, n = 6
14e (n + 1 pairs)
$B_6H_6^{2-}$
$Rh_6(CO)_{16}$

NIDO, n = 5
14e (n + 2 pairs)
$C_5H_5^+$
B_5H_9

ARACHNO, n = 4
14e (n + 3 pairs)
B_4H_{10} $C_4H_4^{2-}$

HYPHO, n = 3
14e (n + 4 pairs)
C_3H_8

electrons produce more open architectures, e.g., nido, arachno, and hypho frameworks that have sep values of $n + 2$, $n + 3$, and $n + 4$, respectively. In the other direction, clusters having sep = n usually adopt capped-closo structures. Chart 3 presents examples of molecules having sep = 7, illustrating the general utility of this approach that extends well beyond polyhedral boranes. Indeed, Wade's Rules are now standard fare in undergraduate and graduate texts and have been applied to clusters of many types including Zintl ions, carbocations, and some metal cluster systems (22); they are not, however, generally useful in dealing with high-density arrays such as those found in bulk metals and many large metal clusters.

A particularly important class of molecules to which these ideas are directly applicable is the nonclassical, or hypercoordinate, hydrocarbons (23). Chart 4 shows two series of molecules that are isoelectronic and isostructural

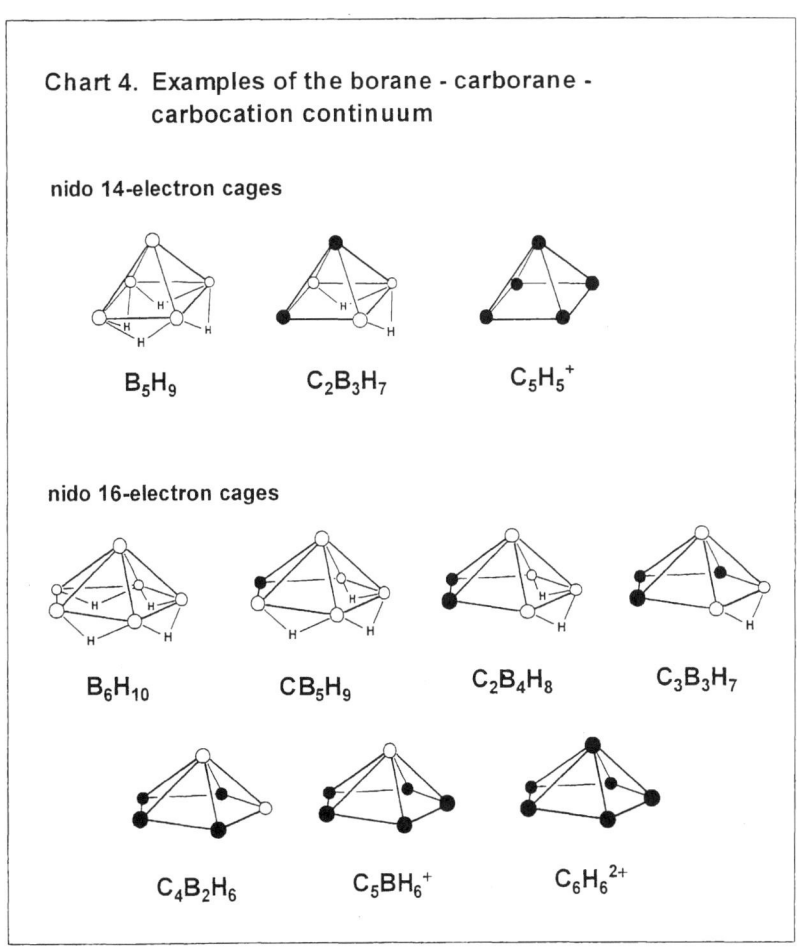

Chart 4. Examples of the borane - carborane - carbocation continuum

nido 14-electron cages

B_5H_9 $C_2B_3H_7$ $C_5H_5^+$

nido 16-electron cages

B_6H_{10} CB_5H_9 $C_2B_4H_8$ $C_3B_3H_7$

$C_4B_2H_6$ $C_5BH_6^+$ $C_6H_6^{2+}$

analogues of, respectively, B_5H_9 and B_6H_{10}, and that graphically illustrate the direct connection between boron hydrides, carboranes, and hydrocarbons. *All of the species shown are known and characterized either in parent form or as alkyl derivatives* (23). In some cases, external substituents can profoundly affect the cage structure; for example, in contrast to $C_4B_2H_6$, $H_4C_4B_2F_2$ has a classical 1,4-diborabenzene ring structure (24).

The "ripple effect" of the original Lipscombian structural ideas was soon manifested in a number of other ways. For example, the calculations of charge distribution in boranes by Eberhardt et. al. (14c), and Lipscomb's prediction that these hydrides would undergo electrophilic substitution (25), inspired an important early NMR study (26) of the electronic effects of halogen substitution in boranes by the prolific synthetic borane chemist Riley Schaeffer (a modern-day Alfred Stock whose own work considerably extended the field). The unexpected stability of mass spectroscopically-generated odd-electron borane cations observed by Greenwood and coworkers (27) were interpreted in terms of the delocalized bonding model and detailed calculations by Lipscomb (14a,28). Parry and Edwards drew on the structural and theoretical contributions of the Lipscomb laboratory to formulate their own proposals for systematizing the chemistry of the lower boranes in terms of symmetric vs. nonsymmetric cleavage (29).

In the area of synthesis, the structural and theoretical framework contributed by Lipscomb provided a foundation for the monumental synthetic work of M. F. Hawthorne and his students, who transformed boron cluster chemistry first by their synthesis of the polyhedral borane anions $B_{10}H_{10}^{2-}$ and $B_{12}H_{12}^{2-}$ (30) and later by the synthesis of metallacarboranes (31). In our own laboratory, considerations of charge distribution in borane anions (14) prompted us to explore reactions of transition metal ions with small borane and carborane anions (32). This, in turn, led to a number of advances including the first closed polyhedral metallaboranes (33), the discovery of metal-induced oxidative cage fusion (34), and the synthesis of $B_{12}H_{16}$, the only neutral B_{12} hydride, via metal-promoted fusion of $B_6H_9^-$ (35)

The development of structure-bonding theory in the boranes was accompanied in the 1960s by the publication of a monograph on NMR spectroscopy of boron hydrides by Eaton and Lipscomb (36) which not only provided a comprehensive source of data in this rapidly advancing field (37) but also was instrumental in advancing the use of NMR as a structural tool for boranes. This was certainly the case in our later development of two-dimensional correlated ^{11}B NMR spectroscopy (COSY), in which we explored both heteronuclear ^{11}B-^1H (38a) and homonuclear ^{11}B-^{11}B coupling (38b,c) in the first reported use of two-dimensional NMR involving a quadrupolar nucleus. This technique, which affords direct insight into electronic structures of borane frameworks of all kinds, allowed us to confirm experimentally some important earlier calculations of the Lipscomb group, e.g., that direct coupling between boron nuclei involved in B-H-B bridges is very small (39).

"A slow sort of country!" said the Queen. "Now here, you see, it takes all the running you can do, to keep in the same place. If you want to get somewhere else, you must run twice as fast as that!"

-- Lewis Carroll, *Alice Through the Looking-Glass* (1)

The Contemporary Scene and a Look Ahead

Fast-forwarding to the present, I have selected a few examples of current research activity in boron cluster science that in one way or another build on the fundamental discoveries and bonding models of polyhedral borane structures that originated a half century ago with Lipscomb and his compatriots. While current publications in these areas may or may not explicitly acknowledge this original work, it is in fact the foundation upon which modern studies of polyhedral boranes and related clusters rest.

Macropolyhedral Condensed Boranes and Supra-icosahedral Clusters

The intrinsic stability of the B_{12} icosahedron, manifested in various forms of elemental boron and in the $B_{12}H_{12}^{2-}$ ion, suggests that this structural principle might be extended indefinitely via synthesis of condensed structures in which icosahedral B_{12} units are fused together with some borons shared between two units. An early example is the hydride $B_{20}H_{16}$, prepared independently by the Lipscomb and Muetterties groups (40,41), which consists of two $B_{10}H_{14}$-like *nido*-B_{10} baskets joined face-to-face. Efforts by Kennedy and coworkers to develop this idea as a synthetic approach have led to the preparation of a series of high-nuclearity polyboranes and metallaboranes consisting of fused icosahedral and icosahedral-fragment units (42). Although this work has generated some quite large polycluster metallaboranes, as yet no species having a supra-icosahedral cage (vide infra) has been isolated. Another major challenge is to develop controllable, predictable synthetic pathways that would allow the construction of specific macropolyhedral target structures, as opposed to serendipitous products. In this connection, Fehlner and coworkers have made major advances in the designed synthesis of metal-rich boron clusters via stepwise insertion of monoboron units to metal cluster scaffolds (43). While their work has concentrated on smaller cage systems, the greater control afforded by this approach offers considerable promise for extension to large systems. The selection of synthetic targets in the macroborane area may be aided by a recent contribution of Balakrishnarajan and Jemmis (44), who developed a set of electron-counting rules for condensed polyboranes that represent a direct extension of the original Lipscomb and Wade formulations. In closely related work, Tillard-Charbonnel, et al. have explored condensed-icosahedral structures in intermetallic compounds, especially those of the group 12 and 13 elements (45).

A question of fundamental theoretical significance that has been probed over the years by Lipscomb and others is whether closed polyhedral boron cages exceeding 12 vertices can have a stable existence. Metallacarboranes of 13 vertices are well known (46), and a few 14-vertex complexes have been prepared (47) although the only crystallographically characterized examples are (η^5-$C_5H_5)_2Fe_2(Me_4C_4B_8H_8)$ isomers (47c,d). Nevertheless, despite calculations (48) suggesting the possible existence of *closo*-polyhedra as large as 30 vertices (or more), no all-boron cluster larger than icosahedral $B_{12}H_{12}^{2-}$ has been isolated, and this remains a formidable challenge for synthetic borane chemists. Interesting hints that the so-called "icosahedral barrier" might yet be overcome are, however, to be found. For example, the cobaltacarborane **A** shown in Chart 5 contains a very open 12-vertex $Co_2C_4B_6$ framework (49) that is, in fact, a

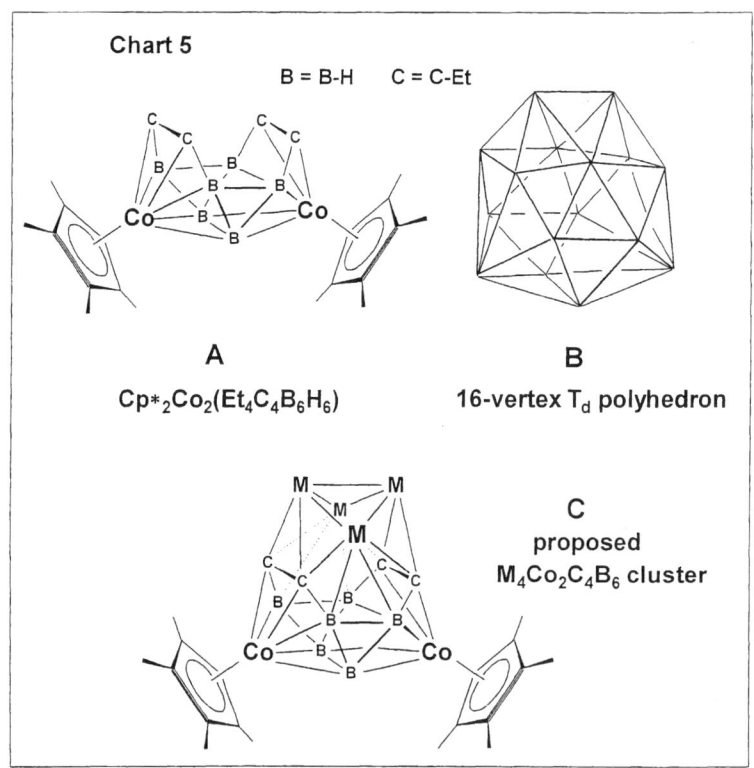

Chart 5

B = B-H   C = C-Et

A
Cp*$_2$Co$_2$(Et$_4$C$_4$B$_6$H$_6$)

B
16-vertex T$_d$ polyhedron

C
proposed M$_4$Co$_2$C$_4$B$_6$ cluster

fragment of a hypothetical 16-vertex polyhedron (B) of T_d symmetry — the same structure predicted by Bicerano, et al. (48c) for a $B_{16}H_{16}$ or $B_{16}H_{16}^{2-}$ hydride. No such borane has been made, although an In_{16} cluster having this same geometry is present in the solid-state structure of $Na_7In_{11.8}$ (50). It may be possible to construct 16-vertex metallacarboranes (C) via addition of four metal vertices to the $Co_2C_4B_6$ cage shown (which itself is in gross "violation" of Wade's electron-counting rules (51)).

Polyhedral Borane Frameworks in Organic Chemistry

The analogy between electron-delocalized borane polyhedra and aromatic hydrocarbons has long been recognized, and in particular it is notable that icosahedral clusters such as $B_{12}H_{12}^{2-}$ and $C_2B_{10}H_{12}$ occupy about the same volume as a benzene ring spinning on one of its 2-fold axes; these polyhedral cages with their 26 skeletal electrons thus truly are three-dimensional aromatic systems. The combination of extremely high thermodynamic stability and structural rigidity makes the icosahedral carboranes very attractive candidates as scaffolds or templates on which to build architecturally novel organic systems and reagents. In effect, such chemistry brings polyhedral boranes into the fold of organic synthesis and thereby creates a new borderline area where myriad new

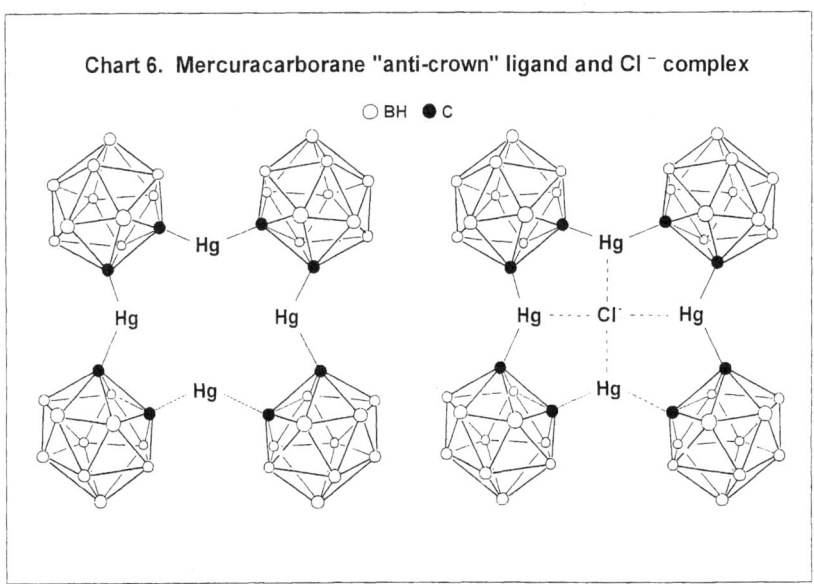

Chart 6. Mercuracarborane "anti-crown" ligand and Cl^- complex

possibilities are opened for exploitation. A major contributor in this area is M. F. Hawthorne, who has demonstrated the powerful potential that carboranes offer as "organic" building-block units (53). Many different examples can be given, but one or two will have to suffice here. The special steric properties of 1,2-$C_2B_{10}H_{12}$ (ortho-carborane) have been used to create a family of "anticrowns" such as $Hg_4(C_2B_{10}H_{10})_4$ (Chart 6) that function as Lewis acids (in contest to familiar crown ethers, which are basic), and hence can be employed to trap halide and other anions (54). In a different application, the 1,12-$C_2B_{10}H_{12}$ (para-carborane) nucleus has been used by two different groups to create rigid linear oligomers or "carborods" [Chart 7(**A**)] that may serve as building block units for rigid rod polymers (55,56). Another concept under current exploration is that of peralkylated "camouflaged" carboranes, whose cage surfaces are protected by a sheath of alkyl groups and are thus effectively spherical hydrocarbons [Chart 7(**C**)] (57). The cyclic macromolecule **B** (58) affords an interesting example of the use of the 1,7-C_2B_{10} (*meta*-carborane) cage geometry in construction.

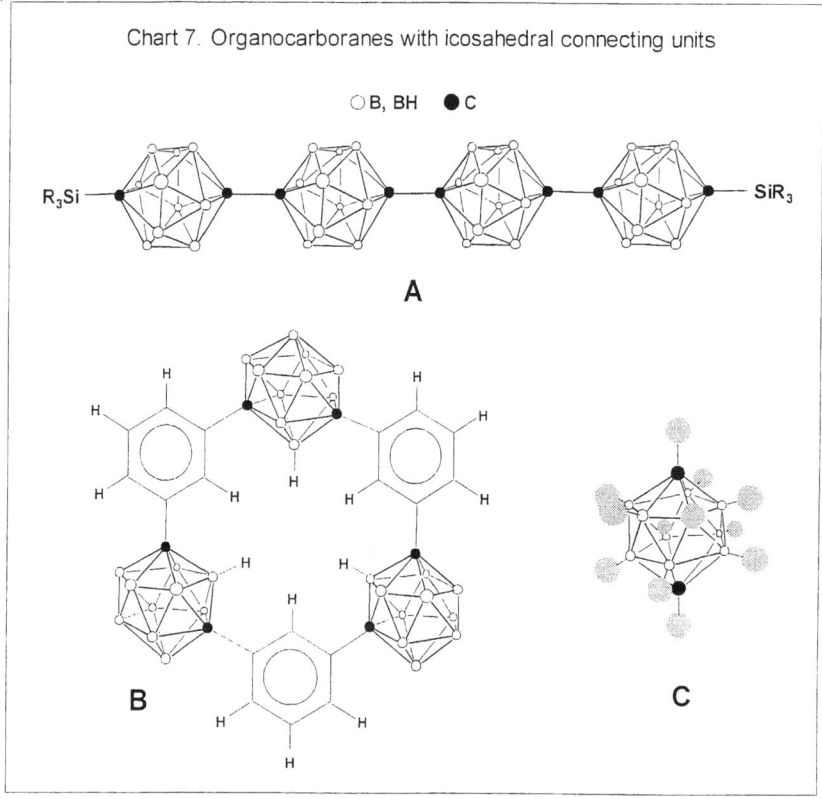

Chart 7. Organocarboranes with icosahedral connecting units

In a kind of inversion of this concept, hydrocarbons can also serve as platforms for polycluster systems: the tri- and tetrametallic complexes shown in Chart 8 (**A** and **B**) were recently prepared in our laboratory and characterized via X-ray crystallography at the University of Heidelberg (59). Compounds of this type are prototypes for extended oligomeric and polymeric networks such as that proposed in Chart 8(**C**), that are expected to exhibit extensive electron-delocalization between the metal centers (60). Systems of this kind, together with alkynyl-and phenyl-linked polyhedra, and multidecker sandwich-based arrays, are current targets of our research (61).

Chart 8. Benzene-Anchored Polynuclear Metallacarboranes

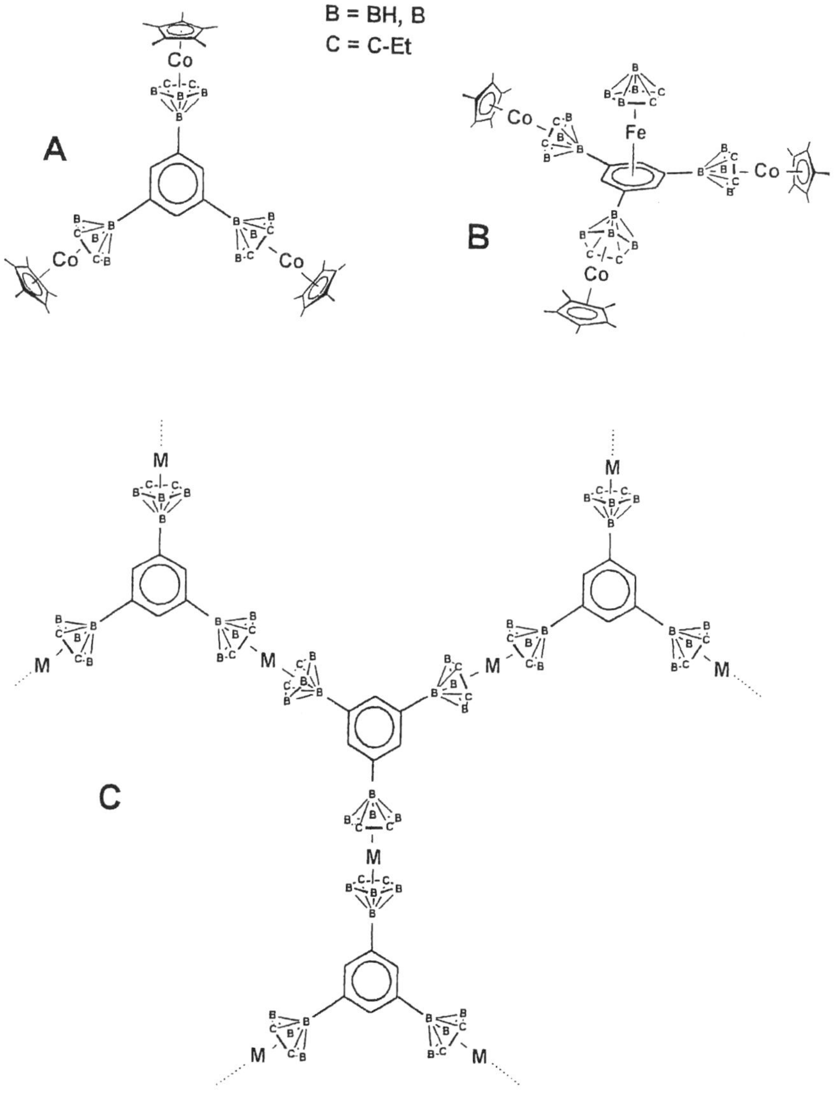

If a nation expects to be ignorant and free, in a state of civilization, it expects what never was and never will be.

--Thomas Jefferson, letter to Colonel Charles Yancey, 1816

On the Borane-Hydrocarbon Border

One particularly interesting facet of carborane structures is their hybrid character, which is most evident in the more carbon-rich species such as those shown in Chart 9 (62-64). Given their intermediate position between the classical hydrocarbons and the polyhedral boranes, these compounds can be expected to (and do) reflect structural characteristics of both classes (23) and are of special interest in theoretical studies of covalent bonding. Here again, Lipscomb's contributions over several decades have markedly influenced the development of

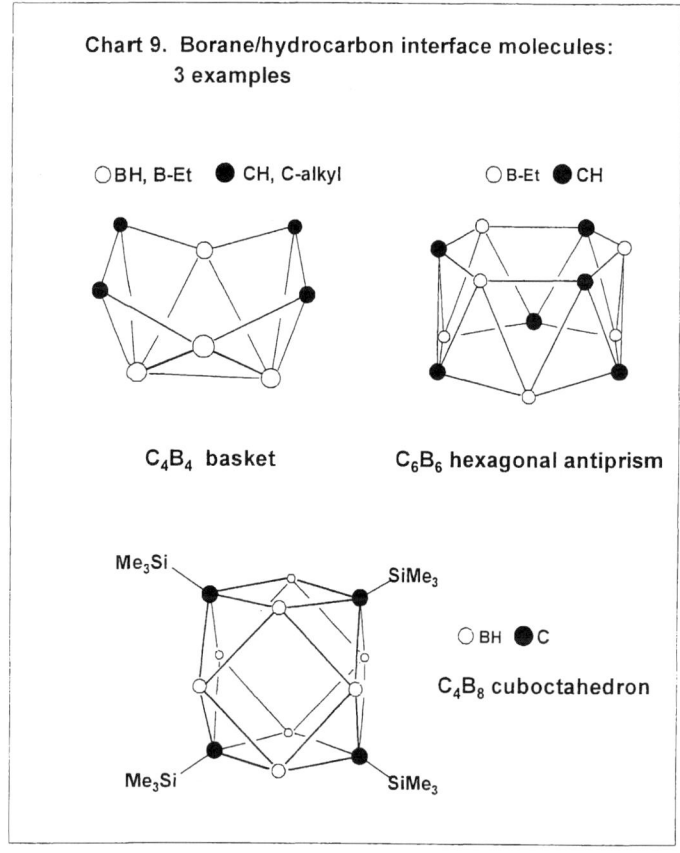

this area. Although there are no general methods for preparing such compounds, the three examples selected here have turned up in serendipitous syntheses under a variety of conditions. In the case of $C_6B_6H_{12}$, it is noteworthy that although the hexa-B-ethyl derivative has the drum shape shown in Chart 9 (63a), an X-ray study of an isomer of the unsubstituted molecule revealed an ethylene-bridged C_4B_6 basket (63b).

The C_4B_8 carboranes furnish an interesting class of organic-inorganic hybrids. Originally prepared accidentally in our laboratory via the transition metal-induced oxidative fusion of $R_2C_2B_4H_4^{2-}$ ligands (34,65), they are 28-electron, 12-vertex (sep = n + 2) systems having two electrons in excess of the 26 required for a closed icosahedron. The resultant cage-opening, however, takes many forms and varies with the choice of R and other factors; moreover, when R is methyl, ethyl, or propyl, the cage framework is fluxional in solution, equilibrating between two different geometries (34). The cuboctahedral tetrakis(C-trimethylsilyl) derivative, whose X-ray structure determination was recently reported by reported by Hosmane, Lipscomb, and their coworkers (Chart 9, bottom) (64), is of special interest in light of Lipscomb's decades-old suggestion of a cuboctahedral intermediate in the proposed diamond-square-diamond (dsd) rearrangement of 1,2- to 1,7-$C_2B_{10}H_{12}$ (66). While there is no direct connection between this C_4B_8 species and the latter mechanism, it does provide the first established example of cuboctahedral geometry in boron cluster chemistry.

Current Applications of Polyhedral Boranes and Carboranes

The evolution of modern borane chemistry that began in the late 1940s has had consequences not only in basic research, but in practical realms as well. A half-century ago, virtually the only known practical use for molecular boron clusters centered on their potential as fuels or fuel additives, exploiting the large energy release that accompanies their combustion. Today this has changed most dramatically. Compounds of this class are finding direct commercial, military, medical, or other uses, or are under active investigation, in a number of areas:

• Heat- and cold-resistant polymers (space and military; GLC liquid phases)

• BNCT (boron neutron capture therapy) cancer treatment

• Radionuclide complexes for tumor imaging

• Recovery of metals from radioactive waste

- Homogeneous catalysis in hydrogenation, hydrosilylation, and isomerization of unsaturated organics

- Weakly coordinating anions (olefin polymerization, lithium batteries)

- BC and BSi refractory coatings

- Lightweight neutron shields

- Airbag propellants

- Reagents for organic synthesis (olefin and alkyne polymerization catalysis, etc.)

- Anticancer agents

- Electronic/magnetic/optical materials (conducting polymers, nonlinear optical materials, etc.)

Many of these projected applications (the last three of which are under active investigation in our laboratory) have been reviewed elsewhere (67); here I briefly call attention to just three areas that are subjects of current research.

New Specialty Materials

Polyhedral boranes and their carbon- and metal- containing relatives offer a unique combination of attributes — thermodynamic and redox stability, electron-delocalized structures, three-dimensional cage geometry, and (to a degree) tailorability via attachment of functional groups — that might be used to advantage in constructing novel materials having useful electronic, optical, or magnetic properties. Several types of materials that come to mind are low-dimensional conducting polymers containing, for example, "carborods" (Chart 7), metallacarborane sandwich polymers (60), or metallacarborane-stabilized alkynyl-linked "molecular wires" that might be used to construct nanoscale electronic circuits (67b,68); dendrimers incorporating carborane or metallacarborane units (69); and nonlinear optical (NLO) materials that can be "tuned" by controlling polarizability via selection of metal centers and substituents (70).

Extraction of Metal Ions from Radioactive Waste

The bis(dicarbollyl)cobalt complex anion 3-Co(1,2-$C_2B_9H_{11}$)$_2^-$ (Chart 10) and related species (31a) have remarkably high solubility in nonaqueous solvents, and have long been known as scavengers *par excellence* for many types of metal cations, including those of cesium, strontium, potassium, and the lanthanide elements (67a). In the Czech Republic (formerly Czechoslovakia) complexes of this class have been employed for over a quarter-century for the extraction of radioactive $^{137}Cs^+$ and $^{90}Sr^{2+}$ from nuclear waste, and as sensors for these and other metal ions in body fluids. Interestingly, while Na$^+$[3-Co(1,2-$C_2B_9H_{11}$)$_2^-$] is comparable to NaCl as an electrolyte, diethyl ether completely extracts the cobalt species from an equal volume of an 0.5 M aqueous solution of the complex! For a time, the use of these compounds was hampered by their decomposition under highly acidic conditions, but the discovery that 3-Co(1,2-$C_2B_9H_5Cl_6$)$_2^-$ withstands 3 M nitric acid allows its use for large-scale recovery of $^{137}Cs^+$ and $^{90}Sr^{2+}$ (67a,71). At present this is the most important practical metallacarborane application, although emerging medical and pharmacological developments may change this picture before long (see below).

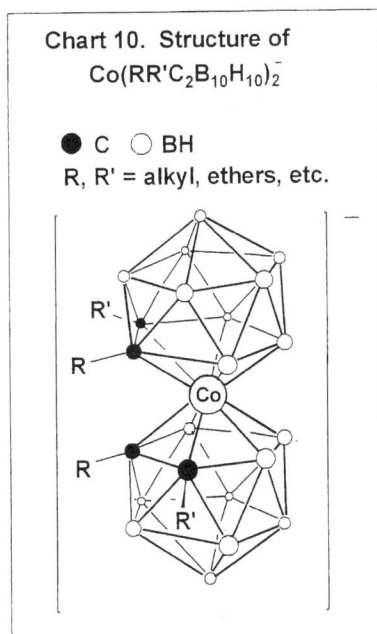

Chart 10. Structure of Co(RR'$C_2B_{10}H_{10}$)$_2^-$

● C   ○ BH
R, R' = alkyl, ethers, etc.

Boranes and Carboranes in Medicine

At the present time, polyhedral boron cluster compounds are under investigation for use in boron neutron capture therapy (BNCT) and as anticancer agents. The first of these applications is far more advanced, having just entered the second stage of clinical trials in the U.S. and western Europe, and is described in detail in several recent reviews (72). This technique relies on a long-known property of the ^{10}B nucleus, which accounts for 20% of boron in Nature, namely, its very high propensity to react with low-energy neutrons to generate an alpha particle (^{4}He^{2+}) and a ^{7}Li^{3+} ion having a total energy of ~2.4 MeV, close to optimum for destruction of one cell radius. Thus, exposure of tumor cells enriched in ^{10}B to a beam of slow neutrons (which otherwise have only minimal effect on tissue) produces a localized lethal effect on those cells. Although the concept of exploiting this property as a method of tumor therapy originated many years ago, the absence of sufficiently stable boron-rich compounds was a hindrance until the exceedingly stable polyhedral borane anions $B_{10}H_{10}^{2-}$ and $B_{12}H_{12}^{2-}$ and their carborane analogues were discovered. Following years of successful use of BNCT in Japan using $Na_2B_{12}H_{11}SH$ to treat glioblastomas (72e), the current clinical trials employ other derivatives including tumor-specific nucleosides containing attached $B_{12}H_{11}$ clusters. The method is of greatest interest in respect to inoperable brain tumors, since it employs two separately benign agents, low-toxicity boron compounds and low-energy (thermal) neutrons, that when combined can selectively destroy the target cells and leave healthy tissue unaffected.

A more recently opened area of interest centers on metallacarboranes as anticancer pharmacological agents. Studies by Prof. Iris Hall and her students have demonstrated that ferratricarbaborane complexes of the type $CpFe^{II}(MeC_3B_7H_9)$ and its analogous Fe^{III} cation exhibit cytotoxicity against several lines of tumor cells (73). These compounds (Chart 11**A**), prepared in Sneddon's group, are electronic analogues of ferricinium cation and other metallocenes that are known anticancer agents (74). Small metallacarboranes bearing C_2B_4 or C_2B_3 ligands are also effective; nearly two dozen such complexes of tantalum, zirconium, iron, cobalt, molybdenum, and tungsten that

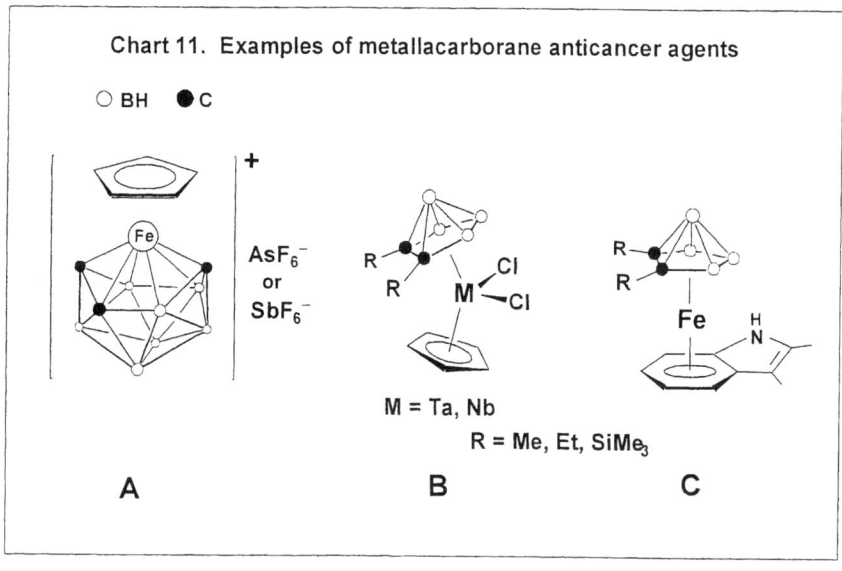

Chart 11. Examples of metallacarborane anticancer agents

were synthesized in our laboratory (e.g., **B** and **C** in Chart 11) are cytotoxic agents toward murine 1210 lymphoid leukemia, murine P388 lymphocytic growth, Tmolt$_3$ leukemia, HeLa-S^3 human uterine carcinoma, Sk-2 melanoma, and Mcf-t breast effusion growth (75). Several of these complexes, interestingly, are more effective against certain glioma and breast cancer cell lines than are the standard clinical drugs. Mode of action studies on selected metallacarborane complexes have shown that the cytotoxicity results from interference with RNA and DNA synthesis (73,75).

These investigations are continuing, with attention now directed to the question of how the efficacy of these compounds might be improved by modifying their their molecular structures and compositions, for example, via the introduction of organic functional groups to the cage, metal, or hydrocarbon ligand. In this projected application — as in others involving boron cage compounds, such as Hawthorne's fascinating Venus flytrap complexes that can serve as vehicles for radiotracer elements (76) — a singular advantage is afforded by the hydrolytic stability of these clusters and their general inertness toward biological systems.

Conclusion

The connection of some of the topics discussed in this paper to the polyhedral bonding theory and other research contributions of Lipscomb and his collaborators in the boron hydride area, are perhaps not always in the forefront of the consciousness of present-day investigators of boron and other cluster types. Nevertheless, the fact is that molecular cluster science in the year 2000 owes a very large debt to the Colonel for giving it an intellectual jump-start fifty-odd years ago, for imaginatively moving it forward at critical points, and for unwaveringly insisting over the decades -- during both good and lean years of funding by federal agencies -- that this is a fundamentally important area of study in which many more exciting discoveries and advances are yet to come.

We have even ventured a few predictions, knowing that if we must join the ranks of boron hydride predictors later proved wrong, we shall be in the best of company

-- W. H. Eberhardt, B. Crawford Jr., and W. N. Lipscomb, "The Valence Structures of the Boron Hydrides", *J. Chem. Phys.* **1954**, *22*, 989

References

1. From *Alice Through the Looking-Glass*, 1872.
2. Grimes, R. N. *Science* **1976**, *194*, 709-710.
3. Lipscomb, W. N., *Advances in Inorganic and Radiochem.* **1959**, 1, 117; see footnote 1.
4. Stock, A., *Hydrides of Boron and Silicon,* Cornell Univ. Press, 1933.
5. Although X-ray crystallography was pioneered by the Braggs as early as 1914, it was not generally applied to molecular structure until the development of digital computers in the 1950s (moreover, its use for determining the structures of low-melting, volatile compounds was totally beyond the state of the art in Stock's day).
6. Frey, P. R., *College Chemistry*, Prentice-Hall, New York, 1952, p. 492.
7. Lewis, J. R., *First- Year College Chemistry* , Barnes and Noble, New York, 1951.
8. Moeller, T., *Inorganic Chemistry*, Wiley, New York, 1952.
9. Pitzer, K. S. *J. Am. Chem. Soc.* **1945**, *67*, 1126.
10. Dulmage, W. J.; Lipscomb, W. N. *J. Am. Chem. Soc.* **1951**, *73*, 3539.
11. Hedberg, K.; Jones, M. E.; Schomaker V. *J. Am. Chem. Soc.* **1951**, *73*, 3538.
12. Kasper, J. S.; Lucht, C. M.; Harker, D. *J. Am. Chem. Soc.* **1948**, *70*, 881.; *Acta Cryst.* **1950**, *3*, 436.
13. (a) Longuet-Higgins, H. C.; de V. Roberts, M. Proc. Roy. Soc. **1954**, *A224*, 336; **1955**, *A230*, 110. (b) Longuet-Higgins, H. C. *J. Chim. Phys.* **1949**, *46*, 268.
14. (a) Lipscomb, W. N. *Boron Hydrides*, Benjamin, New York, 1963. (b) Lipscomb, W. N. *Science* **1977**, *196*, 1047, and references therein. (c) Eberhardt, W. H.; Crawford, B. Jr., and Lipscomb, W. N. *J. Chem. Phys.* **1954**, *22*, 989. (d) Dickerson, R. E.; Lipscomb, W. N., *J. Chem. Phys.* **1957**, *27*, 212. (e) Hoffmann, R.; Lipscomb, W. N. *J. Chem. Phys.* **1962**, *36*, 3489.
15. A neutral borane B_pH_{p+q} can be described as a collection of p BH units and q additional hydrogens. If there are s B-H-B bridges, x extra hydrogens, t 3-center B-B-B bonds, and y 2-center B-B bonds, the limitations imposed by numbers of available electrons and valence orbitals, expressed as the "equations of balance", are that $s + x = q$; $s + t = p$; and $t + y = p - q/2$. Thus, for B_6H_{10} p = 6 and q = 4, which gives 3 solutions in which the styx values are respectively 4220, 3311, or 2402; as the known structure has four bridging hydrogens (s = 4) and no BH_2 groups (x = 0), the first solution is correct. The equations are easily modified for ionic boranes (see ref. 14a, p. 47).
16. (a) Wade, K. *Chem. Commun.* **1971**, 792. (b) Wade, K. *Adv. Inorg. Chem. Radiochem.* **1976**, *18*, 1, and references therein.
17. (a) Rudolph, R. W.; Pretzer, W. R. *Inorg. Chem.* **1972**, *8*, 1974. (b) Rudolph, R. W. *Acc. Chem. Res.* **1976**, *9*, 446.

18. Mingos, D. M. P. *Nature (London) Phys. Sci.* **1972**, *236*, 99.
19. Williams, R. E. *Adv. Inorg. Chem. Radiochem.* **1976**, *18*, 67 and references therein.
20. Grimes, R. N. *Ann. N. Y. Acad. Sci.* **1974**, *239*, 180.
21. King, R. B.; Rouvray, D. H., *J. Am. Chem. Soc.* **1977**, *99*, 24.
22. Mingos, D. M. P.; Wales, D. J. *Introduction to Cluster Chemistry*, Prentice Hall, Englewood Cliffs, NJ, 1990.
23. (a) Olah, G. A.; Prakash, G. K. S.; Williams, R. E.; Field, L. D.; Wade, K. *Hypercarbon Chemistry*, Wiley , New York, 1987. (b) *The Borane, Carborane, Carbocation Continuum*, Casanova, J., Ed., Wiley Interscience, New York, 1998. (c) Grimes, R. N. *Carboranes*, Academic Press, New York, 1970.; and references therein.
24. (a) Timms, P. L. *J. Am. Chem. Soc.* **1968**, *90*, 4585. (b) Maddren, P. S.; Modinos, A.; Timms, P. L.; Woodward, P. *J. Chem. Soc., Dalton Trans.* **1975**, 1272.
25. Lipscomb, W. N., *J. Chim. Phys.* **1956**, *53*, 515.
26. Schaeffer, R.; Shoolery, J. N.; Jones, R., *J. Am. Chem. Soc.* **1958**, *80*, 2670.
27. Greenwood, N. N.; Spalding, T. R.; Taylorson, D. *J. Inorg. Nucl. Chem.* **1980**, *42*, 317.
28. Pepperburg, I. M.; Halgren, T. A.; Lipscomb, W. N. *Inorg. Chem.* **1977**, *16*, 363.
29. Parry, R. W.; Edwards, L. J. *J. Am. Chem. Soc.* **1959**, *81*, 3554.
30. (a) Lipscomb, W. N.; Pitochelli, A. R.; Hawthorne, M. F. *J. Am. Chem. Soc.* **1959**, *81*, 5833. (b) Pitochelli, A. R.; Hawthorne, M. F. *J. Am. Chem. Soc.* **1960**, *82*, 3228.
31. (a) Hawthorne, M. F.; Young, D. C.; Wegner, P. A. *J. Am. Chem. Soc.* **1965**, *87*, 1818. (b) Hawthorne, M. F. *J. Organomet. Chem.* **1975**, *100*, 97.
32. (a) Grimes, R. N., *Accounts Chem. Res.* **1978**, *11*, 420. (b) Grimes, R. N., *Accounts Chem. Res.* **1983**, *16*, 22.
33. (a) Miller, V. R.; Grimes, R. N., *J. Am. Chem. Soc.* **1973**, *95*, 5078. (b) Miller, V. R.; Grimes, R. N., *J. Am. Chem. Soc.* **1976**, *98*, 1600. (c) Miller, V. R.; Weiss, R.; Grimes, R. N., *J. Am. Chem. Soc.* **1977**, *99*, 5646.
34. (a) Maxwell, W. M.; Miller, V. R.; Grimes, R. N., *J. Am. Chem. Soc.* **1974**, *96*, 7116. (b) Grimes, R. N., *Adv. Inorg. Chem. Radiochem.* **1983**, *26*, 55 (c) Grimes, R. N., *Coord. Chem. Rev.* **1995**, *143*, 71.
35. Brewer, C. T.; Swisher, R. G.; Sinn, E.; Grimes, R. N., *J. Am. Chem. Soc.* **1985**, *107*, 3558.
36. Eaton, G. R.; Lipscomb, W. N. *NMR Studies of Boron Hydrides and Related Compounds*, Benjamin, New York, 1969.
37. (a) Todd, L. J.; Siedle, A. R. *Prog. Nucl. Magn. Reson. Spectrosc.* **1979**, *13*, 87. (b) Siedle, A. R. *Annu. Rep. NMR. Spectrosc.* **1982**, *12*, 177.

38. (a) Finster, D. C.; Hutton, W. C.; Grimes, R. N., (b) Venable, T. L.; Hutton, W. C.; Grimes, R. N., *J. Am. Chem. Soc.* **1982**, *104*, 4716. (b) Venable, T. L.; Hutton, W. C.; Grimes, R. N., *J. Am. Chem. Soc.* **1984**, *106*, 29.
39. (a) Switkes, E.; Stevens, R. M.; Lipscomb, W. N., *J. Chem. Phys.* **1969**, *51*, 2085. (b) Switkes, E.; Epstein, I. R.; Tossell, J. A.; Stevens, R. M.; Lipscomb, W. N., *J. Am. Chem. Soc.* **1970**, *92*, 3837. (c) Laws, E. A.; Stevens, R. M.; Lipscomb, W. N., *J. Am. Chem. Soc.* **1972**, *94*, 4467.
40. Dobrott, R. D.; Friedman L. B.; Lipscomb, W. N. *J. Chem. Phys.* **1964**, *40*, 866.
41. Miller, N. R.; Forstener, J. A.; Muetterties, E. L. *Inorg. Chem.* **1964**, *3*, 1690.
42. (a) Kennedy, J. D. *Advances in Boron Chemistry*, Siebert, W., Ed.; Royal Society of Chemistry, Cambridge, U. K., 1997, 451, and references therein. (b) Bould, J.; Clegg, W.; Teat, S. J.; Barton, L.; Rath, N. P.; Thornton-Pett, M.; Kennedy, J. D., *Inorg. Chim. Acta* **1999**, *289*, 95. (c) Bould, J.; Clegg, W.; Kennedy, J. D.; Teat, S. J.; Thornton-Pett, M., *J. Chem. Soc., Dalton Trans.* **1997**, 2005.
43. (a) Fehlner, T. P. *Advances in Boron Chemistry*, Siebert, W., Ed.; Royal Society of Chemistry, Cambridge, U. K., 1997, 463, and references therein. (b) Weller, A. S.; Aldridge, S.; Fehlner, T. P., *Inorg. Chim. Acta* **1999**, *289*, 85. (c) Lei, X.; Shang, M.; Fehlner, T. P., *Organometallics* **2000**, *19*, 118.
44. Balakrishnarajan, M. M.; Jemmis, E. D. *J. Am. Chem. Soc.* **2000**, *122*, 4516.
45. Tillard-Charbonnel, M.; Manteghetti, A.; Belin, C. *Inorg. Chem.* **2000**, *39*, 1684.
46. (a) Grimes, R. N., In "Comprehensive Organometallic Chemistry"; Wilkinson, G., Stone, F. G. A.; Abel, E., Eds.; Pergamon Press: Oxford, England, 1982; Chapter 5.5. (b) Grimes, R. N., In "Comprehensive Organometallic Chemistry II"; Abel, E., Eds.; Wilkinson, G., Stone, F. G. A.; Pergamon Press: Oxford, England, 1995; Volume 1, Chapter 9, pp. 373-430.
47. (a) Evans, W. J.; Hawthorne, M. F., *J. Chem. Soc., Chem. Commun.* **1974**, 38. (b) Maxwell, W. M.; Bryan, R. F.; Sinn, E.; Grimes, R. N., *J. Am. Chem. Soc.* **1977**, *99*, 4008. (c) Maxwell, W. M.; Weiss, R.; Sinn, E.; Grimes, R. N., *J. Am. Chem. Soc.* **1977**, *99*, 4016. (d) Pipal, J. R.; Grimes, R. N., *Inorg. Chem.* **1978**, *17*, 6.
48. (a) Brown, L. D.; Lipscomb, W. N., *Inorg. Chem.* **1977**, *16*, 2989. (b) Bicerano, J.; Marynick, D. S.; Lipscomb, W. N., *Inorg. Chem.* **1978**, *17*, 2041. (c) Bicerano, J.; Marynick, D. S.; Lipscomb, W. N., *Inorg. Chem.* **1978**, *17*, 3443. (d) Schleyer, P. v. R.; Najafian, K.; Mebel, A. M., *Inorg. Chem.* **1998**, *37*, 6765.
49. Piepgrass, K. W.; Curtis, M. A.; Wang, X.; Meng, X.; Sabat, M.; Grimes, R. N., *Inorg. Chem.* **1993**, *32*, 2156.

50. Sevov, S. C.; Corbett, J. D., *Inorg. Chem.* **1992**, *31*, 1895.
51. The $Cp^*_2Co_2Et_4C_4B_6H_6$ cluster has 28 skeletal electrons, derived from contributions of 2e for each BH and CpCo unit and 3e for each CEt group, and hence has n + 2 electron pairs since n = 12. Instead of the predicted nido structure corresponding to a closo system with one missing vertex, its X-ray structure (52a) reveals the *klado* geometry shown in Chart 5, which as noted in the text is a closo 14-vertex framework with 4 missing vertices; such a system would normally correlate with n + 5 electron pairs! It should be noted that two other isomers of $Cp^*_2Co_2Et_4C_4B_6H_6$ have been characterized (52b), one of which is, in fact, a nido system. These findings underline the fact that the electron-counting rules apply to thermodynamically favored structures; for clusters generated under mild conditions, the mechanism of formation may lead to very different kinetically stabilized geometries.
52. (a) Wong, K-S.; Bowser, J. R.; Pipal, J. R.; Grimes, R. N., *J. Am. Chem. Soc.* **1978**, *100*, 5045. (b) Pipal, J. R.; Grimes, R. N., *Inorg. Chem.* **1979**, *18*, 1936.
53. Hawthorne, M. F.; Mortimer, M. D. "Building to Order", *Chemistry in Britain* **1996**, *32*, 32.
54. For leading references, see: (a) Yang, X.; Knobler, C. B.; Zheng, Z.; Hawthorne, M. F., *J. Am. Chem. Soc.* **1994**, *116*, 7142. (b) Zheng, Z.; Knobler, C. B.; Hawthorne, M. F., *J. Am. Chem. Soc.* **1995**, *117*, 5105. (c) Zheng, Z.; Knobler, C. B.; Mortimer, M. D.; Kong, G.; Hawthorne, M. F., *Inorg. Chem.* **1996**, *35*, 1235. Also see: (d) Grimes, R. N. in *Organic Synthesis Highlights III*, J. Mulzer and H. Waldmann, Eds., Wiley-VCH, **1998**, 406.
55. (a) Yang, X.; Jiang, W.; Knobler, C. B.; Hawthorne, M. F., *J. Am. Chem. Soc.* **1992**, *114*, 9719. (b) Jiang, W.; Harwell, D. E.; Mortimer, M. D.; Knobler, C. B.; Hawthorne, M. F., *Inorg. Chem.* **1996**, *35*, 4355.
56. Mueller, J.; Base, K.; Magnera, T. F.; Michl, J., *J. Am. Chem. Soc.* **1992**, *114*, 9721.
57. Jiang, W.; Knobler, C. B.; Mortimer, M. D.; Hawthorne, M. F., *Angew. Chem. Int. Edit. Engl.* **1995**, *34*, 1332.
58. Clegg, W.; Gill, W. R.; MacBride, J. A. H.; Wade, K., *Angew. Chem. Int. Edit. Engl.* **1993**, *32*, 1328.
59. Bluhm, M.; Pritzkow, H.; Siebert, W.; Grimes, R. N., *Angew. Chem. Int. Edit.* **2000**, *39*, 4562.
60. (a) Chin, T. T.; Grimes, R. N.; Geiger, W. E., *Inorg. Chem.* **1999**, *38*, 93. (b) Pipal, J. R.; Grimes, R. N., *Organometallics* **1993**, *12*, 4452; *Ibid.*, 4459. (c) Merkert, J. M.; Davis, J. H., Jr.; Geiger, W. E.; Grimes, R. N., *J. Am. Chem. Soc.* **1992**, *114*, 9846. (d) Stephan, M.; Mueller, P.; Zenneck, U.; Pritzkow, H.; Siebert, W.; Grimes, R. N.; *Inorg. Chem.* **1995**, *34*, 2058.

61. (a) Grimes, R. N., *Applied Organomet. Chem.* **1996**, *10*, 209. (b) Grimes, R. N. *Advances in Boron Chemistry*, Siebert, W., Ed.; Royal Society of Chemistry, Cambridge, U. K., **1997**, 321, and references therein.
62. $R_4C_4B_4H_4$: (a) Fehlner, T. P., *J. Am. Chem. Soc.* **1977**, *99*, 8355. (b) Siebert, W.; El-Essawi, M. E. M., *Chem. Ber.* **1979**, *112*, 1480. (c) Mirabelli, M. G. L.; Sneddon, L. G. *Organometallics* **1986**, *5*, 1510. (d) Mirabelli, M. G. L.; Carroll, P. J.; Sneddon, L. G. *J. Am. Chem. Soc.* **1989**, *111*, 592. (e) Boring, E. A.; Sabat, M.; Finn, M. G.; Grimes, R. N., *Organometallics* **1998**, *17*, 3865.
63. (a) $H_6C_6B_6Et_6$: Wrackmeyer, B.; Schanz, H-J.; Hofmann, M.; Schleyer, P. v. R., *Angew. Chem. Int. Edit.* **1998**, *37*, 1245. (b) $H_6C_6B_6H_6$: Gruener, B.; Jelinek, T.; Plzak, Z.; Kennedy, J. D.; Ormsby, D. L.; Greatrex, R.; Stibr, B., *Angew. Chem. Int. Edit.* **1999**, *38*, 1806.
64. Cuboctahedral $(SiMe_4)_4C_4B_8H_8$: Hosmane, N. S.; Colacot, T. J.; Zhang, H.; Yang, J.; Maguire, J. A.; Wang, Y.; Ezhova, M. B.; Franken, A.; Demissie, T.; Lu, K-J.; Zhu, D.; Thomas, J. L. C.; Collins, J. D.; Gray, T. G.; Hosmane, S. N.; Lipscomb, W. N., *Organometallics* **1998**, *17*, 5294. A number of other other $R_4C_4B_8H_8$ carboranes with varying structures have been structurally characterized: see ref. 34, and references therein.
65. Maxwell, W. M.; Miller, V. R.; Grimes, R. N., *J. Am. Chem. Soc.* **1976**, *98*, 4818.
66. Lipscomb, W. N. *Science* **1966**, *153*, 373.
67. (a) Plesek, J. *Chem. Rev.* **1992**, *92*, 269. (b) Grimes, R. N., *Coord. Chem. Rev.* **2000**, *200-202*, 773.
68. Malaba, D.; Sabat, M.; Grimes, R. N., *Eur. J. Inorg. Chem.* **2001**, 2557.
69. Armspach, D.; Cattalini, M.; Constable, E. C.; Housecroft, C. E.; Phillips, D., *Chem. Commun.* **1996**, 1823.
70. (a) Base, K.; Tierney, M. T.; Fort, A.; Muller, J.; Grinstaff, M. W., *Inorg. Chem.* **1999**, *38*, 287, and references therein. (b) Abe, J.; Nemoto, N.; Nagase, Yu; Shirai, Y.; Iyoda, T., *Inorg. Chem.* **1998**, *37*, 172. (c) Murphy, D. M.; Mingos, D. M. P.; Forward, J. M., *J. Materials Chem.* **1993**, *3*, 67. (d) McKinney, J. D.; McQuillan, F. S.; Chen, H.; Hamor, T. A.; Jones, C. J.; Slaski, M.; Cross, G. H.; Harding, C. J., J. Organomet. Chem. 1997, 547, 253.
71. Hurlburt, P. K.; Miller, R. L.; Abney, K. D.; Foreman, T. M.; Butcher, R. J.; Kinkead, S. A., *Inorg. Chem.* **1995**, *34*, 5215.
72. See, for example: (a) Soloway, A. H.; Zhuo, J.-C.; Rong, F.-G.; Lunato, A. J.; Ives, D. H.; Barth, R. F.; Anisuzzaman, A. K. M.; Barth, C. D.; Barnum, B. A., *J. Organometal. Chem.* **1999**, *581*, 150. (b) Hawthorne, M. F., *Molecular Medicine Today* **1998**, *4*, 174. (c) Soloway, A. H.; Tjarks, W.; Barnum, B. A.; Rong, F.-G.; Barth, R. F.; Codogni, I. M.; Wilson, J. G., *Chem. Rev.* **1998**, *98*, 1515. (d) Kabalka, G. W.; *Expert Opinion on*

Therapeutic Patients **1998**, *8*, 545. (e) Hawthorne, M. F. *Angew. Chem. Int. ed. Engl.* **1993**, *32*, 950.
73. (a) Wasczcak, M. D.; Lee, C. C.; Hall, I. H.; Carroll, P. J.; Sneddon, L. G., *Angew. Chem. Int. Edit. Engl.* **1997**, *36*, 2228. (b) Hall, I. H.; Warren, A. E.; Lee, C. C.; Wasczcak, M. D.; Sneddon, L. G., *Anticancer Research* **1998**, *18*, 951.
74. Leading references: (a) Köpf-Maier, R.; Köpf, H.; Neuse, E. W., *Angew. Chem.* **1984**, *23*, 456. (b) Von Köpf-Maier, P; Köpf, H.; Neuse, E. W., *J. Cancer Res. Clin. Oncol.* **1984**, *108*, 336. (c) Neuse, E. W.; Kanzawa, F., *Applied Organometal. Chem.* **1990**, *4*, 19. (d) Neuse, E. W.; Mojapelo, B. S., *Transition Met. Chem.* **1985**, *10*, 135.
75. (a) Hall, I. H.; Tolmie, C. E.; Barnes, B. J.; Curtis, M. A.; Russell, J. M.; Finn, M. G.; Grimes, R. N., *Appl. Organometal. Chem.* **2000**, *14*, 108. (b) Hall, I. H.; Lackey, C. B.; Kistler, T. D.; Durham, R. W. Jr., Russell, J. M.; Grimes, R. N., *Anticancer Research* **2000**, *20*, 2345.
76. Paxton, R. J.; Beatty, B. G.; Hawthorne, M. F.; Varadarajan, A.; Williams, L. E.; Curtis, F. L., Knobler, C. B.; Beatty, J. D.; Shively, J. E., *Proc. Natl. Acad. USA* **1991**, *88*, 3387.
77. Pitzer, K. S., *J. Am. Chem. Soc.* **1945**, *67*, 1126.

Chapter 3

Recent Developments in the Chemistry of Main Group Metallacarboranes of the C_2B_4-Carborane Ligands

Narayan S. Hosmane[1] and John A. Maguire[2]

[1]Department of Chemistry and Biochemistry, The Michael Faraday Laboratories, Northern Illinois University, DeKalb, IL 60115–2862
[2]Department of Chemistry, Southern Methodist University, Dallas, TX 75275–0314

This account describes some recent findings in the area of small cage, metallacarboranes. The results on the following compounds are discussed. (1) A class of C_4B_8-carboranes that act as restricted electron-acceptors by removing only the valence electrons of a group 1 or group 2 metals. (2) Some main group metallacarboranes that demonstrate their importance beyond their use as versatile synthons for the production of *d*- and *f*-block metallacarboranes. (3) The half- and full-sandwich gallacarboranes that raise the possibility of using such compounds as precursors in the formation of conducting materials. (4) Electron-deficient cage compounds that can be transformed into electron-precise heterocyclic rings.

There has been extensive research reported on the chemical and structural properties of the metallacarboranes in the pentagonal bipyramidal (MC_2B_4) and the icosahedral (MC_2B_9) cage systems (1,2). These complexes are generally synthesized by the reaction of the mono- or dianions of the nido-C_2B_4 or C_2B_9 carboranes with suitable metal reagents, usually in the form of their halides. Much of the emphasis for these studies comes from the fact that the two nido-carboranes have 6 π-electrons, delocalized in orbitals centered on their open pentagonal face that are very similar to those found in the cyclopentadienide ligand (Cp), $[C_5H_5]^-$. Because of this, there is a parallel between the two ligand systems. Our work has involved synthetic, structural, reactivity and theoretical studies on the full- and half-sandwich metallacarboranes derived from the interactions of $[nido\text{-}2\text{-}(SiMe_3)\text{-}n\text{-}(R)\text{-}2,n\text{-}C_2B_4H_4]^{2-}$ [n = 3, 4; R = $SiMe_3$, Me, H] with main group (2), d-group (3-5), and f-group metals (6). Here we report herein some of our latest results in this fascinating area of chemistry.

Syntheses of C_2B_4 and C_4B_8-Carboranes

The reaction of $[nido\text{-}2\text{-}(SiMe_3)\text{-}n\text{-}(R)\text{-}2,n\text{-}C_2B_4H_4]^{2-}$ with $NiCl_2$ in hexane did not give the expected nickelacarborane products, but instead the carboranes underwent oxidative cage closure reactions to give the corresponding closo-1-$(SiMe_3)$-2-(R)-1,2-$C_2B_4H_4$ and nickel metal (8), as shown in **Scheme I**.

Oxidative cage fusion also accompanied the cage closure reactions, leading to the formation of the tetracarbon carborane coproducts, nido-2,4,x,y-$(SiMe_3)_4$-2,4,x,y-$C_4B_8H_8$ (x = 7, y = 9, (**III** in Scheme I); x = 6, y = 12, (**IV** in Scheme I) (6). Since all cage carbon atoms in these tetracarbon carboranes are separated by at least one boron atom, they will subsequently be referred to as the "carbons apart" isomers. Grimes and co-workers had earlier reported on the syntheses, structures and reactivities of several "carbons adjacent" tetracarbon carboranes, in which the cage carbons were localized on one side of very distorted icosahedral cages (8-14). In one of these isomers the carbon atoms were bonded contiguously in a Z-shaped pattern (12), while the other had a more open structure in which the middle C–C bond was no longer present (13). However, the proximate locations of the cage carbons were maintained in all isomers. These carbons adjacent compounds were obtained as the products from the mild air oxidation of the metal-hydride complexes, $(R_2C_2B_4H_4)_2MH_x$ (M = Fe (x = 2), Co (x =1); R = CH_3, C_2H_5, $n\text{-}C_3H_7$ and $CH_2C_6H_5$) (9-11,14). On the other hand, the carbons apart tetracarbon carboranes were obtained directly as one of the oxidation products from the reaction of $[nido\text{-}2\text{-}(SiMe_3)\text{-}3\text{-}(R)\text{-}2,3\text{-}C_2B_4H_4]^{2-}$ (R = $SiMe_3$, n-Bu or t-Bu) and $NiCl_2$ (8). One of these, the nido-2,4,7,9-$(SiMe_3)_4$–2,4,7,9-$C_4B_8H_8$, shown in **Figure 1(a)**, is of particular interest in that it is a 12-vertex cage whose structure is based more on a cuboctahedron than an icosahedron (8). A cuboctahedral structure was proposed by Lipscomb for the key intermediate in the diamond-square-diamond (DSD) mechanism for the rearrangement of closo-1,2-$C_2B_{10}H_{12}$ to closo-1,7-$C_2B_{10}H_{12}$ (15). Interestingly, nido-2,4,7,9-$(SiMe_3)_4$–2,4,7,9-$C_4B_8H_8$, which has D_{2h} cage symmetry, is fluxional with the structure of the proposed transition state for atom mixing being that of a cuboctahedron having D_{4h} cage symmetry. It seems

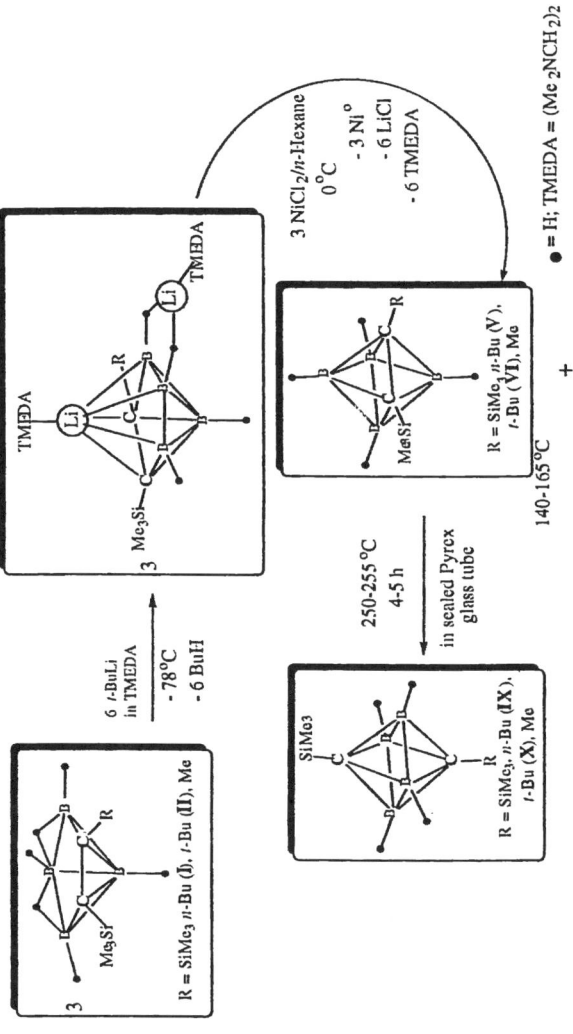

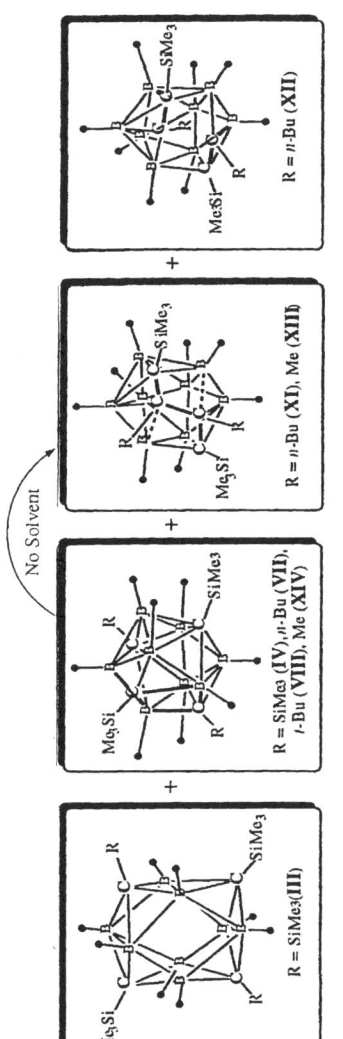

Scheme I. Syntheses of *closo*-C_2B_4- and *nido*-C_4B_8-Carboranes

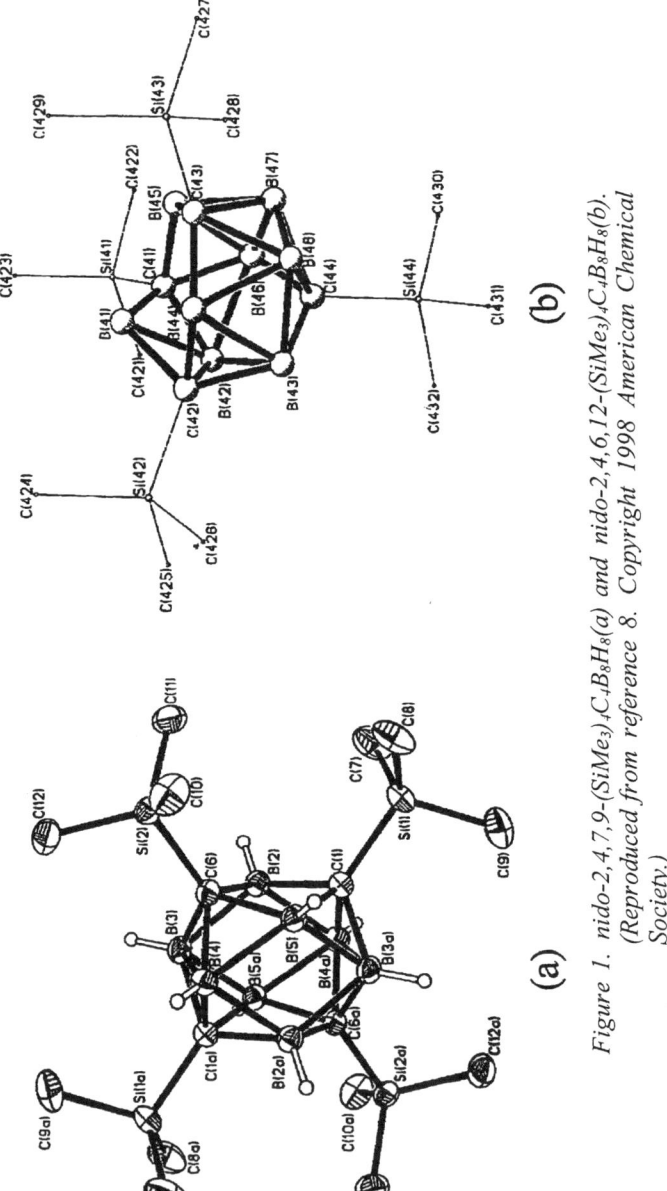

Figure 1. nido-2,4,7,9-(SiMe$_3$)$_4$C$_4$B$_8$H$_8$(a) and nido-2,4,6,12-(SiMe$_3$)$_4$C$_4$B$_8$H$_8$(b). (Reproduced from reference 8. Copyright 1998 American Chemical Society.)

therefore that the cuboctahedron is an important structure in the intermolecular rearrangements of both the dicarba- and tetracarba- twelve vertex carboranes. The other isomer, exemplified by *nido*-2,4,6,12-(SiMe$_3$)$_4$-2,4,6,12-C$_4$B$_8$H$_8$, shown in Figure 1(b) had a more traditional nido-cage structure with an open C$_3$B$_3$ face surmounting a B$_5$ ring and an apical cage carbon (8). This isomer also exhibits nonrigid stereochemistry, but goes through a transition state having C$_{2v}$ symmetry. When two of the SiMe$_3$ moieties were substituted by less bulky *n*-butyl groups, both carbons adjacent and carbons apart isomers of C$_4$B$_8$-carborane were isolated. The cage geometries of these carbons adjacent isomers, shown in **Figure 2**, are identical to those observed earlier by Grimes. These results indicate that steric effects play an important, but not exclusive, role in dictating the geometry of the tetracarbon carboranes.

Reactivities of "Carbons Apart" C$_4$B$_8$-Carboranes

All of the C$_4$B$_8$-carboranes can be reduced to give dianionic cages, which on metallation produce, at least formally, 13-vertex metallacarboranes. Grimes and coworkers have reported on the metallacarboranes in the carbons adjacent system (1,15-19). Multiple structures were found for a particular carborane and metal group, with the yield of any particular metallacarborane being low. These were thought to arise by the metal trapping some of the numerous isomers of the C$_4$B$_8$ dianions that were present in equilibrium in reaction solutions. While the reports on the metallacarboranes derived from the carbons adjacent carboranes are fairly extensive, those on the carbons apart system are more limited and restricted to some group 1 and group 2 metals (20-22). **Scheme II** outlines the general procedures for the syntheses of the known carbons apart compounds.

According to this scheme, the reactions between Mg and the group 1 metals go by different paths. The Mg reaction seems to be a straight forward two-electron transfer to give the magnesacarboranes. However, each group 1 metal reaction goes through a short lived paramagnetic intermediate, which reacts with an additional metal atom to produce either monoprotonated ML$_n$[(SiMe$_3$)$_4$C$_4$B$_8$H$_9$] (M = Li, Na, K; L = THF; n = 4 or L = TMEDA, n = 2) or in the case of Cs the novel polymeric compound, [*exo*-[(_-H) $_2$Cs(TMEDA)]-1-Cs-2,4,7,9-(SiMe$_3$)$_4$-2,4,7,9-C$_4$B$_8$H$_8$]$_n$ (21,22). The cesiacarborane is the first example of a cesium compound in which the metal interacts with a carborane cage to form repeating units held together by -metal-carborane-metal- linkages (21). The X-ray diffraction study on the cesium complex confirmed its polymeric structure in which each C$_4$B$_8$- carborane fragment serves as a ligand to two Cs atoms, one bonded through an open six-membered face and to the other *via* upper- and lower-belt M-H-E (where E= B or C) interactions (**Figure 3**) (21,22). This

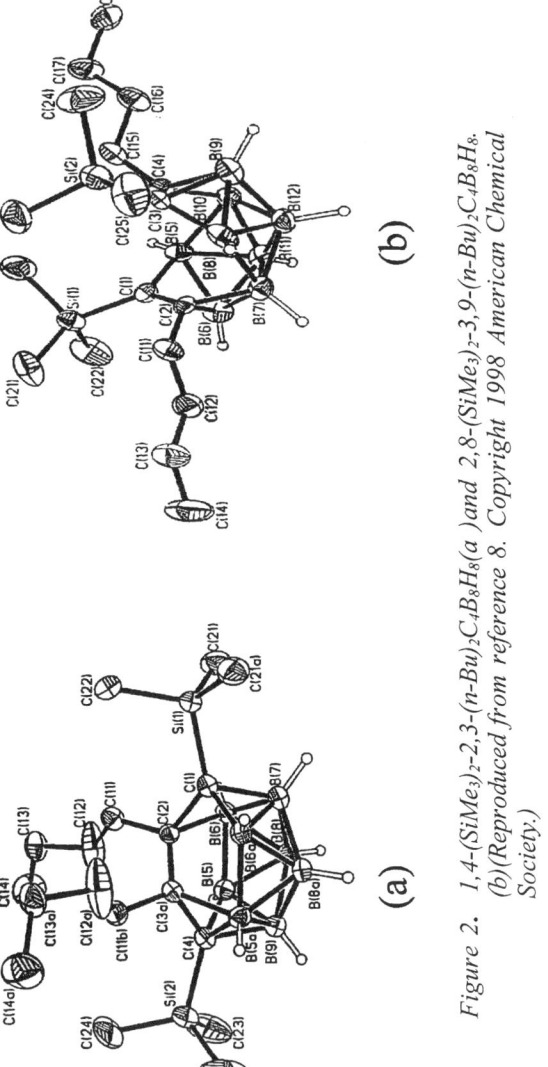

Figure 2. 1,4-$(SiMe_3)_2$-2,3-$(n$-$Bu)_2C_4B_8H_8$ (a) and 2,8-$(SiMe_3)_2$-3,9-$(n$-$Bu)_2C_4B_8H_8$. (b) (Reproduced from reference 8. Copyright 1998 American Chemical Society.)

structural feature is somewhat similar to that of the Sr complex of the $[C_2B_{10}H_{12}]^{2-}$ ligand reported by Hawthorne and coworkers (23). However, the interatomic distances of Cs to the carborane cage atoms are such that it could be regarded as a cesium-carborane complex in which some degree of interaction exists between the metal and the π-electron density on the carborane cage. Since this cesium compound can also be prepared by an ion-exchange reaction directly from the lighter group 1 salts of the C_4B_8-cage, further study of this and related compounds in solvent extraction of radioactive cesium metal (^{137}Cs) from nuclear waste is underway.

Scheme II shows that the carbons apart tetracarbon carborane, *nido*-2,6-(R)$_2$-4,12-(SiMe$_3$)$_2$-2,4,6,12-$C_4B_8H_8$ (R = SiMe$_3$, *n*-butyl), and several of its B-alkylated derivatives, react with Mg metal in THF to produce magnesacarboranes (20,22). Two types of cages were found, one in (THF)$_2$Mg(SiMe$_3$)$_4$(B-Me)$C_4B_7H_7$, shown in **Figure 4(a)**, and the other in (L)$_2$Mg(SiMe$_3$)$_2$(R)$_2$(B-Y)$C_4B_7H_7$ (L = THF, R = SiMe$_3$, Y = *t*-Bu; L = THF, R = SiMe$_3$, Y = H; (L)$_2$ = TMEDA, R = *n*-Bu, Y = H), shown in Figure 3(b). Both magnesacarboranes showed the presence of electron precise carbon and boron atoms, as well as electron deficient cage fragments. The presence of both types of carbon and boron atoms in formally non-exopolyhedral positions is probably dictated by the sometimes conflicting tendencies of the carbon atoms to occupy both nonadjacent and low coordination sites.

Reactivity of Carbons Adjacent C_4B_8-Carboranes

A carbons adjacent magnesacarborane, exo-(μ-H)$_3$Mg(THF)$_3$(SiMe)$_2$(Me)$_2C_4B_8H_8$ was synthesized, in 81% yield, by the reaction of the metal with the (SiMe$_3$)$_2$(Me)$_2C_4B_8H_8$ precursor. Single crystal X-ray diffraction studies showed the compound to be composed of an exo-polyhedral [Mg(THF)$_3$]$^{2+}$ cation that is loosely bound to a [(SiMe$_3$)$_2$(Me)$_2C_4B_8H_8$]$^{2-}$ cage. This carborane polyhedron is best described as ten-vertex *arachno*-(SiMe$_3$)$_2C_2B_8H_8$ cage that subtends an electron precise MeC=CMe fragment (**Figure 5**) (22).

Conversion of a Carborane into an Electron-Precise Heterocyclic Ring

It has long been recognized that the dominant structural patterns found in alkyl and aromatic hydrocarbons are reflections of fragments of the structures of elemental carbon (diamond or graphite) where hydrogen atoms have replaced C-C bonds to give "electron precise-carbon hydrides". In a similar way the structures of boron hydrides are reflections of the icosahedral units found in elemental boron and, with the replacement of a (BH)⁻ unit by an isoelectronic and isolobal CH or CR unit, gives the structures of the corresponding "electron

54

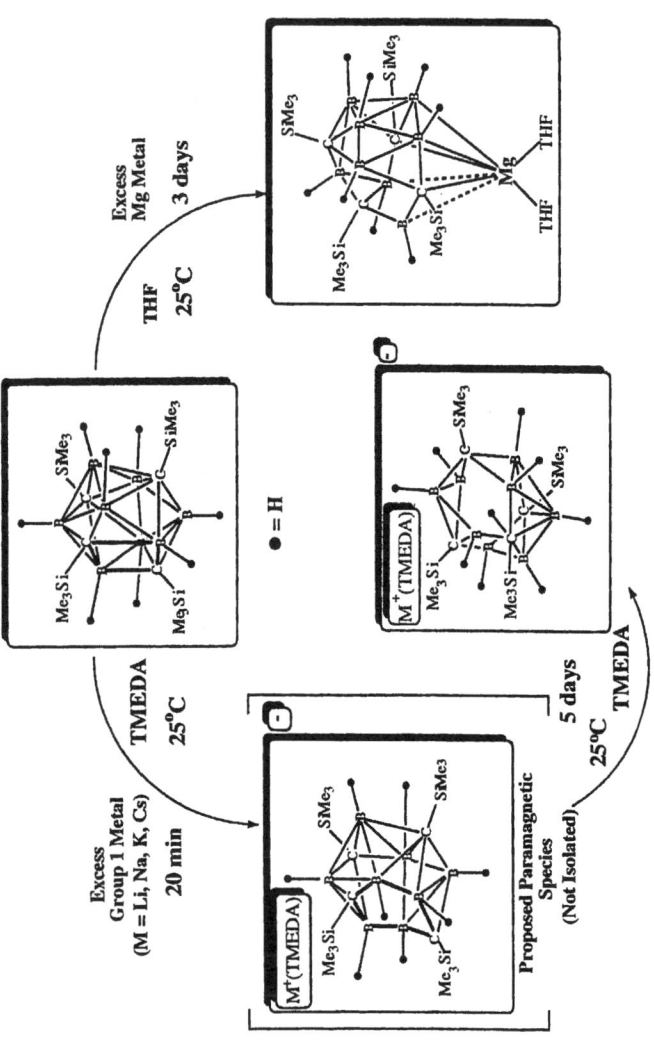

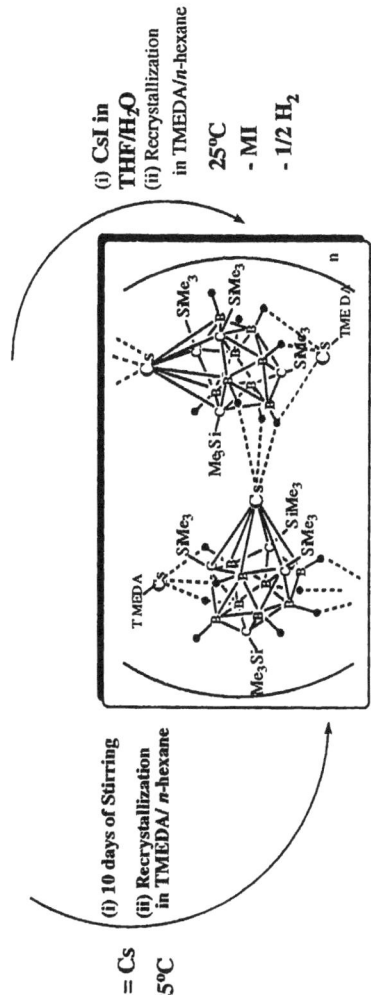

Scheme II. Syntheses of Tetracarbon-carborane Compounds of Group 1 and Group 2 Metals

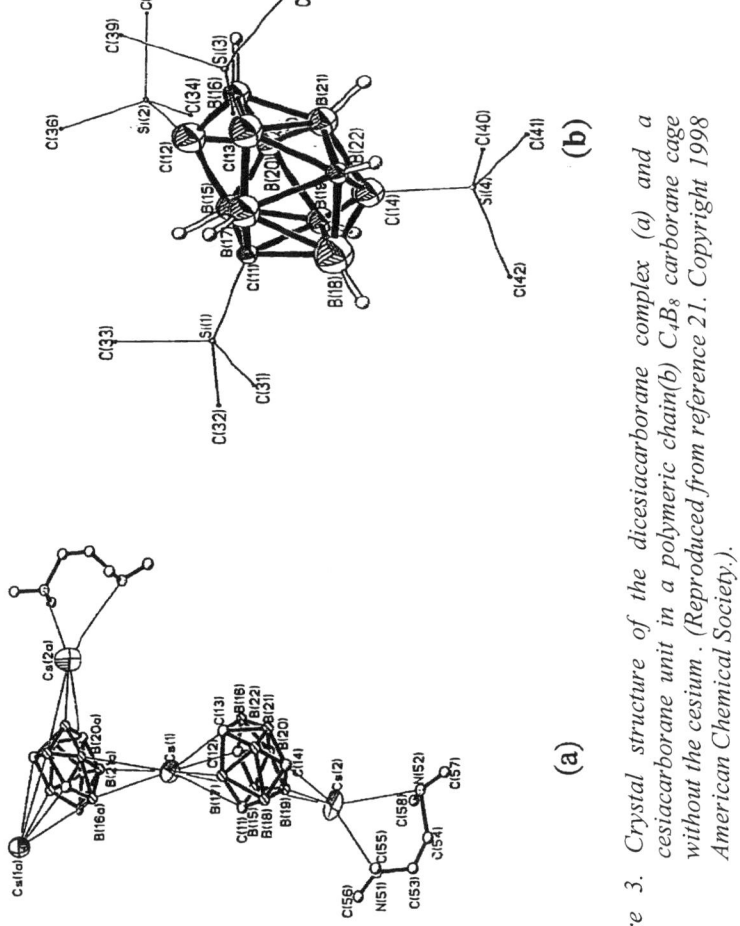

Figure 3. Crystal structure of the dicesiacarborane complex (a) and a cesiacarborane unit in a polymeric chain(b) C_4B_8 carborane cage without the cesium. (Reproduced from reference 21. Copyright 1998 American Chemical Society.).

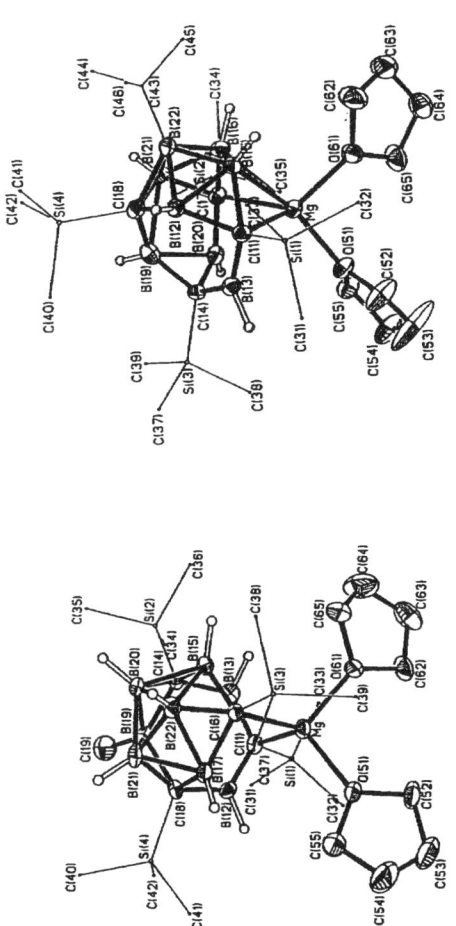

(a) (THF)$_2$Mg(SiMe$_3$)$_4$(B-Me)C$_4$B$_7$H$_7$ (b) (THF)$_2$Mg(SiMe$_3$)$_4$(B-t-Bu)C$_4$B$_7$H$_7$

Figure 4. Crystal Structures of Magnesacarboranes, of the C$_4$B$_8$-Cage Systems. (Reproduced from reference 22. Copyright 2000 American Chemical Society.).

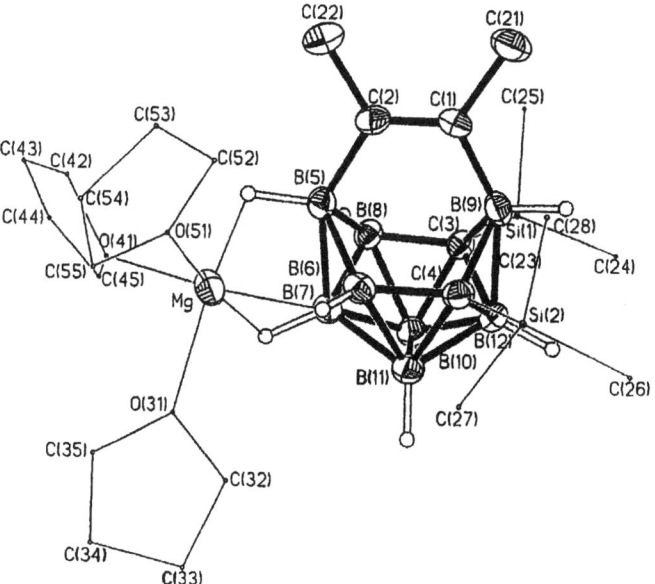

Figure 5. Crystal Structure of exo-(μ-H)$_3$Mg(THF)$_3$(SiMe$_3$)$_2$(Me)$_2$C$_4$B$_8$H$_8$ (Reproduced from reference 22. Copyright 2000 American Chemical Society.)

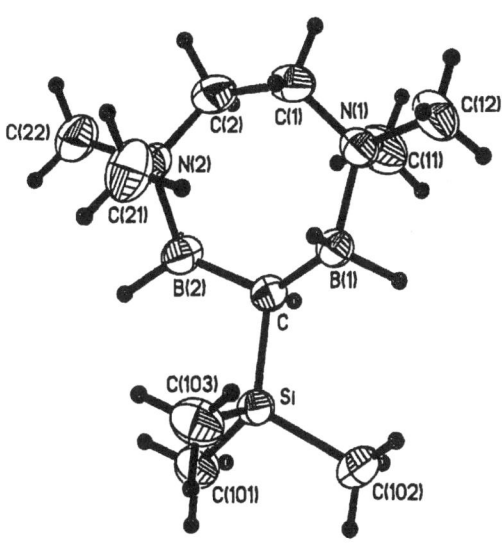

Figure 6. Electron-Precise Heterocyclic Ring

deficient" carborane derivatives (24-26). Although the magnesacarboranes in Figures 4 and 5 consist of both electron-precise atoms and electron-deficient carborane units in a single cage framework, there have been no reports on the complete conversion of an electron-deficient carborane cage into a totally electron-precise B-C heterocyclic ring. As part of our exploration of new methodologies in this area, a neat sample of *nido*-2,3-(SiMe$_3$)$_2$-2,3-C$_2$B$_4$H$_6$ was reacted with a large excess of wet *N,N,N,N*-tetramethylethylenediamine (TMEDA) *in vacuo* with constant stirring at room temperature for 7 days to

2,3-(SiMe$_3$)$_2$-2,3-C$_2$B$_4$H$_6$ + 2 H$_2$O + 2 (Me$_2$NCH$_2$)$_2$ → 2 (SiMe$_3$)(Me)$_4$N$_2$C$_3$B$_2$ + O$_2$ (1)

yield a seven-membered "electron-precise" heterocyclic ring, 1-trimethylsilyl-3,6-dimethyl-3,6-diaza-1,4,5-tricarbaheptaborane, whose solid state geometry was confirmed by X-ray crystallography (see **Figure 6** and eq. 1) (27). Although the mechanism for this transformation is not known, the net result of the reaction can be described as the production of two TMEDA-trapped (SiMe$_3$)CB$_2$H$_5$ fragments for each starting carborane. No evidence exists as yet as to the order of the coordination/redox steps in the reaction sequence. However, the study of reactions such as that shown in eq. 1 could provide clues as to the mechanisms of the decompositions of boranes and carboranes.

Reactivity of C$_2$B$_4$-Carboranes with Group 13 Elements

There has been a number of reports on the syntheses, structures and reactivities of half-sandwich alkylgallacarboranes of both the carbons adjacent and carbons apart C_2B_4-carborane systems (2). Those reports demonstrate that the reactive site of the molecule is the apical gallium metal that acts as a Lewis acid and reacts with bases, such as 2,2'-bipyridine, 2,2'-bipyrimidine, and 2,2':6',2"-terpyridine, to form donor-acceptor complexes. The base-gallacarborane complexes have distorted geometries in that the apical gallium atoms are dislocated, or slipped, away from the centroidal positions above the C_2B_3 faces. In all cases the direction of slippage is away from the cage carbons towards the adjacently bonded boron atoms; in the carbons apart (base)-gallacarborane complexes this leads to η^2- coordination of the metal, while in the analogous carbons adjacent complexes, η^3- coordination is found (28). This distortion pattern is not unique to the gallacarboranes, but is also found in other groups 13 and 14 base-metallacarborane compounds (1,2). As shown in **Scheme III**, depending on the relative amounts of $GaCl_3$ and the C_2B_4 dianions, high yields of both the half- and full-sandwich complexes could be obtained (28,29). The crystal structures of the full-sandwich compounds, [*commo*-1,1'-Ga(2,n-(SiMe_3)_2-1,2,n-GaC_2B_4H_4)_2]⁻, (n = 3, 4) are shown in **Figure 7**. When a 1:1 carborane:$GaCl_3$ molar ratio was used, the corresponding half-sandwich chlorogallacarborane, *closo*-1-(TMEDA)-1-(Cl)-2,n-(SiMe_3)_2-1,2,n-GaC_2B_4H_4 was obtained (29) which could be further converted to the corresponding *closo*-gallacarborane hydride and alkyl derivatives (30). The structures of the gallacarboranes given in **Figure 8**, show the same type of slip distortions found in the other base-gallacarborane complexes. Slip distortions are also found in the full-sandwich gallacarboranes, shown in Figure 7, but are much less than those in the half-sandwich complexes. For example, the maximum differences in the Ga-C_2B_3 atom distances were 0.34 Å and 0.45 Å, for the carbons apart and carbons adjacent full-sandwich complexes, respectively, shown in Figure 7. These can be compared to differences of from 0.88 Å to 0.76 Å for the base-gallacarborane half-sandwich.complexes (2,28,29).

The reaction of [Ga(*t*-Bu)Cl_2]_2 with the disodium compound, *closo-exo*-5,6-Na(THF)_2-1-Na(THF)_2-2,4-(SiMe_3)_2-2,4-$C_2B_4H_4$, in a 1:2 molar ratio, produced the expected half-sandwich gallacarborane, 1-Ga(*t*-Bu)-2,4-(SiMe_3)_2-2,4-$C_2B_4H_4$ and the surprising Ga(II)-Ga(II)-linked digallacarborane, shown in **Figure 9** (31). This digallacarborane is of special interest in that the Ga-Ga bond distance is 2.340(2) Å, which can be compared to Ga-Ga distances of 2.343(2) Å (32) and 2.301 Å (33), respectively, found in the radical anions, [(*i*-Pr)_3C_6H_2Ga]_2•⁻ and [((Me_3Si)_2HC)_2Ga]_2•⁻ and the value of 2.319 Å in Na_2[{((*i*-Pr)_3C_6H_2)_2C_6H_3}Ga]_2 (34). The former two compounds are thought to possess 1-electron π bonds to give gallene-type compounds and the latter compound is reputed to be the first example of a gallyne (35,36). Despite the short Ga-Ga bond distance in the digallacarborane shown in Figure 9, all evidence points to a metal-metal single bond. At least formally, its structure can be thought of as arising from the

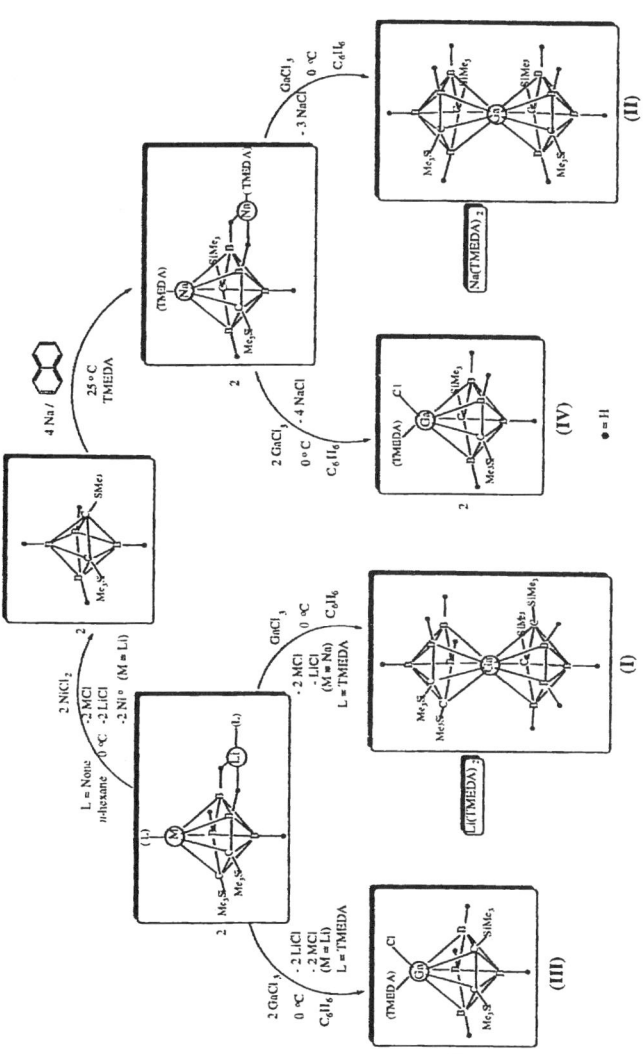

Scheme III. Syntheses of Half- and Full-Sandwich Gallacarboranes of the C$_2$B$_4$-Cage Systems

62

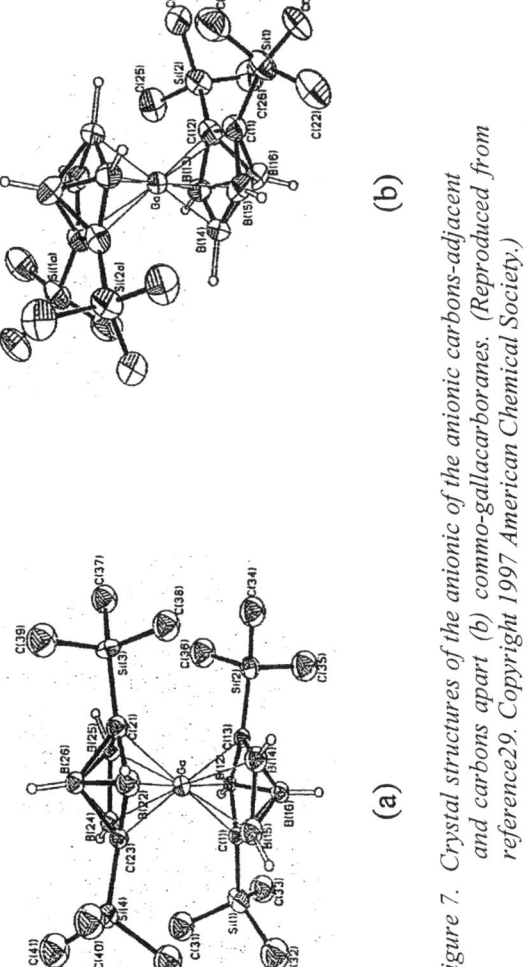

Figure 7. Crystal structures of the anionic carbons-adjacent and carbons apart (b) commo-gallacarboranes. (Reproduced from reference29. Copyright 1997 American Chemical Society.)

63

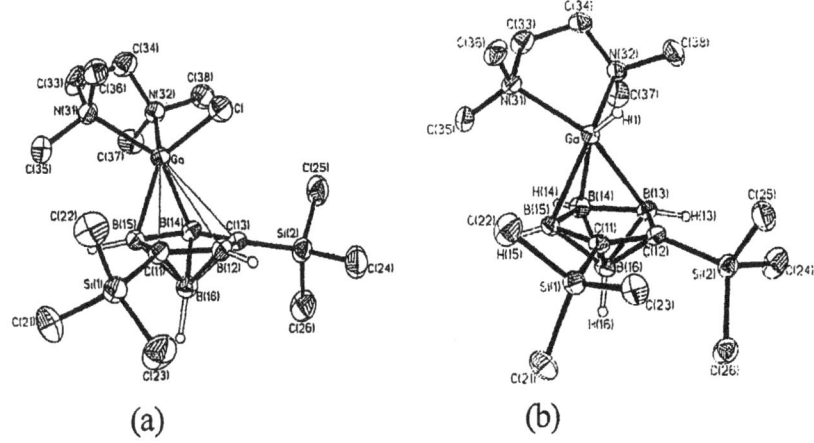

Figure 8. Crystal structure of the carbons apart chlorogallacarborane (a) and the carbons adjacent hydridogallacarborane (b).

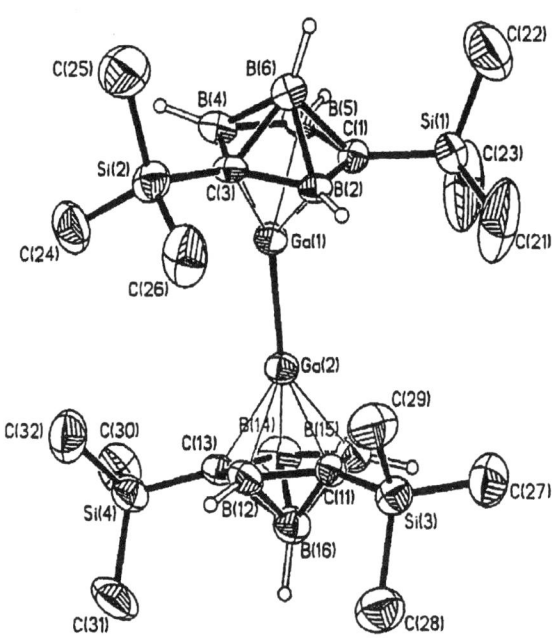

Figure 9. Crystal structure of a Ga(II)-Ga(II)-linked digallacarborane. (Reproduced from reference31. Copyright VCH Verlagsgellschaft mbH.)

substitution of a $(R)_2C_2B_4H_4Ga-$ group for a $t\text{-}(CH_3)_3C-$ group in the previously characterized 1-$(CH_3)_3$C-2,4-$(SiMe_3)_2$-1,2,3-$GaC_2B_4H_4$ (2). Although the mechanism is not known, the dependence of the reaction on the nature of the group 1 metal in the precursor carborane indicates that the digallacarborane is probably not the result of a simple elimination of t-Bu groups during the isolation of the half-sandwich gallacarborane *via* vacuum distillation (31).

Acknowledgment. This work was supported by grants from the donors of the Petroleum Research Fund, administered by the American Chemical Society, National Institute of Health, the National Science Foundation (Grant No. CHE-9988045) and the R. A. Welch Foundation (Grant No. N-1322). The perseverance of numerous undergraduate students, postdoctoral associates, visiting scholars and other co-workers in many of these studies is gratefully acknowledged.

References

1. For general references see: (a) *Comprehensive Organometallic Chemistry, II,* Abel, E. W.; Stone, F. G. A.; Wilkinson, G., Eds.; Elsevier Science Ltd.: Oxford, **1995**; Volume 1, Chapters 6-8 9.
2. Saxena, A. K.; Maguire, J. A.; Hosmane, N. S. *Chem. Rev.* **1997**, *97,* 2421 and references therein.
3. Hosmane, N. S.; Wang, Y.; Zhang, H.; Lu, K.-J.; Maguire, J. A.; Gray, T. G.; Brooks, K. A.; Waldhör, E.; Kaim, W.; Kremer, R. K. *Organometallics,* **1997**, *16,* 1365-1377.
4. Zhang, H.; Wang, Y.; Saxena, A. K.; Oki. A. R.; Maguire, J. A.; Hosmane, N. S. *Organometallics,* **1993**, *12,* 3933.
5. Thomas, C. J.; Lei, J.; Zhang, H.; Siriwardane, U.; Maguire, J. A.; Weiss, V. P.; Brooks, K. A.; Hosmane, N. S. *Organometallics,* **1995**, *14,* 1365.
6. Hosmane, N.S.; Wang, Y.; Oki. A. R.; Zhang, H.; Zhu, D.; McDonald, E. M.; Maguire, J. A. *Phosphorus, Sulfur, and Silicon,* **1994**, *93-94,* 253 and references therein.
7. Hosmane, N. S.; Saxena, A. K.; Barreto, R. D.; Zhang, H.; Maguire, J. A.; Jia, L.; Wang, Y.; Oki, A. R.; Grover, K. V.; Whitten, S. J.; Dawson, K.; Tolle, M. A.; Siriwardane, U.; Demissie, T.; Fagner, J. S. *Organometallics,* **1993**, *12,* 3001.
8. Hosmane, N. S.; Colacot, T. J.; Zhang, H.; Yang, J.; Maguiire, J. A.; Wang, Y.; Ezhova, M. B.; Franken, A.; Demissie, T.; Lu, K.-J.; Zhu, D.; Thomas, J. L. C.; Collins, J.; Gray, T. G.; Hosmane, S. N.; Lipscomb, W. N., *Organometallics,* **1998**, *17,* 5294.
9. Grimes, R. N. *Adv. Inorg. Chem. Radiochem.* **1983**, *26,* 55.
10. Maynard, R. B.; Grimes, R. N. *J. Am. Chem. Soc.,* **1982**, *104,* 5983.
11. Maxwell, W. M.; Miller, V. R.; Grimes, R. N. *Inorg. Chem.,* **1976**, *15,* 1343.
12. Freyberg, D. P.; Weiss, R.; Sinn, E.; Grimes, R. N. *Inorg. Chem.,* **1977**, *16,* 1847.
13. Venable, T. L.; Maynard, R. B.; Grimes, R. N. *J. Am. Chem. Soc.,* **1984**, *106,* 6187.
14. Spencer, J. T.; Pourian, M. R.; Butcher, R. J.; Sinn, E.; Grimes, R. N. *Organometallics,* **1987**, *6,* 335.
15. Lipscomb, W. N. *Science,* **1966**, *153,* 3734, and references therein.
16. Maxwell, W. M.; Bryan, R. F.; Grimes, R. n. *J. Am. Chem. Soc.* **1977**, *99,* 4008;
17. Maxwell, W. M.; Weiss, R.; Sinn, E.; Grimes, R. N. *J. Am. Chem. Soc.* **1977**, *99,* 4016.
18. Grimes, R. N.; Pipal, J. R.; Sinn, E. *J. Am. Chem. Soc.* **1979**, *101,* 4172
19. Maxwell W. M.; Grimes, R. N. *Inorg. Chem.* **1979**, *18,* 2174.
20. Hosmane, N. S.; Zhang,H.; Wang, Y.; Lu, K.-J.; Thomas, C. J.; Ezhova, M. B.; Helfert, S. C.; Collins, J. D.; Maguire, J. A.; Gray, T. G. *Organometallics* **1996**, *15,* 2425.

21. Hosmane, N.S.; Demissie, T.; Zhang, H.; Maguire, J. A.; Lipscomb, W. N.; Baumann, F.; Kaim, W. *Organometallics* **1998** *17*, 293.
22. Hosmane, N. S.; Zhang, H.; Maguire, J. A.; Wang, Y.; Demissie, T.; Colacot, T. J.; Ezhova, M. B.; Lu, K. –J.; Zhu, D.; Gray, T. G.; Helfert, S. C.; Hosmane, S. N.; Collins, J. D.; Baumann, F.; Kaim, W.; Lipscomb, W. N. *Organometallics,* **2000** *19*, 497.
23. Khattar, R.; Knobler, C. B.; Hawthorne, M. F. *Inorg. Chem.,* **1990**, *29*, 2191.
24. Lipscomb, W. N. In *"Boron Hydrides"*, Benjamin: New York, 1963, p. 89.
25. Muetterties, E. L.; Knoth, W. H. In *"Polyhedral Boranes"*, Marcel Decker: New York, 1968, p.40.
26. Parry, R. W. In *"The Borane, Carborane, Carbocation Continuum"*, Casanova, J. Ed., Wiley: New York,1998, p.191.
27. Zheng, C.; Hosmane, N. S. *Acta Cryst.,* **1999**, *C55*, 000.
28. Hosmane, N. S.; Saxena, A. K.; Lu, K.-J.; Maguire, J. A.; Zhang, H.; Wang, Y.; Thomas, C. J.; Zhu, D.; Grover, B. R.; Gray, T. G.; Eintracht, J. E. *Organometallics,* **1995**, *14*, 5104.
29. Hosmane, N. S.; Lu, K. -J.; Zhang H.; Maguire, J. A. *Organometallics,* **1997**, *16*, 5163.
30. Hosmane, N. S. In *"Advances in Boron Chemistry "*, Proceedings of the Ninth International Meeting on Boron Chemistry, held on 14-18 July 1996, in Heidelberg, Germany; Siebert, W., Ed.; Royal Society of Chemistry, London, UK (1997)349.
31. Saxena, A. K.; Zhang, H.; Maguire, J. A.; Hosmane, N. S.; Cowley, A. H. *Angew. Chem., Int. Ed. Engl.,* **1995**, *34*, 332.
32. Xe, H.; Bartlett, R. A.; Olmstead, M. M.; Ruhlandt-Senge, K.; Sturgeon, B. E.; Power, P. P. *Angew. Chem. Int. Ed. Engl.* **1993**, *32*, 717.
33. Uhl, W.; Schütz, W.; Kiam, W.; Waldhör, E. *J. Organomet. Chem.* **1995**, *501*, 79.
34. Su, J.; Li, X.-W.; Crittendon, R. C.; Robinson, G. H. *J. Am. Chem. Soc.* **1997**, *119*, 5471.
35. Xie, Y.; Grev, R. S.; Gu, J.; Scheafer, H. F. III, Schleyer, P. v. R. *J. Am. Chem. Soc.* **1998**, *120* 3773.
36. The is not universal agreement about the existence of Ga̅Ga bond, for an alternative view, see: Cotton, F. A.; Cowley, A. H.; Feng, X. *J. Am. Chem. Soc.* **1998**, *120*, 1795.

Chapter 4

Formation of Nanostructured Phases of Fe, Co, and Ni by a Freeze-Out Technique

Kimloan T. Nguyen, Alfred A. Zinn, and Herbert D. Kaesz

Department of Chemistry and Biochemistry, University of California, Los Angeles, CA 90095–1569

> Thermal decomposition of $Fe(CO)_5$, $Co_2(CO)_8$, and $Ni(CO)_4$ in molten phenanthrene or pyrene in the temperature range of 250 to 160 °C, respectively, leads to spherical nanoclusters in the range of 4-19 nm for Fe and 11-23 nm for Co, and nanocrystallites of an average size of 46 nm of Ni. The carbonyls are introduced in hexadecane solution under conditions leading to rapid cooling below the freezing point of the phenanthrene or pyrene shortly after the injection and decomposition of the precursor. Formation of the solid prevents accretion of the nanoparticles and also protects them from air-oxidation. The suspensions are stable for at least six months without deterioration.

Currently, nanoparticle phases which are of interest for increased information storage (1), are generated using metal vapor condensation methods (2) or chemical and electrochemical reduction of metal salts (3). The former allows the generation of products with exceptional purity, but requires ultra high vacuum and very high temperatures for metal evaporation and is applicable for producing only small amounts of material. The various reduction processes, on the other hand, have the disadvantage of showing high levels of impurities (4).

Since nanophase metal powders are pyrophoric, they need to be stabilized and protected from oxidation. This has been achieved by covering the particle surface with coordinating ligands or surfactants which also inhibit further particle growth. Alternatively, they have been embedded in a polymer matrix (5). In most cases, however, the protecting groups react with the surface atoms of the nanoparticles forming a strongly bonded layer of organic groups which causes unpredictable changes in magnetic properties.

---

a Dedicated with respect and affection to W.N. Lipscomb, Jr.

For our investigations, we chose the thermal decomposition of the metal carbonyls $Fe(CO)_5$, $Co_2(CO)_8$, and $Ni(CO)_4$ to generate nanophases of Fe, Co, and Ni. These carbonyl compounds are commercially available, show good solubility in non-polar organic solvents, and are known to decompose relatively cleanly into the respective metal and CO (6), equations (1), (2) and (3)

$$Fe(CO)_5 \xrightarrow{150\ °C} Fe + 5\ CO \qquad 1$$

$$Co_2(CO)_8 \xrightarrow{50\text{-}100\ °C} 2\ Co + 8\ CO \qquad 2$$

$$Ni(CO)_4 \xrightarrow{170\text{-}180\ °C} Ni + 4\ CO \qquad 3$$

The carbonyls are placed in hexadecane solution under argon. The solutions are injected into a bath of molten phenanthrene (m.p.: 101°; b.p.: 340°) or pyrene (m.p.: 149°; b.p.: 410°), whose freezing points are somewhat below the temperature required for the decomposition of the metal carbonyl complexes. The temperature drop is accomplished by injection of the solution containing the precursors which solutions are typically at room temperature or lower. The temperature drop is controlled by the volume of solution introduced (10-20 ml). This internal quenching process is delicately balanced to permit decomposition of the carbonyls before the resulting nanoparticles become imbedded in the solid.

In a typical experiment the phenanthrene or pyrene is heated to the decomposition temperature of the organometallic precursor (160° for $Co_2(CO)_8$, 170° for $Ni(CO)_4$, or 250 °C for $Fe(CO)_5$). The precursor is dissolved in hexadecane at room temperature and injected into the molten phenanthrene or pyrene respectively. Upon contact with the hot reaction medium, the metal carbonyls instantly decompose, generating CO gas and highly magnetic black metal powders. The volume of the injected solution is chosen in such a way that the reaction temperature drops 40-50 °C within 30 sec and solidifies within 1-2 min. We found that these parameters are sufficient to ensure decomposition of all the introduced precursor material.

The solids are then cooled to room temperature and stored until needed. For analysis the matrix is dissolved in toluene, and the products are isolated by filtration and drying under vacuum. The powders were characterized using X-ray diffraction (XRD), scanning electron microscopy (SEM), and transmission electron microscopy (TEM). The metal powders are all highly magnetic as confirmed qualitatively by holding a permanent magnet to the glass wall of the storage flask containing the dried metal powders. The XRD-patterns of the powders as-obtained show only weak and broad peaks, indicating a very small particle size or amorphous material. However, after annealing in vacuum at 150° C for 48 hrs, the peaks sharpen and match the expected patterns for the pure metals. SEM-pictures show agglomerates of very small particles. These large agglomerates could be resolved by TEM into single nanoparticles which allowed determination of their size distribution. The decomposition of $Fe(CO)_5$ results in the formation of spherical shaped nanoclusters which adhered= to each other, forming large agglomerates, but did not fuse together (Fig. 1 a). Most iron particles show a size distribution in the range of 8-19 nm; however, a number of agglomerates showed particles as small as 4 nm (Fig. 2). Similar results are found for the decomposition products of $Co_2(CO)_8$. The cobalt particles are found to be spherical in shape and did also adhere to each other forming large agglomerates while preserving their nanoparticulate identity (Fig. 1 b). The size distribution, however, was somewhat larger being in the range of 11-24 nm (Fig. 2). By contrast, $Ni(CO)_4$ showed a somewhat different behavior. TEM pictures revealed that the Ni particles exhibited a distinct but irregular crystalline shape and were found to be larger in size averaging about 46 nm; these larger crystallites also displayed a strong tendency to clump together forming larger agglomerates (Fig. 1 c). Close examination of the TEM images revealed that these crystallites must have formed upon fusion of smaller particles, their shapes being clearly distinguishable within the crystallites; their size distribution is found to be in the range of 6-16 nm (Fig. 2). The XRD patterns and size distribution of individual samples of Fe, Co, and Ni powders are presented in an Appendix to this paper.

To isolate the metal particles, the matrix material is easily removed by washing with various solvents. The solid mixture is treated with benzene or toluene and yields metal suspensions which when dried and exposed to air heat up and smoke and sometimes ignite; the powders then consist of mainly metal oxides with loss of ferromagnetic properties. Washing the powders with octanol renders them air stable by forming a protective film over the particles. Excess octanol can be removed by washing the metal powders with methanol, ethanol, or isopropanol. Such washed powders exhibit air sensitivity.

By varying the decomposition temperature we discovered an interesting trend for the observed particle size. Ni exhibited the most dramatic change. An XRD pattern for a sample formed at 180° C consists of sharp peaks with a high count of a macro crystalline cubic Ni phase. The particle size change with a 10 °C temperature increase for Co and Fe, however, is much less pronounced. Co samples made at 170 °C show a medium size increase by a factor of 2-4, while in an iron sample obtained at 260 °C is essentially unchanged from that obtained at 260 °C. The latter observation is remarkable since the reaction is carried out at a much higher temperature than those for Co or Ni. These results might well derive from the increase in the metal-metal bond strengths [$(D°_{298}/kJ\,mol^{-1})$: Fe, 100.0 ±21; Co, 167.0 ±25, and Ni, 203.3 ±1] (7) providing a thermodynamic basis for the observed increase in particle growth rate .

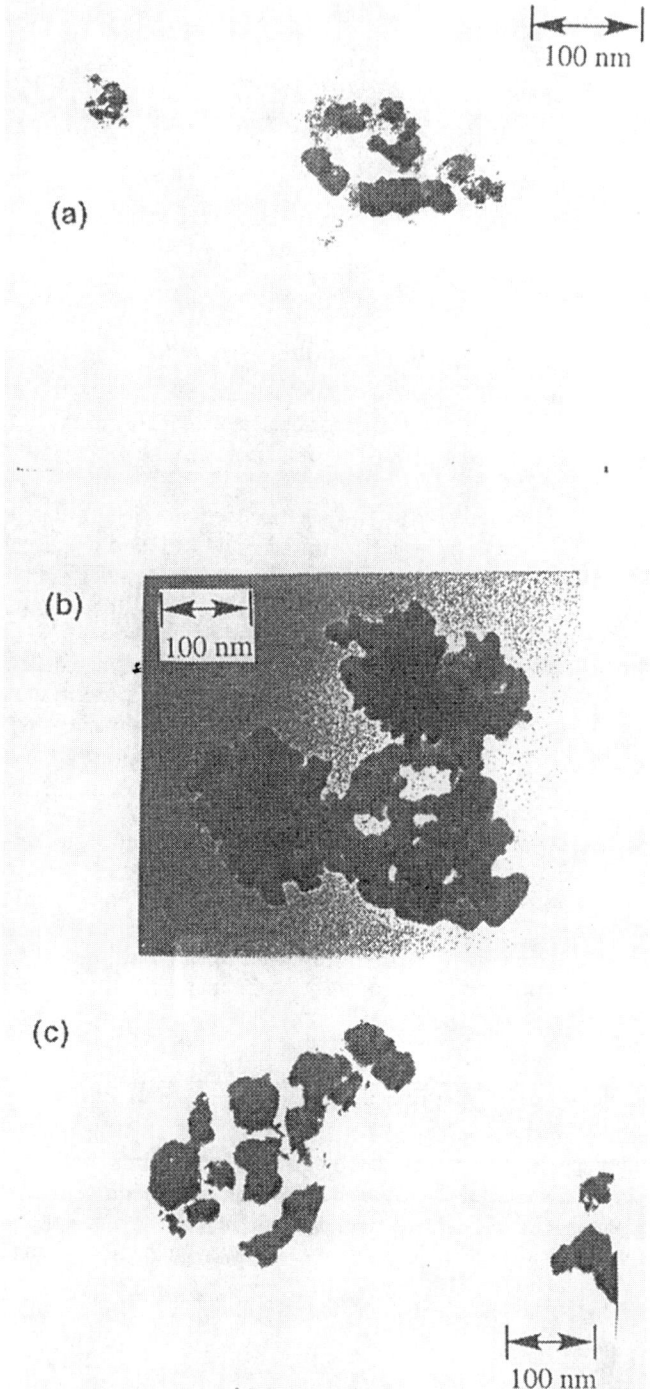

Fig. 1. TEM images of metallic nanoclusters: (a) Fe obtained at 250 °C; (b) Co obtained at 162 °C; (c) Ni obtained at 170 °C.

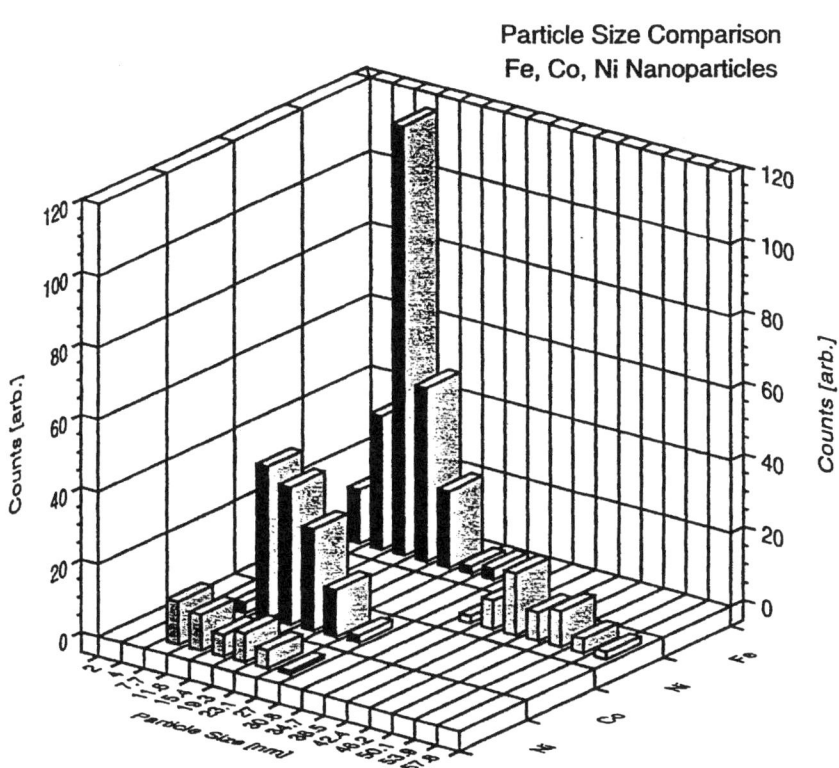

Fig. 2. Composite size distribution of metal nanoparticles (from left to right):
Ni obtained at 170 °C; Co obtained at 162 °C; Ni obtained at 182 °C;
Fe obtained at 250 °C.

References and Notes

(1)   J. C. Mallinson, *The Foundation of Magnetic Recording*, (Boston Academic Press, ed. 2,1993); C. D. Mee and E. D. Daniel, Eds., Magnetic Recording Handbook (McGraw-Hill, New York, 1989).

(2)   R. W. Siegel, Material Science and Engineering, B19, 37 (1993); F. E. Luborsky and P. E. Lawrence, J. All;. Phys. 32, 2315 (1961); M. Kishimoto, S. Kitahata, M. Amemiya, IEEE Trans. Mag. MAG-22, 732 (1986); G. Ziao and C. L. Chien, Appl. Phys. Lett. 51, 1280 (1987).

(3)   R. W. Siegel, NanoStructured Materials, 4, 121 (1994).

(4)   R. E. Treece et al., Inorg. Chem. 32, 2745 (1993); S. Gangopadhyay, G. C. Hadjipanayis, C. M. Sorensen, K. J. Klabunde, Mat. Res. Soc. Symp. Proc. 206, 55 (1991); L. E. Brus, J. Phys. Chem., 90, 255 (1986); L. E. Brus, New J. Chem., 11, 123 (1987).

(5)   J. S. Bradley, J. M. Millar, E. W. Hill, K. Am. Chem. Soc. 113, 4016 (1991); Y. Wang, N. Herron, J. Phys. Chem., 91, 257 (1987); P. Llanos, J. Kl. Thomas, Chem. Phys. Lettl, 125, 299 (1986); M. Meyer, C. Wallberg, K. Kurihara, J. H. Fendler, J. Chem. Soc., Chem. Commun., 90 (1984).

(6)   T. Kodas, M. Hampden-Smith, Eds., The Chemistry of Metal CVD (VCH, Weinheim, Germany, 1994).

(7)   D. A. Lide, Ed., Handbook of Chemistry and Physics, (CRC Press Inc., ed. 74,1993-1994).

(8)   We thank Dr. Richard Lysse for his help with the SEM and TEM measurements. This work was supported by the Department of Chemistry and Biochemistry at UCLA and by National Science Foundation Grant CHE 9208398 under the Materials Synthesis and Processing initiative.

Appendix

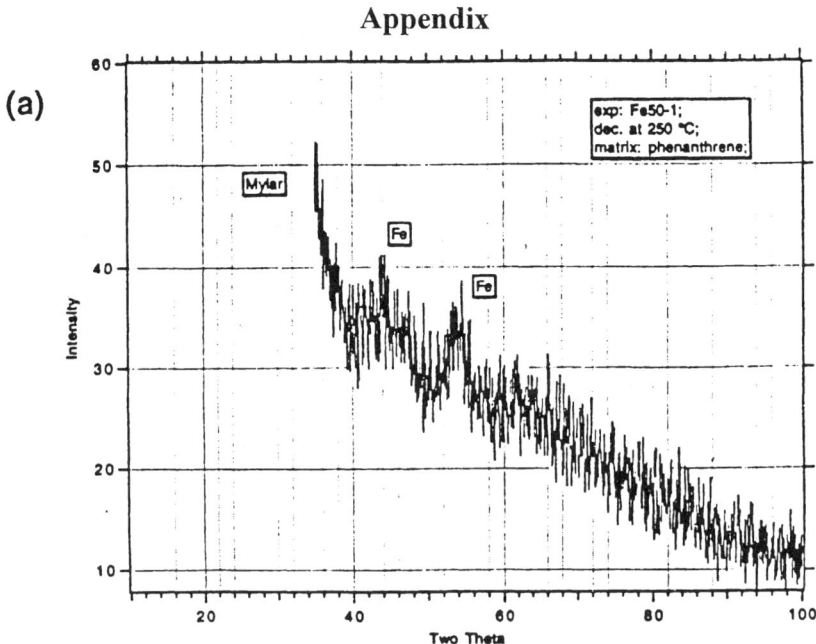

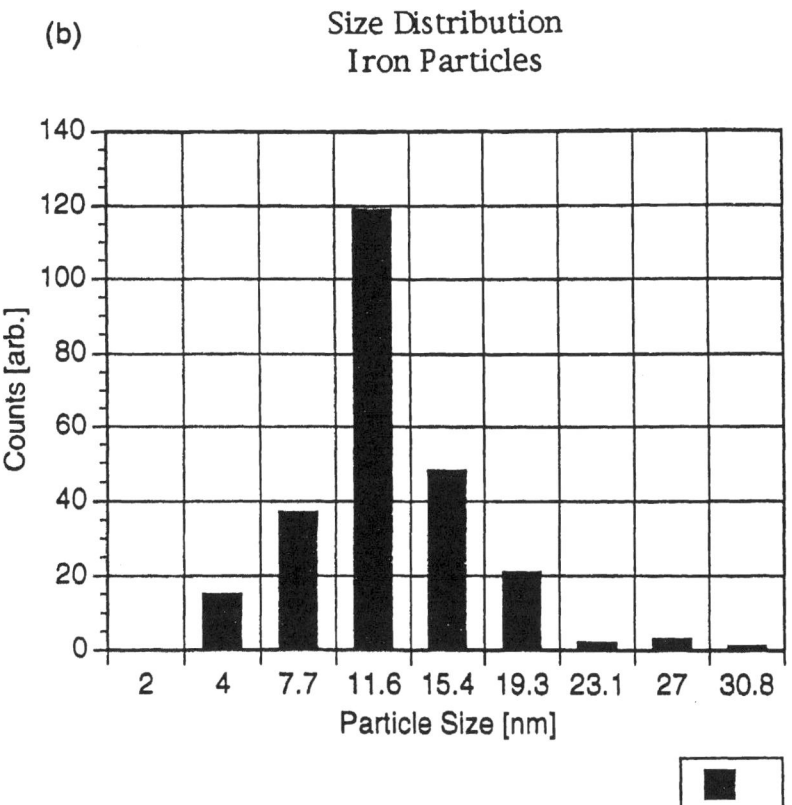

Fig. S-1. Fe powder as obtained at 250 °C. (a) XRD-pattern. (b) Size distribution.

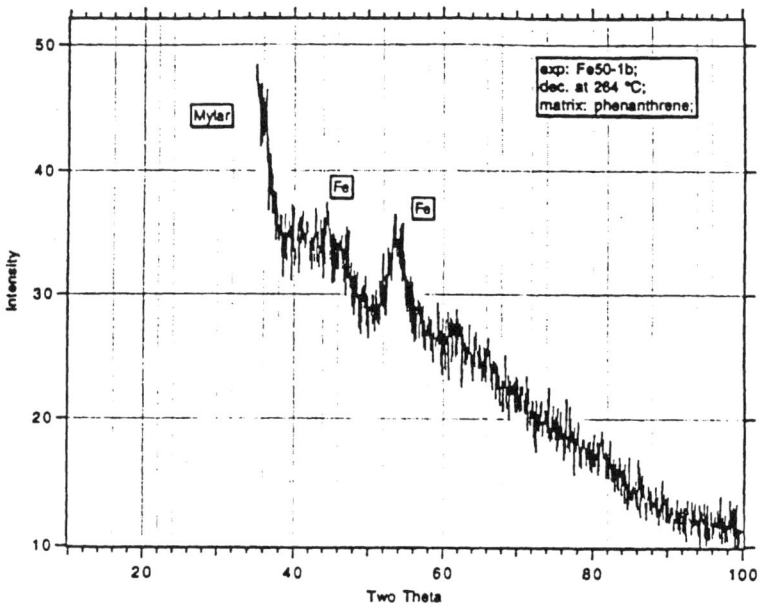

Fig. S-2. XRD-pattern of the Fe powder as obtained at 264 °C; size distribution is comparable to Fe powder obtained at 250 °C

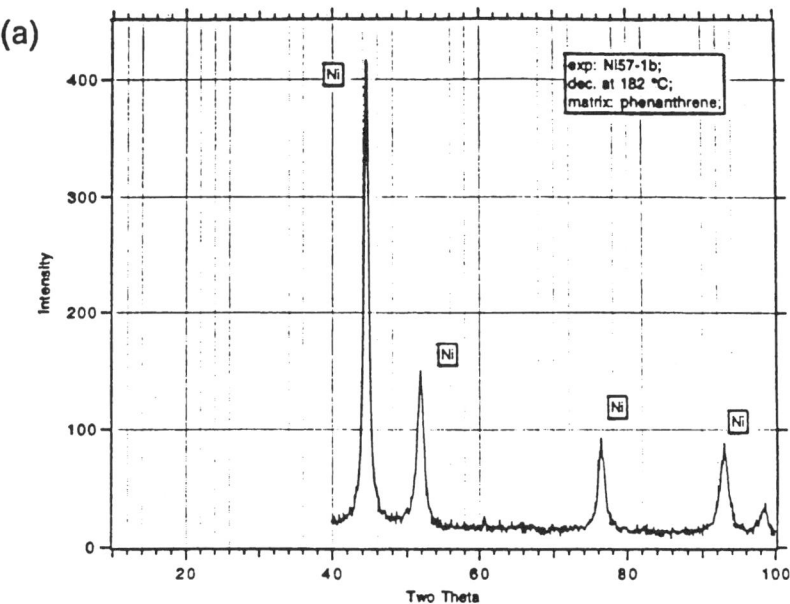

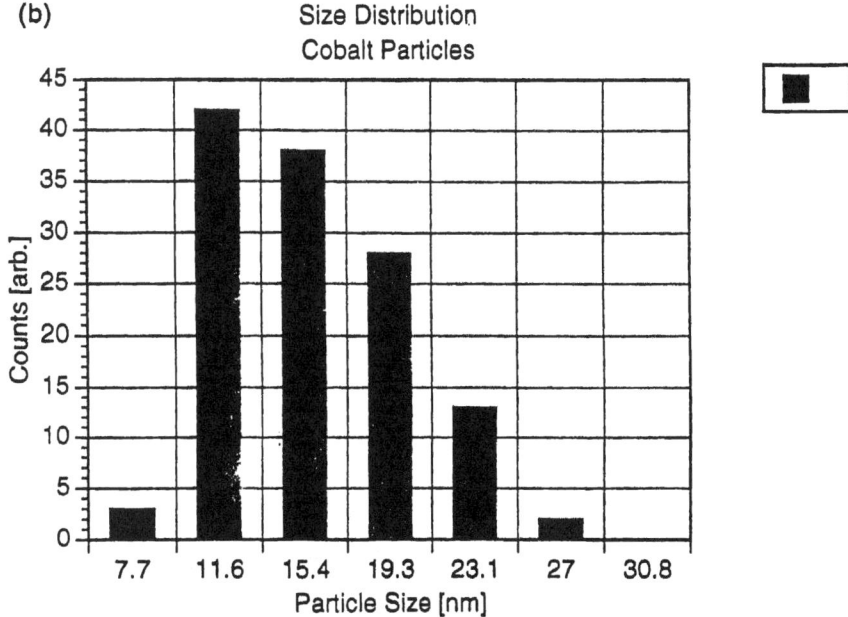

Fig. S-3. Co powder as obtained at 162 °C.
(a) XRD-pattern. (b) Size distribution.

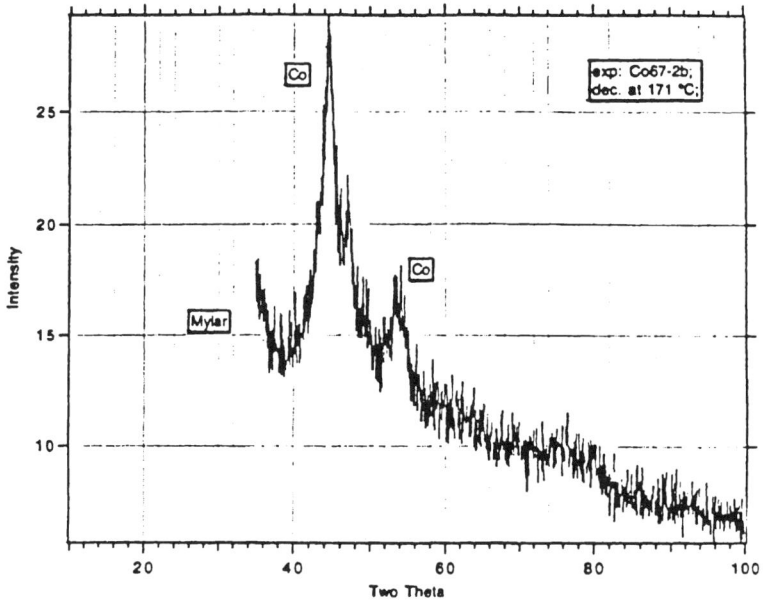

Fig. S-4. XRD-pattern of the Co powder as obtained at 171 °C; these show an average size increase by a factor of 2-4 as compared to Co powder obtained at 162 °C.

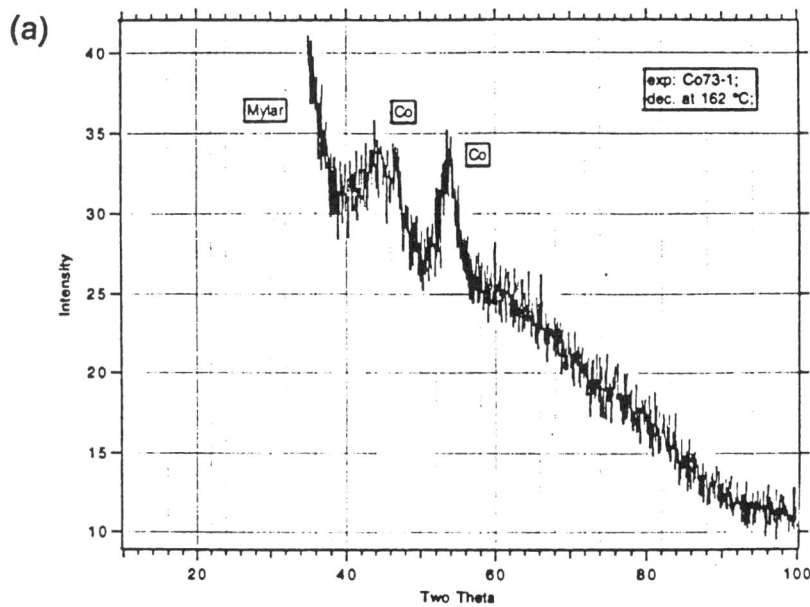

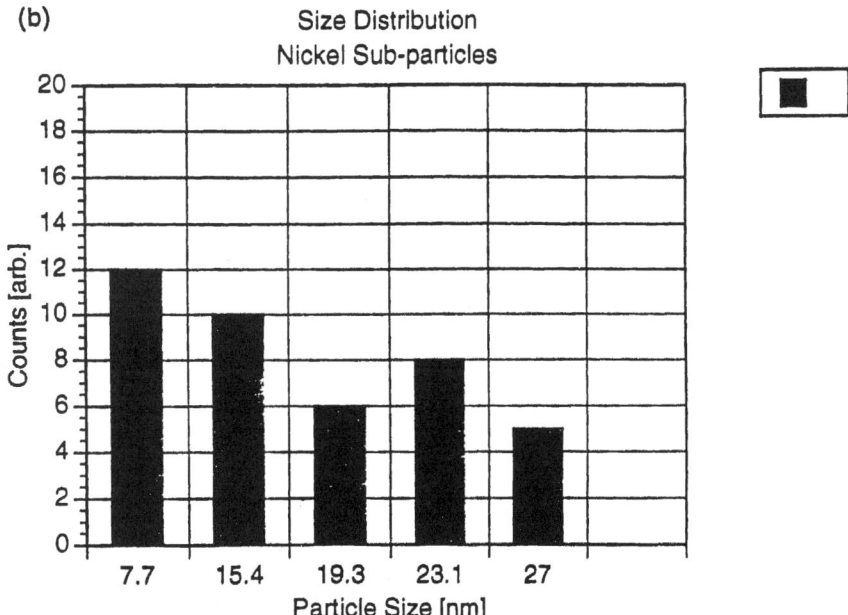

Fig. S-5. Ni powder as obtained at 170 °C. (a) XRD-pattern.
(b) Size distribution.

(a)

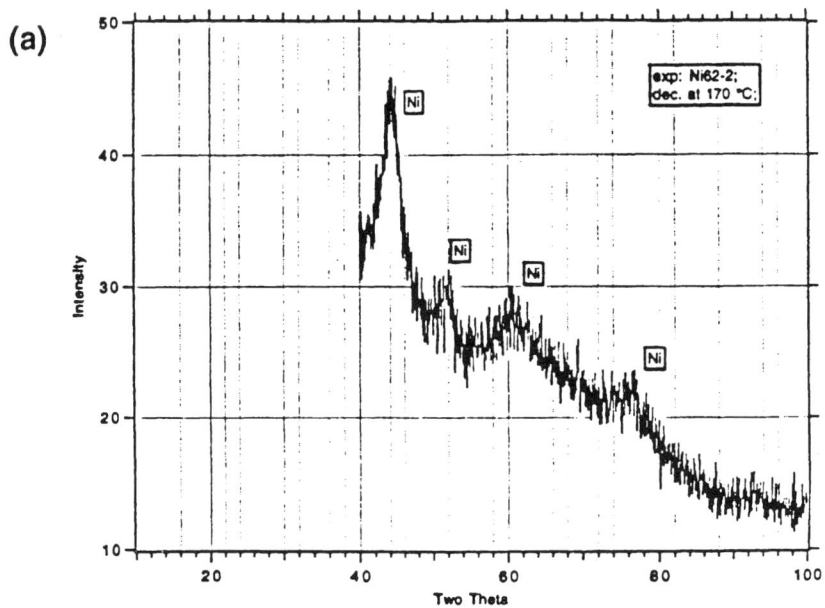

(b)

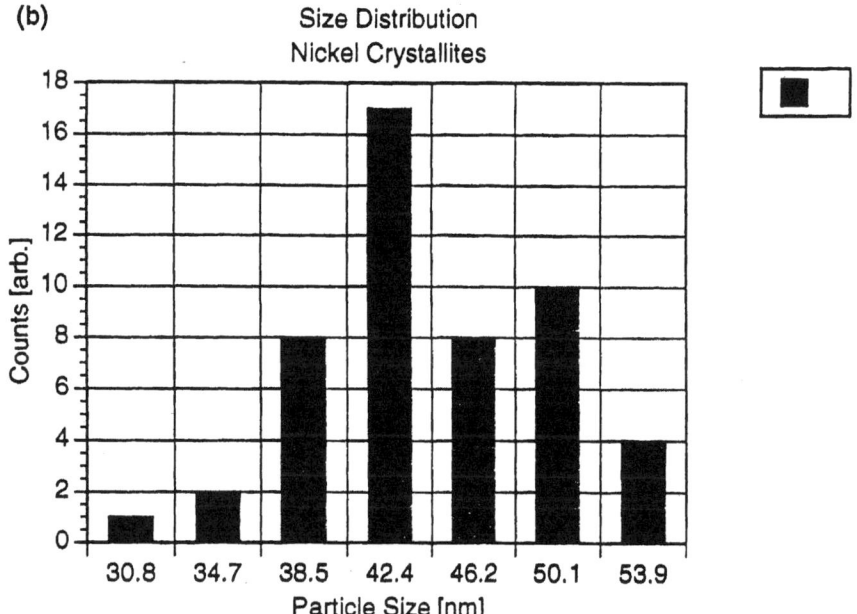

S-6. Ni powder as obtained at 182 °C. (a) XRD-pattern . (b) Size distribution. A high count of a macro crystalline cubic Ni phase is observed as compared to Ni powder as obtained at 170 °C

Chapter 5

Proposed New Materials: Boron Fullerenes, Nanotubes, and Nanotori

Vladimir Dadashev[1], Asta Gindulyte[1,2], William N. Lipscomb[3], Lou Massa[1,2], and Richard Squire[4]

[1]Department of Chemistry, Hunter College, 695 Park Avenue, New York, NY 10021
[2]The Graduate School, City University of New York, 365 Fifth Avenue, New York, NY 10016
[3]Gibbs Chemical Laboratory, Harvard University, Cambridge, MA 02138
[4]Division of Physical Sciences, Marshall University, 901 West Dupont Avenue, Belle, WV 25015

Actually existing carbon fullerenes, nanotubes, and nanotori are of interest, for reasons including their mechanical and electromagnetic properties. Because of a geometrical duality between these carbon compounds and their boron analogs, equally interesting proposed boron fullerenes, nanotubes, and nanotori are immediately suggested in a natural application of the Descartes-Euler formula, and its generalization the Euler-Poincare formula. There is a unique correspondence between the carbon and boron duals.

[†]His co-authors dedicate their contribution in this paper to Professor W. N. Lipscomb, on the occasion of his 80th birthday.

Introduction

For the description of polyhedra there exists a relationship which could easily have been known to the ancient Greeks. As it happened it was left to Descartes, and 100 years later independently to Euler, to discover that

$$P - C + F = 2 \qquad (1)$$

where P is the number of points (vertices), C is the number of connections (edges), and F is the number of faces in an arbitrary polyhedron. Equation 1 is the Descartes-Euler formula [1]. One may quickly test its validity against the Platonic solids, which are pictured in Figure 1.

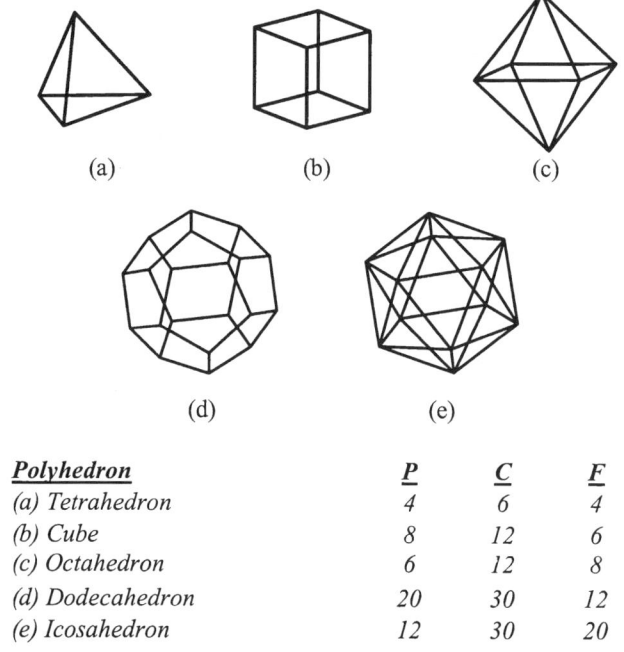

Polyhedron	*P*	*C*	*F*
(a) Tetrahedron	*4*	*6*	*4*
(b) Cube	*8*	*12*	*6*
(c) Octahedron	*6*	*12*	*8*
(d) Dodecahedron	*20*	*30*	*12*
(e) Icosahedron	*12*	*30*	*20*

Figure 1. Regular Polyhedra. While the present paper suggests the possibility of new materials, the Platonic solids stand for old materials, indeed. To ancient Greeks these were shapes of atoms of fire, earth, air, quintessence, and water, respectively.

For example, for a cube $P - C + F = 8 - 12 + 6 = 2$, as required. Similarly for an octahedron $P - C + F = 6 - 12 + 8 = 2$. The octahedron, moreover, is a geometrical dual of the cube, because the role of P and F are interchanged in the two structures. Thus the 8 points of the cube correspond to the 8 faces of the octahedron, and the 6 faces of the cube correspond to the 6 points of the octahedron. The duals share exactly the same value C (=12), and exactly, the same point group symmetry (O_h). For any case where the Descartes-Euler formula applies, duals are defined by the "interchange" of P and F, C held constant. Thus, Figure 1 indicates the dodecahedron and icosahedron are duals, and the tetrahedron is self-dual.

A new class of molecules that have created interest in recent years are (see Figure 2) carbon fullerenes [2]. They are known to have highly unusual and useful mechanical and electromagnetic properties. For example, they can be induced to display "high temperature" superconductivity. Were it possible to create boron duals to the carbon fullerenes we suggest they too would have useful mechanical and electromagnetic properties. This because they would have the same number of chemical bonds and the same point group symmetry as their carbon duals. And yet, there would possibly be valuable property differences, as for example due to the electron deficient bonds of boron, and vibrational effective mass differences between boron and carbon normal modes. And for some applications boron fullerenes would be uniquely useful, as in medical neutron capture therapy, based upon the high neutron cross section of the boron nucleus.

With the aid of the Descartes-Euler formula it has been shown that every single-cage carbon fullerene corresponds to a boron fullerene dual [3]. The archtype carbon fullerene is C_{60} (a buckeyball) shown in Figure 3 together with its boron fullerene dual $B_{32}H_{32}^{2-}$, whose probable stability had been suggested [4] prior to the experimental discovery of fullerenes [5]. The P and F values of C_{60} are "interchanged" with those of $B_{32}H_{32}^{2-}$, and both molecules have equal numbers of bond "connections", and the same icosahedral point group symmetry. A quantum chemical molecular orbital study shows that $B_{32}H_{32}^{2-}$ has a geometrical energy minimum and HOMO-LUMO energy gap comparable to $B_{12}H_{12}^{2-}$, and therefore too ought to be a stable molecule if it could be synthesized. Multi-cage carbon fullerene structures, which may be pictured as formed from single fullerenes sharing one or more common faces, are also possible. Here too, the Descartes-Euler formula may be used to construct the corresponding multi-cage boron fullerene duals. Again molecular orbital studies show they are of comparable stability.

It has proved possible to synthesize cylindrical single walled nanotubes (SWNT) of pure carbon [2]. Their geometry is related to the cage-like carbon fullerenes. Like them too, much is expected of their mechanical and electrical properties. Thus, it may be that cables made from SWNT's (see Figure 4) might be a hundred times stronger and six times lighter than a steel cable of equal diameter. The electrical conductivity of SWNT's has been a subject of intense study. SWNT's placed across external connections (see Figure 5) have been used to study their conductivity properties, which appear to depend upon tube radius and the angle of chiral twist about the tube axis [6-7]. SWNT's are likely candidates to form molecule size electronic devices.

Figure 2. The concept of a beautifully symmetric soccerball molecule has captured the public imagination. The glamor and cache of these molecules is symbolized by their representation as "molecule of the year" on the cover of the journal Science. C_{60} is the archetype example of the class of pure carbon cage molecules called fullerenes. Reprinted with permission from Science, 20 Dec. 1991. Copyright [1991] American Association for the Advancement of Science.

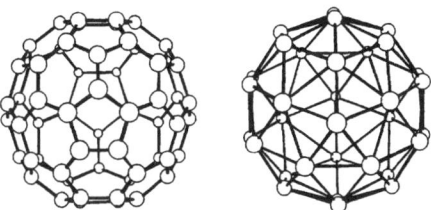

Figure 3. The carbon fullerene C_{60} and its geometrical dual $B_{32}H_{32}^{2-}$, which had been suggested long ago [4].

Figure 4. An artist rendition of a satellite tethered to the surface of the earth by a cable of carbon single wall nanotubes. Such a cable is expected to be a hundred times stronger than a steel cable of comparable cross section, and six times as light. The conception pictured here lends credibility to the notion of an "elevator to the sky". Reprinted with permission, American Scientist, July-August 1997. Copyright [1997] D. W. Miller.

Figure 5. A carbon single wall nanotube across gold external connections allowing the study of their conductivity properties. Reprinted by permission from Nature, 3 April 1997. Copyright [1997] Macmillan Magazines Ltd.

The geometry of open SWNT's is qualitatively different than that of the single-cage and multi-cage fullerenes. In the latter cases, each of the fullerenes may be topologically "deformed" onto the surface of a sphere. But, an open cylinder nanotube may not be so deformed. The Descartes-Euler formula does not hold for an open-ended nanotube. What then becomes of the dual concept, based upon the Descartes-Euler formula, when we come to open nanotube geometry? Evidently a generalization of the Descartes-Euler formula is required since it occurs that

$$P - C + F = 0 \qquad (2)$$

for open nanotubes. Using this equation, it has been shown the idea of boron nanotube duals may be saved [8]. Why the right hand side of equation (2) is zero, instead of two, we mention later.

Quite remarkably it occurs that in the same type of experiments which generate carbon SWNT's, as much as one percent of the total production may be in the form of nanotori, which spontaneously self assemble [9]. These are SWNT's which have been "bent around" and joined at their formerly open ends, creating a ("doughnut" shape) molecular torus (see Figure 6). It is expected that carbon nanotori will prove to be technologically interesting molecules. They might be used to form mechanical components of molecular machines. They might afford the possibility of forming a closed loop of molecular current and a concomitant molecular magnetic field. They might afford a molecular scale test of the Bohm-Aharonov effect.

The existence of carbon nanotori immediately provokes the question of whether or not boron nanotori, which are their duals, might have meaning. We suggest for the first time, the concept of geometrical duals, which has been applied to cage-like fullerenes, and nanotubes, can be extended to nanotori. Whereas, the Descartes-Euler formula has been used to define the dual relationship between carbon and boron cage-like fullerenes, here we discuss a generalization called the Euler-Poincare formula which defines the dual relationship between carbon and boron nanotori. Why the Euler characteristic, $P - C + F$, equals zero for an open nanotube, becomes evident. We shall see that nanotori, nanotubes, and planer sheet networks share a topological connection.

The Euler-Poincare formula invokes the use of Betti numbers [10] which may be calculated as the count of the number of critical points, of various types, associated with the geometrical structure of nanotori. The theory of Morse functions [11] relates critical points to topological structure. We shall show, an alternating sum of Betti numbers defines the Euler characteristic of a torus to be zero. This connects the topology of a nanotorus, nanotube, and planar sheet, which have the same Euler characteristic. We show that for every possible carbon nanotorus there is a geometrical dual boron nanotorus.

In what follows, firstly, we review how the simple idea of geometrical duals is immediately suggestive of the possible existence of boron fullerenes, given their carbon fullerene analogs. Molecular orbital calculations are supportive of boron fullerene stability. Also the way in which the concept of geometrical duals may be extended to nanotubes is reviewed. Secondly, we apply concepts of Morse theory and topology to the case of nanotori, to show how the notion of geometrical duals is generalized to this case. Finally, we summarize and discuss our results.

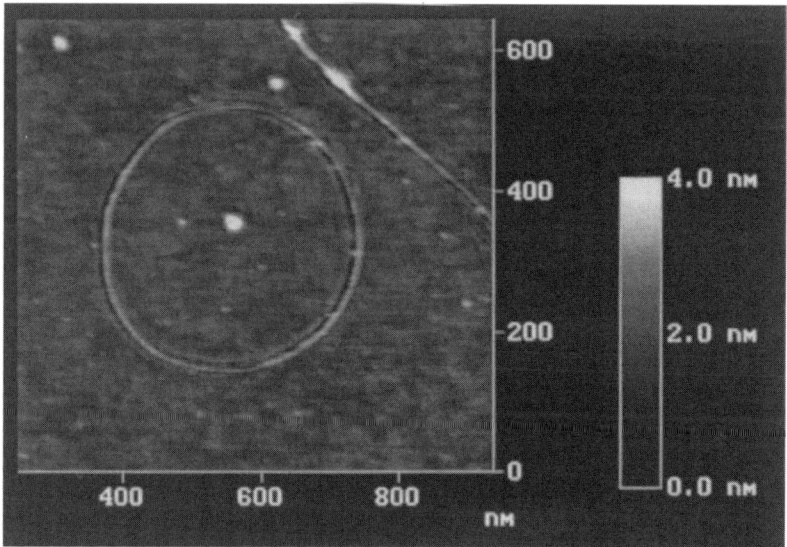

Figure 6. A pure carbon nanotorus. When first observed by Smalley and co-workers they called them "crop circles" as an expression of their skepticism that these were in fact true molecular tori. Recall "flying saucers" have been invoked in the popular press to explain "crop circles", as in a field of wheat. Reprinted by permission from Nature, J. Liu et. al., Nature, vol. 385, 780-781, 27 Feb 1997. Copyright [1997] Macmillan Magazines Ltd.

Boron Fullerenes and Nanotubes

In Table I we list a few examples of the geometrical correspondence which prevails between closo boron hydrides and the carbon fullerenes [3]. The geometrical structure of Buckminsterfullerene maps into that of the 32-vertex closo boron hydride, $B_{32}H_{32}$. Both molecules of symmetry I_h display correspondence of the geometrical centers of the 32 carbon (polygon) faces to the 32 boron vertices, the 60 boron faces to the 60 carbon vertices, and the 90 carbon contacts to the 90 boron contacts. In accordance with the Descartes-Euler formula, for both molecules the sum of the vertices and faces exceeds by 2 the number of contacts. A cursory review of Table I indicates that similar correspondences prevail for all the examples listed. Clearly the list of examples may be readily extended.

Table I. Single Cage Carbon Compounds and Their Boron Analogs.

Diagram	Formula	P	F	C+2	Sym	P	F	Formula	Diagram
	C_{20}	20	12	32	I_h	12	20	$B_{12}H_{12}$	
	C_{28}	28	16	44	T_d	16	28	$B_{16}H_{16}$	
	C_{60}	60	32	92	I_h	32	60	$B_{32}H_{32}$	
	C_{70}	70	37	107	D_{5h}	37	70	$B_{37}H_{37}$	
	C_{80}	80	42	122	I_h	42	80	$B_{42}H_{42}$	

We mention that the Descartes-Euler formula may be used to predict which closo boron hydride is the analog of each experimentally known fullerene. Thus, for a given fullerene, if the number of carbons is multiplied by 3/2 (a formal carbon contact number) to give the total number of carbon contacts, one may then calculate the number of carbon faces whose equality with the number of boron vertices yields immediately the molecular formula desired.

We review quantum chemical calculations [12] relating to geometry, charge state, and stability of boron fullerenes which should prove to be of interest for attempts at synthesis of these molecules. The methods ab initio, PRDDO, AM1, and LDF have been used to study boron fullerenes B_NH_N (N=1, 12, 16, 32, 37, 42). The different methods largely agree in their predictions of geometry, charge state and relative stability of the boron fullerenes, thus mutually reinforcing their individual predictions of these properties.

The geometries were optimized using the approximate ab initio method of partial retention of diatomic differential overlap (PRDDO) [13] and the semiempirical AM1 [14] method. Energies were then evaluated at the PRDDO optimal geometries by performing ab initio calculations with Gaussian 92 [15] at the 3-21G basis set level, and local density functional (LDF) calculations using the Perdew-Zunger exchange and correlation functionals [16]; the Gaussian basis sets were 11s/7p/1d contracted 5/3/1 for B and 6s/1p contracted 3/1 for H.

The geometries were constrained to have the generally assumed point group symmetries of the carbon fullerenes which are the geometrical duals of boron fullerenes. The independent bond distances for each of the boron fullerene molecular structures have been calculated. In general there is good agreement between the PRDDO and AM1 predictions of bond distances. The magnitudes obtained are not very different from standard distances used for known closo boron hydrides [4]. The molecules have preferred charge state -2, except for $B_{16}H_{16}$ which is predicted to have a charge of zero. Knowledge of the charge states is thought to be important for plans to synthesize these compounds. Thus, synthesis of the negatively charged boron fullerenes would presumably become stabilized in the presence of charge compensating positive ions.

In Table II the properties indicative of energetic stability are shown. For each molecule B_NH_N we calculated the quantities E/N the total energy per BH unit. In Table II we report the magnitude of the relative total energy per B-H unit compared with that of $B_{12}H_{12}$, denoted $(E/N)_r$.

We judge stability by examining $(E/N)_r$ for each of the molecules. The molecules are stable relative to separated BH groups. Known to be an extremely stable molecule, $B_{12}H_{12}^{2-}$ yields the largest stabilization per BH unit and the largest HOMO-LUMO gap.

Thus, the ab initio method applied to the proposed boron fullerenes yields results predicting geometry, charge state, and stability. These quantities are thought to be useful for the possible synthesis of these molecules whose properties might prove to be interesting, as has occurred with carbon fullerenes.

In a modest extension to less spherical molecules, which do not have internal bonding contacts, we also applied the Descartes-Euler formula to suggest the probable geometries of multicage boron fullerenes [17-18]. Such molecules were studied with molecular orbital calculations [19] to establish their energetic stability, bond lengths, and ionic charge state. In our earlier study of highly spherical single cage boron fullerenes we employed a variety of quantum methods which included HF/3-21G, PRDDO, LDF, and AM1 calculations. The different methods agreed fairly closely in their predictions of geometries, charge states and relative stabilities of the boron fullerenes. Here we report only ab initio (HF/STO-3G) calculations. Even for calculations which employ more sophisticated basis functions and higher level methods, we anticipate results will remain similar to those reported here.

Table II. Stability Factors for the Boron Fullerenes B_nH_n, based on HF/3-21G calculations [12].

N	SYM	$(E/N)_r$
12	I_h	1.0000
16	T_d	0.9995
32	I_h	0.9990
37	D_{5h}	0.9994
42	I_h	0.9995

In Table III we highlight the geometrical correspondence between multicage carbon fullerenes and their geometrical dual multicage boron fullerene analogues consistent with the Descartes-Euler formula. The carbon species C_{20}, C_{35}, and C_{47} are like polyhedra suggested [20] for Zr_8C_{12}, $Zr_{13}C_{22}$, and $Zr_{18}C_{29}$. Our C_{56} molecule of symmetry T_d, differs from the known analogue of C_{57} which has internal bonds and does not yield in an obvious manner a dual polyhedron satisfying the Descartes-Euler formula. This, of course, does not speak against the carbon 57-vertex structure as a reasonable one. The carbon species C_{40}, C_{56} (D_3), and C_{68} all arise as duals of their boron analogues obtained from boron "diamond" geometries, as in $B_{22}H_{22}^{2-}$ (D_{5d}), $B_{30}H_{30}^{2-}$ (D_3), $B_{36}H_{36}^{2-}$ (T), which in turn are obtained from the transition state "square" geometries as in $B_{22}H_{22}^{2-}$ (D_{5h}), $B_{30}H_{30}^{2-}$ (D_{3h}), $B_{36}H_{36}^{2-}$ (T_d). Such diamond-square-diamond (DSD) rearrangement mechanisms had been predicted earlier [21] and have been verified numerically in the cases suggested above.

Table III. Multicage Carbon Compounds and Their Boron Analogs.

Diagram	Formula	P	F	C+2	Sym	P	F	Formula	Diagram
	C_{20}	20	12	32	I_h[a]	12	20	$B_{12}H_{12}$	
	C_{35}	35	22	57	D_{5h}	22	35	$B_{22}H_{22}$	
	C_{40}	40	22	62	D_{5d}	22	40	$B_{22}H_{22}$	
	C_{47}	47	30	77	D_{3h}	30	47	$B_{30}H_{30}$	
	C_{56}	56	30	86	D_3	30	56	$B_{30}H_{30}$	
	C_{56}	56	36	92	T_d	36	56	$B_{36}H_{36}$	
	C_{68}	68	36	104	T	36	68	$B_{36}H_{36}$	

[a]NOTE: As a result of Jahn-Teller distortion C_{20} has a C_1 symmetry.

The geometries were optimized by minimizing of the STO-3G Hartree-Fock energies of all boron species listed in Table III. The geometries were constrained to have the assumed point group symmetries of the carbon fullerenes which are duals of the boron fullerenes studied. We leave for future studies a complete search of the entire molecular energy surface releasing all symmetry constraints. When the symmetry constraint for the "square" boron geometries was lifted, they spontaneously reverted to the "diamond" geometries of reduced overall molecular symmetry. The independent bond lengths associated with all multicage boron fullerenes of Table III have been calculated. The BH bonds and triangular BB bonds are not very different from standard distances previously used for known closo boron hydrides [4]. Also calculated were variations in boron atomic charges, derived from Mulliken population analysis. Although the qualitative trends in atomic charge analysis were probably good, we would expect a much improved basis to accentuate atomic charge differences among the boron atoms, and to give improved quantitative atomic charges.

In Table IV we report energetic stability factors for the multicage boron fullerenes. For each molecule we list $(E/N)_r$, the relative total energy per BH unit compared against the case of $B_{12}H_{12}^{2-}$, which is the most stable case. All these molecules are stable relative to separated BH groups. The stability of the multicage boron fullerenes is comparable to that of the single cage highly spherical boron fullerenes reported earlier.

Table IV. Stability Factors for the Multicage Boron Fullerenes $B_nH_n^{2-}$, based on HF/STO-3G calculations [19].

N	SYM	$(E/N)_r$
12	I_h	1.0000
22	D_{5h}	0.9994
22	D_{5d}	0.9997
30	D_{3h}	0.9991
30	D_3	0.9996
36	T_d	0.9990
36	T	0.9995

The preferred charge state for each of the boron multicage fullerenes was assumed to be 2- for all calculations reported here. This conforms with most earlier calculations for the single cage boron fullerenes [12] and with AM1 semiempirical calculations [22] of the multicage boron fullerenes. Knowledge of the preferred charge states may well be important to plans for synthesis of these compounds, which we would like to encourage.

Carbon nanotubes are cylindrical structures related to carbon fullerene structures. Indeed, carbon nanotube cylinders are often "capped" at their ends with "hemispherical" carbon fullerenes, illustrating the close relation of the two types of structure. Nanotube structures are of great interest because of their mechanical and one-dimensional electrical properties [2]. In our discussion of boron nanotubes [8], the duals of carbon nanotubes, we employ a generalization of Descartes-Euler formula, viz., the Euler-Poincare formula for a cylinder,

$$P - C + F = 0 \qquad (2)$$

where P, C, and F retain their previous meanings. Why the right side of the above equation has the value zero is apparent later.

Of course there is any number of different possible nanotubes. We restrict our discussion to a few types. They are sufficiently general that the analysis shown here can be adapted afterward to any other case. In Figure 7 we show selected examples of carbon nanotubes and their boron duals derived from the Euler-Poincare formula. The boron duals may be imagined as arising from a correspondence between a boron atom and the center of the carbon faces in the carbon nanotubes. In the figure the hydrogen atoms, which extend radially outward from each boron atom, are not shown in order to simplify the drawing. Moreover, it may be possible to synthesize boron nanotubes (and boron fullerenes) as pure boron compounds, i.e., without bonds to hydrogen atoms. We would expect the basic boron geometrical structure to be the same with or without the hydrogen atoms attached. But, we recognize the electronic structure associated with the boron geometry will be greatly modified by the formation of BH bonds.

In Table V we display a count of P, C, and F for each of the molecules displayed in Figure 7. It may be seen how the Euler-Poincare formula is satisfied

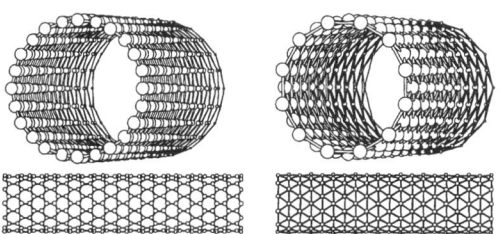

Figure 7a. Carbon (left) and boron (right) nanotubes, shown as end-on views (top) and side views (bottom). The case shown is the "zigzag" geometry corresponding to (n,m) = (12,0).

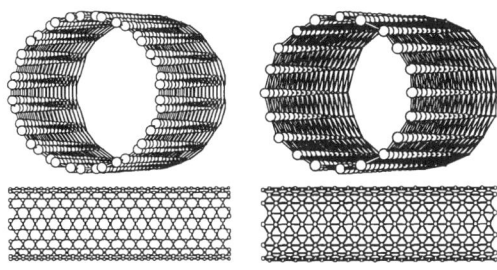

Figure 7b. Carbon and boron nanotubes viewed as in Figure 7a; the "armchair" geometry corresponding to (n,m) = (10,10).

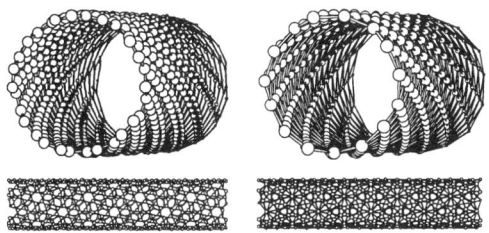

Figure 7c. Carbon and boron nanotubes viewed as in Figure 7a; an example chiral geometry corresponding to (n,m) = (10,5).

in each case. Notice in particular how the roles of P and F are interchanged in going from a carbon nanotube to its dual boron nanotube partner. Notice also that the number of bonds is common to the carbon and boron nanotube duals. Obtaining the count of the bonds requires consideration of all dangling bonds in both carbon and boron cases. Dangling bonds are associated with dangling faces, and these too must be counted to obtain the values listed in the table. We mention that the nanotube cylinders (including their dangling bonds) are topologically related to a torus, for which the Euler-Poicare formula also applies. The nanotubes also map into a flat sheet lattice structure, illustrated in Figure 8.

One may specify a single wall carbon nanotube "by bisecting a C_{60} molecule at the equator and joining the two resulting hemispheres with a cylindrical tube one monolayer thick and with the same diameter as C_{60}". Bisecting the C_{60} molecule along different directions will yield carbon nanotubes of differing symmetry. Thus, the "zigzag" and "armchair" carbon nanotubes of Figure 8 are related to bisections of C_{60} normal to a 3-fold axis and a 5-fold axis, respectively. Analogous, if "spherical" $B_{32}H_{32}$, the geometrical dual of C_{60}, is bisected normal to a 5-fold and 3-fold axis one specifies thereby the dual "zigzag" and "armchair" boron nanotubes respectively, also shown in Figure 10. Additional chiral nanotubes, an example of which is illustrated in Figure 10, can be defined in relation to more general "hemispherical" caps.

Table V. Carbon and Boron Dual Nanotubes. Listed are the factors of the Euler-Poincare formula for a cylinder, viz., P − C + F = 0.

Formula	P	F	C	(n,m)	Tube Type	P	F	Formula
C_{480}	480	240	720	(12,0)	zigzag	240	480	$B_{240}H_{240}$
C_{720}	720	360	1080	(10,10)	armchair	360	720	$B_{360}H_{360}$
C_{600}	600	300	900	(10,5)	chiral	300	600	$B_{300}H_{300}$

Figure 8 illustrates that a variety of nanotubes may be defined in terms of tube diameter, d, angle ϕ, and wrapping vector $W = na_1 + ma_2$. The "zigzag" and "armchair" nanotubes correspond to angles ϕ equal to 0° and 30° respectively. A general chiral nanotube, such as displayed in Figure 7c corresponds to a chiral angle ϕ falling between the extremes, 0° and 30°.

A mathematical representation of a general nanotube is formed by "rolling up" the plane of Figure 8, joining the two ends of the chiral vector W. Variation of the tube diameter, d, and the chiral angle, ϕ, are considered to control the properties of the various nanotubes formed [2]. That carbon nanotubes are semiconducting or metallic depending on variation of ϕ and d has been confirmed experimentally [6-7]. It is our expectation that if boron nanotubes could be synthesized, their electrical properties, also controlled by variation of ϕ and d parameters, would, like their carbon nanotube duals, be of very great interest.

Morse Theory, Betti Numbers, and Boron Nanotori

The Descartes-Euler formula, equation 1, has been used to define the class of molecules called boron fullerenes as the topological duals of carbon fullerenes. In order to extend the concept of duality to nanotubes, however, the Descartes-Euler formula must be generalized to the Euler-Poincare formula as in Equation 2. One may understand why the right side of Equation 2 is zero by use of Betti numbers [10]. Betti numbers may be calculated as a count of the number of critical points of each type (i.e., minima, saddle points, maxima) associated with the geometrical structure of a molecule. The Euler-Poincare formula [11] may be written in a very general way in terms of Betti numbers as

$$P - C + F = \beta_1 - \beta_2 + \beta_3 \quad (3).$$

The left side of Equation 3 is called the Euler characteristic, and it equals the right side, an "alternating sum" of Betti numbers.

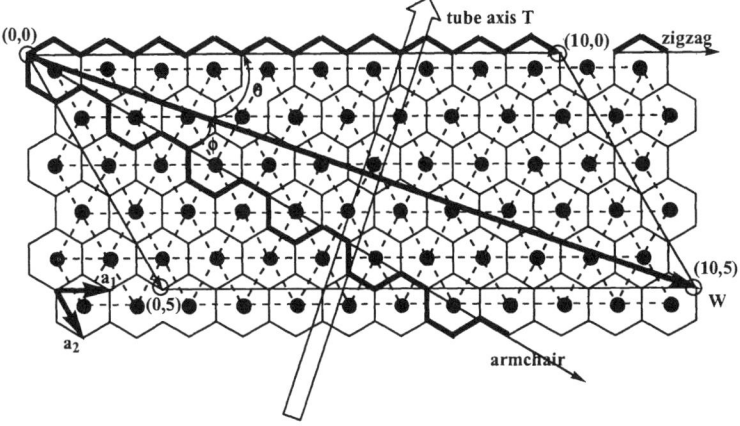

Figure 8. Relation of Nanotube and lattice for carbon (vertices of solid line hexagons) and boron (black dots), illustrating lattice vectors a1 and a2, and the "wrapping vector" W = n a1 + m a2. The extreme cases of "zigzag" (n,0) and "armchair" (n, n) are indicated. The chiral angle, ϕ, is that subtended by the general wrapping vector, W, and the armchair vector. The sum of the angles ϕ and θ is 30°. The tube axis vector, T, is orthogonal to the wrapping vector W. The tube diameter, d = |W|/π. Notice the duality: every boron point is at the center of a carbon face, and the number of boron and carbon bonds is the same.

Morse theory [11] is the study of the relationship between critical points and topology. A Morse function may, for the example of a torus, be defined as the height measured from its minimum as origin. As we go up in height, we go from a minimum, through a saddle point, through another saddle point, to a maximum. The general rule for evaluating the Euler characteristic in terms of Betti numbers is that we alternately add and subtract the number of critical points of increasing index . Thus, one minimum minus two saddle points plus one maximum gives zero, for the Euler characteristic of a torus. These alternating Betti numbers, which are local quantities give us a sum which is a global quantity. One may not find another topological structure with Poincare number zero which cannot be distorted into a torus.

The Euler-Poincare formula implies a duality which we have explored in the context of molecular nanotori. We show that for every possible carbon nanotorus there is a topological dual boron nanotorus. Moreover the alternating sum of Betti numbers is also zero for an open nanotube, and a planar network, which is consistent with the possibility of "converting" a nanotorus into an open nanotube, and thence into a planar network.

First we discuss Critical Points and the Topology of a torus abstractly. Then we apply the discussion to carbon and boron tori duals. Critical points, defined as the first derivative equals zero, $\partial f/\partial x = 0$, are fundamental in calculus in the method used to find minima, maxima, and inflection points of a simple function. This concept may be extended to describe topology. Let a smooth real valued function f represent the height of a point on a torus standing on its edge (see Figure 9), and solve for the critical points of f. Also, require that the second derivatives not be zero. Mathematically, this means that the Hessian matrix, H_f, of second derivatives is invertible, or det $H_f(p) \neq 0$ where $H_f(p) \equiv \{\partial^2 f /\partial x_i \partial x_j\}$.

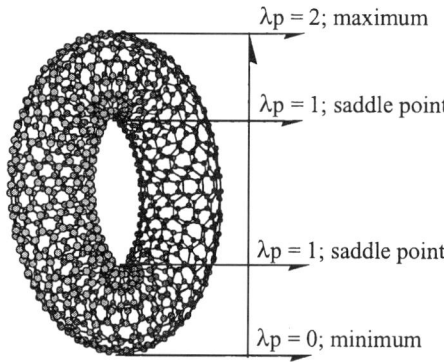

$\lambda p = 2$; maximum

$\lambda p = 1$; saddle point

$\lambda p = 1$; saddle point

$\lambda p = 0$; minimum

Figure 9. The critical points of a torus of index λ_P.

In general we can approximate our height function f in coordinates local to the neighborhood of each critical point by $f(x) = c - x_1^2 - x_2^2 \ldots + \ldots x^{n-1} + x^n$. At $x = p$, these local coordinates vanish $(x_1(p) \ldots x_n(p) = 0)$ and our function equals c, a constant, as it should. (For the torus there are 2 terms.) Each critical point has an index λ_p defined to be the number of negative eigenvalues of $H_f(p)$. The number of negative eigenvalues equals the number of independent directions along which f is decreasing. For example, at a saddlepoint there is one negative eigenvalue. From this index information we can create a polynomial called a Morse series by the following prescription: $M_t(f) = \Sigma_{all\ p}\ t^{\lambda_p}$. "Take t to the exponent of the number of negative eigenvalues for all critical points, p."

We now connect with topology through the Poincare series, $P_t(m)$ constructed from the Betti numbers of M. An intuitive method of getting Betti numbers is to count the holes in 0, 1, and 2 dimensions [23]. A "zero dimensional" hole measures the connectedness of the surface and as such, for the case of a torus, $\beta_0 = 1$. There are two nonhomologous 1-dimensional holes, in a torus, so $\beta_1 = 2$. Finally there is one 2-dimensional hole which is the inside of the torus, so $\beta_2 = 1$. We have then $P_t(m) = \beta_0 - \beta_1 + \beta_2 = 1 - 2 + 1 = 0$. Thus in the case of a torus we have a "perfect" Morse function since, $M_t(f) = P_t(m)$. We can derive local quantities such as the number of critical points by substituting $t = 1$ into $M_t(f)$, obtaining $M_1(f) = 1 + 2 + 1 = 4$. We can also find the Euler-Poincare characteristic χ, by evaluating $M_t(f)$ at $t = -1$. Thus $M_{-1}(f) = 1 - 2 + 1 = 0$.

A carbon torus may be imagined as being formed from an open carbon nanotube which "bends around" into a circle and joins at its open ends. Clearly a carbon graphite sheet, which may be rolled into a carbon nanotube, which in turn may be bent around and joined into a carbon nanotorus all have related geometrical structure. The discussion of boron nanotori, the duals of carbon nanotori, is based upon the Euler-Poincare formula, Equation 3. We have seen in the case of a torus, the Betti numbers of dimension zero, one, and two have respectively, the values 1, 2, and 1, and therefore the alternating sum of Betti numbers equals zero. The boron nanotori duals of carbon nanotori are defined by an interchange in the roles played by P and F in the Euler-Poincare formula which holds for nanotori. Thus the number of points in a carbon nanotorus equals the number of faces in its boron nanotorus dual, while the number of carbon faces equals the number of boron points, and the number of connections are equal. We have constructed computer graphical models of three sets of carbon/boron nanotori duals. The smallest of these duals are pictured in Figure 12. In Table VI we display a count of P, C, and F for all three sets of duals. It may be seen how the Euler-Poincare formula is satisfied in each case. Notice that the role of P and F are interchanged in going from carbon nanotorus to its boron dual nanotorus partner. Notice that the number of contacts is common to the carbon and boron nanotori duals. Now we can understand why nanotori, nanotubes, and planar networks are topologically related. For the latter two structures β_0, β_1, and β_2 are each zero, because there are no true minima, saddle points, or maxima. Therefore their alternating sum of Betti numbers is zero, just as in the case of nanotori. Thus they share a common value of Euler characteristic. For this reason they are deformable into one another. If boron nanotori could be synthesized, their mechanical and electromagnetic properties, like those of their carbon nanotori duals, would be of great interest.

Table VI. Carbon and Boron Dual Nanotori. Listed are the factors of the Euler-Poincare formula for a torus, viz., P − C + F = 0.

Formula	P	F	C	P	F	Formula
C_{960}	960	480	1440	480	960	$B_{480}H_{480}$
C_{1920}	1920	960	2880	960	1920	$B_{960}H_{960}$
C_{3840}	3840	1920	5760	1920	3840	$B_{1920}H_{1920}$

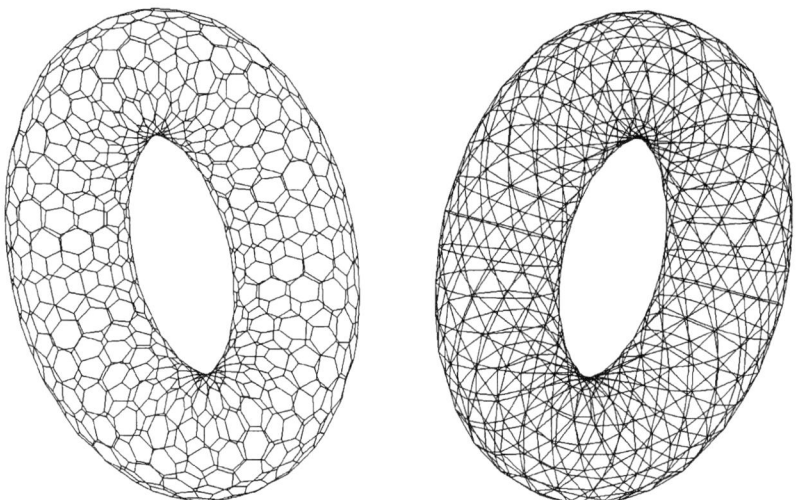

Figure 10. Nanotori geometrical duals, C_{960} (left) and B_{480} (right).

Summary and Discussion

We have shown how a simple idea, the Descartes-Euler formula, defines the boron fullerenes as the geometrical duals of the carbon fullerenes. As the carbon fullerenes are actually existing molecules, the Descartes-Euler formula is immediately suggestive of the possible existence of their boron duals. This is confirmed by quantum chemical molecular orbital calculations which indicate considerable stability for boron fullerene cage and multi-cage geometries. Therefore we encourage attempts at the synthesis of these proposed compounds.

There are many properties of carbon fullerenes that have aroused interest. Because of the similarities inherent in the topological relationship of duals, it is assumed boron fullerenes would be equally interesting. We draw attention to one property in particular, viz., superconductivity. The carbon fullerenes can be induced to display "high temperature" superconductivity. The mechanism for superconductivity in these compounds is unknown. If the BCS mechanism happened to be relevant, then [2]

$$kT_c = (h/2\pi)\omega e^{-1/(N_oV)} \qquad (4)$$

where Tc is the critical temperature for transition to the superconducting state, k is Boltzman's constant, h is Plank's constant, N_o is the density of states at the Fermi level, V is the Cooper pair potential, and ω; is the angular frequency of the normal mode vibration inducing the Cooper pair interaction. T_c is directly proportional to the normal mode frequency ω, given by $\omega = \sqrt{(k/\mu)}$, where k is the Hooke's law "spring constant" for the normal mode, and μ is the effective mass of the normal mode. We point out that since the mass of boron is less than that of carbon, and the number of boron atoms participating in a normal mode, fewer than those of carbon, this suggests that the effective mass of a boron fullerene normal mode should be less than that of an analogous normal mode of a carbon dual. On these grounds alone, it would be implied that, other things similar, a higher frequency ω, and a higher T_c might occur for boron fullerenes.

Open-ended carbon nanotubes have a geometry which certainly is related to fullerenes, but the topology differs. Whereas all single cage and multicage fullerenes are deformable onto a sphere, an open ended cylinder is not. Although for open ended SWNT's the Euler characteristic is no longer two, direct inspection shows it always equals zero. This allows a definition of boron duals of carbon SWNT's to occur, again by swapping the roles played by P and F (points and faces) in the Euler characteristic. Since the mechanical and electrical properties of carbon SWNT's are so compelling, synthesis of their analog boron duals is much to be desired.

Carbon nanotori spontaneously self assemble, in the same type of experiments, which lead to carbon fullerenes and nanotubes. We have shown a generalization of the Descartes-Euler formula, called the Euler-Poincare formula, may be used to define the boron duals of carbon nanotori. The Euler characteristic is obtained as an alternating sum of Betti numbers. The Betti numbers may be calculated as the count of the various types of critical points (minima, saddle points, maxima) associated with a Morse function defined on a molecular geometry. It occurs that the Betti numbers of a nanotorus, open ended nanotube, and planer nano-network deliver Euler characteristics equal to zero. Sharing the same Euler characteristic the structures are transformable into one another by simple cuts (or in the reverse direction by simple joins) which do not alter the overall connectedness of the structure.

Actually produced carbon nanotori typically contain tens of thousands of atoms. At first glance this would discourage an application of quantum mechanical molecular orbital calculations to study their stability, and that of their boron duals. Without discussing here the details, we simply point out a method used in Quantum Crystallography (QCr) exists which makes such a calculation practical [24-29]. A nanotorus can be decomposed into fragments consisting of a kernel of atoms and its neighborhood atoms. The entire nanotorus is a sum of kernels. Because of the axial symmetry of a nanotorus all kernels are identical, and therefore one kernel alone is sufficient to define the density matrix of an entire nanotorus. The near sightedness of the density matrix [31] which is tantamount to $S \to 0$, as R increases (S is an overlap integral of atomic orbitals separated by a distance R) ensures that the number of neighborhood atoms which fill out a fragment are not so numerous as to prevent the practical implementation of a molecular orbital calculation of an entire fragment. Similar calculations could usefully be applied to an open ended nanotube of infinite length (in such a case end effects could be neglected).

Carbon fullerenes, nanotubes, and nanotori actually exist, and their properties are fundamental and technologicaly important. The Descartes-Euler formula (and its generalization, the Euler-Poincare formula) immediately suggests the possible existence of their geometrical duals. We suggest very great interest would be attached to synthesis of any of the proposed dual boron fullerenes, nanotubes, and nanotori. It would seem that experiments quite analogous to those which successfully produce the carbon compounds might well produce their boron duals, too.

Acknowledgement

L. M. acknowledges an IBM Shared University Research (SUR) grant, a CUNY Research Award, a CUNY Collaborative Award, and a NASA JOVE/JAG grant.

References

1. Coxeter, H. S. M. *Introduction to Geometry;* John Wiley & Sons: New York, NY, 1961.
2. Dresselhaus, M. S.; Dresselhaus, G.; Eklund, P. C. *Science of Fullerenes and Carbon Nanotubes;* Academic: San Diego, CA, 1996.
3. Lipscomb, W.N.; Massa, L. *Inorg. Chem.* **1992**, *31*, 2297.
4. Bicerano, J.; Marynick, D. S.; Lipscomb, W. N. *Inorg. Chem.* **1978**, *17*, 3443.
5. Kroto, H.; Heath, J. R.; O'Brian, S. C.; Curl, R. F.; Smalley, R. E. *Nature* **1985**, *318*, 162.
6. Wildoer, J. W. G.; Venema, L. C.; Rinzler, A. G.; Smalley, R. E.; Dekker, C. *Nature* **1998**, *391*, 59.
7. Odom, T. W.; Huang, J.-L.; Kim, P.; Lieber, C. M. *Nature* **1998**, *391*, 62.
8. Gindulyte, A.; Lipscomb, W. N.; Massa, L. *Inorg. Chem.* **1998**; *37*, 6544.
9. Liu, J.; Dai, H.; Hafner, J. H.; Colbert, D. T.; Smalley, R. E. *Nature* **1997**, 385, 780.
10. Alexandroff, P. *Elementary Concepts of Topology;* Dover: New York, NY, 1961.
11. Nash, C.; Sen, S. *Topology and Geometry for Physicists;* Academic Press: New York, NY, 1983.
12. Derecskei-Kovacs, A.; Dunlap, B. I.; Lipscomb, W. N.; Lowrey, A.; Marynick, D. S.; Massa, L. *Inorg. Chem.* **1994**, *33*, 5617.
13. Halgren, T. A.; Lipscomb, W. N. *J. Chem. Phys.* **1973**, *58*, 1569.
14. Dewar, M. J. S.; Zoebisch, E. G.; Healy, E.; Stewart, J. J. P. *J. Am. Chem. Soc.* **1985**, *107*, 3902.
15. Gaussian 92/DFT, Revision G.1, Frisch, M. J.; et. al. Gaussian Inc.: Pittsburgh, PA, 1993.
16. Perdew, J. P.; Zunger, A. *Phys. Rev. B* **1981**, *23*, 5048.
17. Lipscomb, W. N.; Massa, L. *Phosphorus Sulfur* **1994**, *87*, 125.
18. Lipscomb, W. N.; Massa, L. *Inorg. Chem.* **1994**, *33*, 5155.
19. Gindulyte, A.; Krishnamachari, N.; Lipscomb, W. N.; Massa, L. *Inorg. Chem.* **1998**, 37, 6546.
20. Wei, S.; Guo, B. C.; Purnell, J.; Buzza, S.; Castleman, A.W. Jr. *Science* **1992**, *256*, 818.
21. Lipscomb, W. N. *Science* **1966**, *153*, 373.
22. Krishnamachari N., (Unpublished).
23. Hocking J.G., Young G.S., Topology, Dover, New York, 1961.
24. Massa, L.; Huang, L.; Karle, J. *Int. J. Quantum Chem. Symp.* **1995**, *29*, 371.
25. Huang, L.; Massa, L.; Karle, J. *Int. J. Quantum Chem. Symp.* **1996**, *30*, 479.
26. Karle, J.; Huang, L.; Massa, L. *Pure Appl. Chem.* **1998**, *70*, 319.

27. Huang, L.; Massa, L; Karle, J. In *Encyclopedia of Computational Chemistry;* Schleyer, P. v. R., Sr. Ed.; John Wiley & Sons: New York, NY, 1998; pp 1457-1464.
28. Huang, L.; Massa L.; Karle, J. In *Proceedings of the10th Organic Crystal Chemistry Symposium (Rydzyna, August 1997): J. Mol. Struct;* Barnes, J., Ed. (Salford); Eds. of special issue: Jones, D. W. (Bradford); Rychlewska, U. (Poznan), 1999; 474, pp 9-12.
29. Karle, J.; Huang, L.; Massa, L. In *Current Challenges on Large Supramolecular Assemblies;* Tsoucaris, G., Ed.; NATO Science Series C: Mathermatical and Physical Sciences; 1999; Vol. 519, pp 1-5.
30. Kohn, W. *Phys. Rev. Lett.* **1996**, *76*, 3168.

Chapter 6

Oscillations, Waves, and Patterns in Chemistry and Biology

Irving R. Epstein

Department of Chemistry and Volen Center for Complex Systems, MS 015, Brandeis University, Waltham, MA 02454

Chemical systems with complex kinetics exhibit a fascinating range of dynamical phenomena. These include periodic and aperiodic (chaotic) temporal oscillation as well as spatial patterns and waves. Many of these phenomena mimic similar behavior in living systems. With the addition of global feedback in an unstirred medium, the prototype chemical oscillator, the Belousov-Zhabotinsky reaction, gives rise to clusters, i.e., spatial domains that oscillate in phase, but out of phase with other domains in the system. Clusters are also thought to arise in systems of coupled neurons.

Introduction

One of the primary lessons that Bill Lipscomb has imparted to his students is that one should always be on the alert for new, unpredicted phenomena, and that one should not be afraid to tackle them simply because one has not been formally trained to do so. The phenomena that collectively make up the field of nonlinear chemical dynamics lie very far from my work with the Colonel, but my study of them has proved richly rewarding, as I hope to demonstrate in this chapter. I shall first present an overview of some of the most significant phenomena in the field and then focus on a single area of current interest, the creation of clusters generated by applying a global negative feedback to an oscillatory reaction-diffusion system.

Nonlinear Chemical Dynamics

Although some of the fundamental discoveries in nonlinear chemical dynamics were made at the beginning of the twentieth century and arguably even earlier, the field itself did not emerge until the mid-1960's, when Zhabotinsky's development (1) of the oscillatory reaction discovered by Belousov (2) finally convinced a skeptical chemical community that periodic reactions were indeed compatible with the Second Law of Thermodynamics as well as all other known rules of chemistry and physics. Since the discovery of the Belousov-Zhabotinsky (BZ) reaction, nonlinear chemical dynamics has grown rapidly in both breadth and depth (3).

Oscillatory Chemical Reactions

The BZ reaction was not the first to be discovered. Indeed, Bray (4) had found oscillations in the reaction of iodate and hydrogen peroxide over a third of a century before Belousov's work. Both the Bray and BZ oscillators were discovered by serendipity rather than by deliberate design or search, and both were initially scoffed at by the vast majority of chemists as artifacts or even frauds.

In the early 1980's, our group at Brandeis took up the challenge of developing a systematic design algorithm for the synthesis of new oscillating chemical reactions. Our first success, the chlorite-arsenite-iodate reaction (5), was quickly followed by the discovery/design of several dozen new chemical oscillators. The design algorithm, which is discussed in detail elsewhere (6), rests on combining complex, often autocatalytic, kinetics with flow reactor technology to keep the system far from equilibrium. Figure 1 shows a "periodic table" of the known chemical oscillators. In Table I, I show the mechanism of the bromate-chlorite-iodide oscillator (7) to demonstrate that even a "simple" inorganic oscillating reaction is likely to be extremely complex mechanistically. The behavior of these chemical oscillators is suggestive of that of biological clocks, and a major motivation for the study of the former has been the quest for insights into the latter.

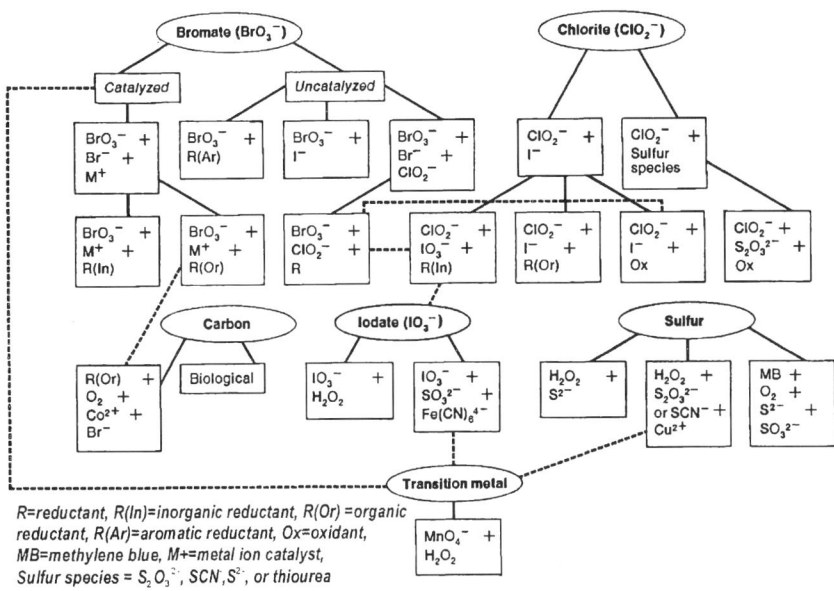

Figure 1. A "taxonomy" of chemical oscillators. Solid lines link related systems within a family of oscillators. Broken lines connect systems with common features.

Patterns and Waves

If an oscillating reaction, like the BZ, is run in an unstirred system, e.g., a petri dish, new phenomena may emerge. The simplest of these patterns consists of a set of concentric rings, a "target pattern," which develops in an initially red dish of BZ reagent by formation of a blue spot that grows to a disk. The inside of the disk turns back to red, leaving a blue ring, inside of which develops another blue spot that eventually becomes a second ring. The process continues as the rings grow and new rings develop. Occasionally, another center will form. When rings from different centers collide, the overlapping portions annihilate, generating complex patterns. If one breaks a ring mechanically, e.g., by passing a stirring rod through it, or if there are imperfections in the petri dish, then

Table I. Mechanism of the Bromate-Chlorite-Iodide Oscillating Reaction

No.	Reaction
(1)	$2H^+ + BrO_3^- + I^- \rightarrow HBrO_2 + HOI$
(2)	$HBrO_2 + HOI \rightarrow HIO_2 + HOBr$
(3)	$I^- + HOI + H^+ \leftrightarrow I_2 + H_2O$
(4)	$BrO_3^- + HOI + H^+ \rightarrow HBrO_2 + HIO_2$
(5)	$BrO_3^- + HIO_2 \rightarrow HBrO_2 + IO_3^-$
(6)	$HOBr + I_2 \leftrightarrow HOI + IBr$
(7)	$IBr + H_2O \leftrightarrow HOI + Br^- + H^+$
(8)	$HBrO_2 + Br^- + H^+ \rightarrow 2HOBr$
(9)	$HOBr + Br^- + H^+ \leftrightarrow Br_2 + H_2O$
(10)	$BrO_3^- + Br^- + 2H^+ \leftrightarrow HBrO_2 + HOBr$
(11)	$HIO_2 + Br^- + H^+ \leftrightarrow HOI + HOBr$
(12)	$HIO_2 + HOBr \rightarrow IO_3^- + Br^- + 2H^+$
(13)	$H^+ + HClO_2 + I^- \rightarrow HOCl + HOI$
(14)	$HClO_2 + HOI \rightarrow HOCl + HIO_2$
(15)	$HOCl + I^- \rightarrow HOI + Cl^-$
(16)	$HIO_2 + I^- + H^+ \leftrightarrow 2HOI$
(17)	$HOI + HOCl \rightarrow HIO_2 + Cl^- + H^+$
(18)	$HOCl + Br^- \rightarrow HOBr + Cl^-$
(19)	$HBrO_2 + HOCl \rightarrow BrO_3^- + Cl^- + 2H^+$
(20)	$HClO_2 + HOBr \rightarrow HBrO_2 + HOCl$

spirals develop rather than rings. In Figure 2a we see a spiral pattern in the BZ reaction, a pattern that bears a striking resemblance to that seen in Figure 2b, which shows the slime mold *Dictyostelium discoideum* in the process of coming together to form spores under conditions of limited nutrient supply.

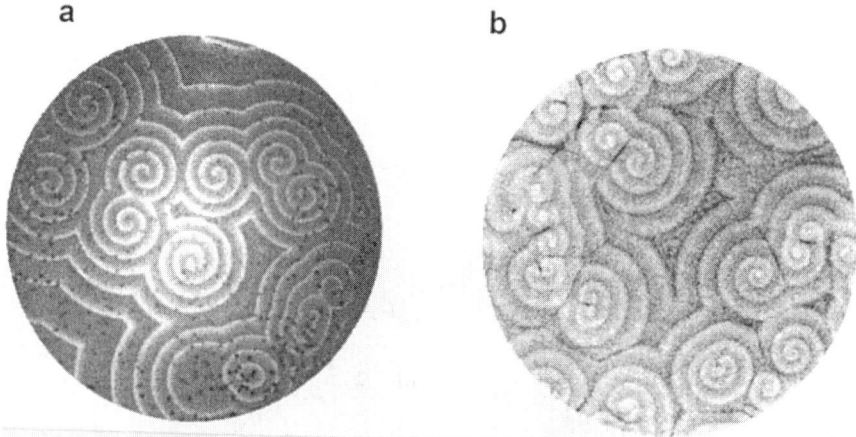

Figure 2. Spiral waves (a) in the BZ reaction, and (b) in the aggregating slime mold Dictyostelium discoideum.

Turing Patterns

The patterns described above consist of *traveling* waves, which move through space. The analogy between such waves in chemical systems and in neural systems was first pointed out nearly a century ago (8). Many patterns of interest in biological systems, such as animal coat patterns or the morphologies of organs, are *stationary*, i.e., they do not change in time (9). In a seminal paper in 1952, Turing (10) suggested that complex chemical reactions coupled with diffusion should be capable of giving rise to stationary patterns. Turing's idea has been the inspiration for a wide range of theoretical work in areas ranging from biology to astrophysics, but it required nearly four decades for the first unambiguous experimental evidence of Turing patterns, a study (11) of the chlorite-iodide-malonic acid oscillating reaction (12).

Turing patterns in the closely related chlorine dioxide-iodine-malonic acid system are shown in Figure 3. As a system parameter, in this case the input concentration of chorine dioxide, is varied, the pattern undergoes a bifurcation from spots to stripes. Comparison of the patterns in Figure 3 with those seen on the male and female tropical fish pictured in Figure 4 is at least suggestive that Turing patterns may play a role in pattern formation in certain species. The bifurcation parameter here is presumably a hormone related to sex determination in the fish.

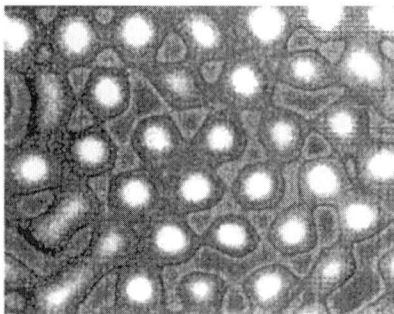

Figure 3. Turing patterns in the chlorine dioxide-iodine-malonic acid reaction in an unstirred gel reactor. Initial concentrations: [malonic acid] = $1x10^{-2}$ M; $[I_2]$ = $8x10^{-4}$ M; $[ClO_2]$ = $1x10^{-3}$ M (left), $1.5x10^{-3}$ M (right).

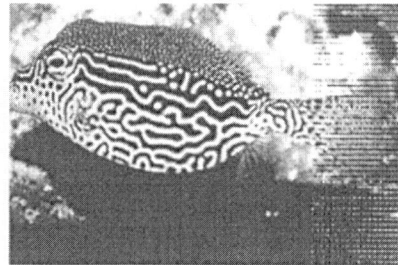

Figure 4. A tropical fish. Male (left) and female (right).

Oscillatory Cluster Patterns in the BZ Reaction with Global Feedback

As chemists have become more sophisticated in their ability to design and understand chemical oscillators, and as they have increasingly sought systems that are relevant to biological processes, oscillatory systems with feedback have become an area of growing interest. We describe here experiments (13) and computer simulations (14) on a photosensitive variant of the BZ reaction, in which the catalyst is a ruthenium bipyridyl complex, Ru(bpy)$_3$ (15).

The reaction is carried out in a thin layer of silica gel (16) with a global negative feedback imposed through illumination (13), as shown in Figure 5.

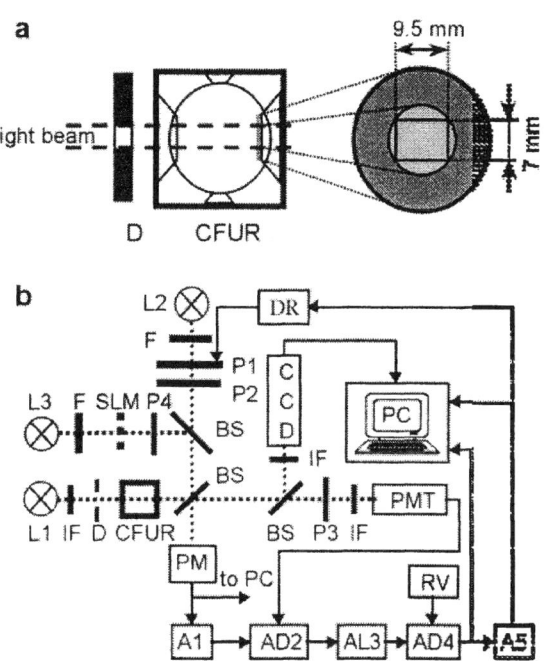

Figure 5. Experimental arrangement for global feedback. (a) Continuously fed unstirred reactor (CFUR) consists of continuously stirred tank reactor (CSTR) and thin layer of silica gel containing immobilized $Ru(bpy)_3$ polymerized on reactor optical window (gray circle). Diaphragm (D) selects illuminated working area of gel (light gray circle). Rectangular frame shows field of view of CCD camera. (b) Low-intensity analyzing light from stabilized 45 W light source (L1) passes through working area of gel and is collected by lens and directed to photomultiplier (PM). AD2 and AD4 are differential amplifiers, AL3 is logarithmic amplifier, A5 is dc amplifier. Driver (DR) rotates polarizer P1 and controls intensity of actinic light from 450 W Xe Arc lamp (L2). P1 – P4 are polarizers, IFs are interference filters, Fs are bandpass filters, BS are beam splitters; lenses and collimators are not shown. 150 W Xe Arc lamp (L2) serves to set patterns of initial conditions. Image of spatial light modulator (SLM) is focused in plane of silica gel. Actinic light intensity is measured by power meter (PMT). Polarizers P2-P3 and P4-P3 are crossed to separate optical channels.

(Reproduced with permission from reference 11. Copyright 2000 Nature (London.)

The average concentration of $Ru(bpy)_3^{3+}$, Z_{av}, taken over the working area of the gel, is employed to control the intensity I of actinic light from lamp L1 by varying the angle between polarizers P1 and P2. The gain of amplifier A5 is used to control the strength of the feedback via the feedback coefficient g. The (negative or inhibitory) feedback acts in such a way that if Z_{av} exceeds a target concentration set by the experimenter, the actinic light intensity is increased, thereby producing bromide ion, which inhibits the oxidation of $Ru(bpy)_3^{2+}$, a key step in the oscillation (17). We investigate how pattern formation depends on g, with the initial reagent concentrations chosen so that without any feedback the system generates bulk oscillations. In Figure 6, we present a "phase diagram" that summarizes the results obtained as g is varied.

Clusters

Various traveling wave patterns are found when g is less than g^*, a value that depends on the initial concentrations of reagents. When the feedback coefficient exceeds g^*, *cluster* patterns can occur. Clusters consist of sets of domains in which nearly all of the elements in a domain oscillate with the same amplitude and phase (18-19). Clusters have been observed in model studies of arrays of coupled neurons (20), but they are rare in chemical systems. In the simplest case, a system consists of two clusters that oscillate 180° out of phase; each cluster can consist of several spatial domains.

When g exceeds 2×10^4 M^{-1}, *standing clusters* arise. Figure 7 depicts one period of oscillation of a pattern that arose from uniform initial conditions. These clusters resemble standing waves, except that they lack a characteristic wavelength, owing to the global nature of the feedback. At $t = 4$ s, Z_{av} reaches a maximum. After this instant, the white domains start to fade and the system gradually evolves to the uniform reduced (dark) state. During the second half-period, the regions that were dark during the first half-period become bright, and at $t = 29$ s the pattern displays another maximum in Z_{av}.

Since the domains oscillate out of phase, the period, T_{av}, of oscillations of Z_{av} is half the period of the local oscillations. Other patterns of standing clusters can be obtained by varying the initial conditions, e.g., by illuminating the system briefly through a mask with the desired pattern.

At larger g, two other types of patterns arise. If the initial reagent concentrations lie far from the boundary of the oscillatory region of the parameter space, we observe *irregular clusters*. Local oscillations in these patterns are aperiodic, but the average concentration, Z_{av}, (and, consequently, I) oscillates approximately periodically with period, T_{av} (Figure 8b). Figure 8a displays snapshots of irregular clusters at intervals of T_{av}.

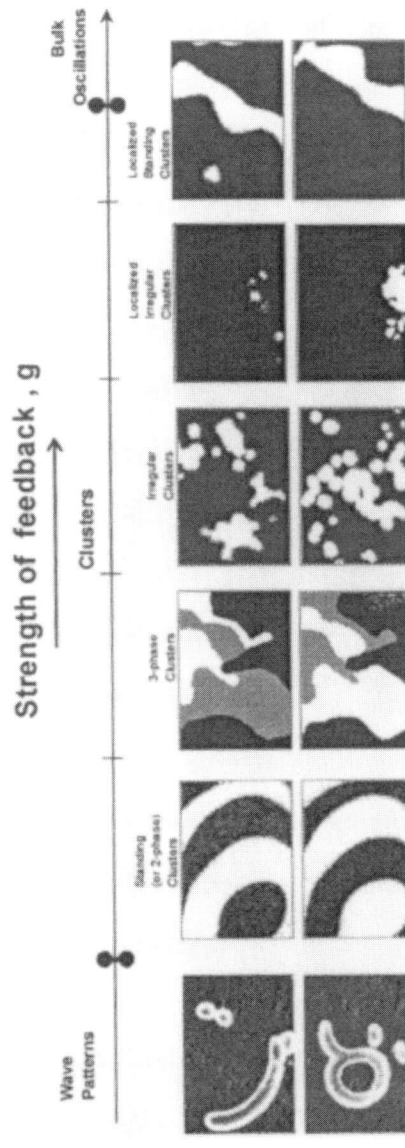

Figure 6 Family of cluster patterns observed in experiments for variable strength g of the global negative feedback. Snapshots of cluster patterns are separated in time by one period of global oscillation.

(Reproduced with permission from reference 11. Copyright 2000 Nature (London).)

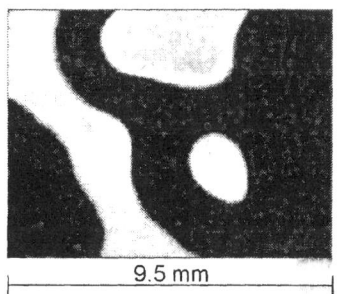 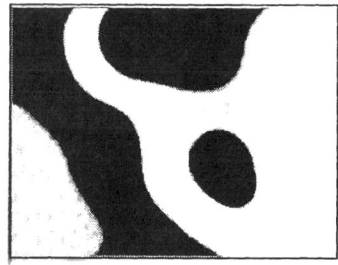

Figure 7. Standing clusters arising from uniform initial conditions. Frames are separated by one-half period.

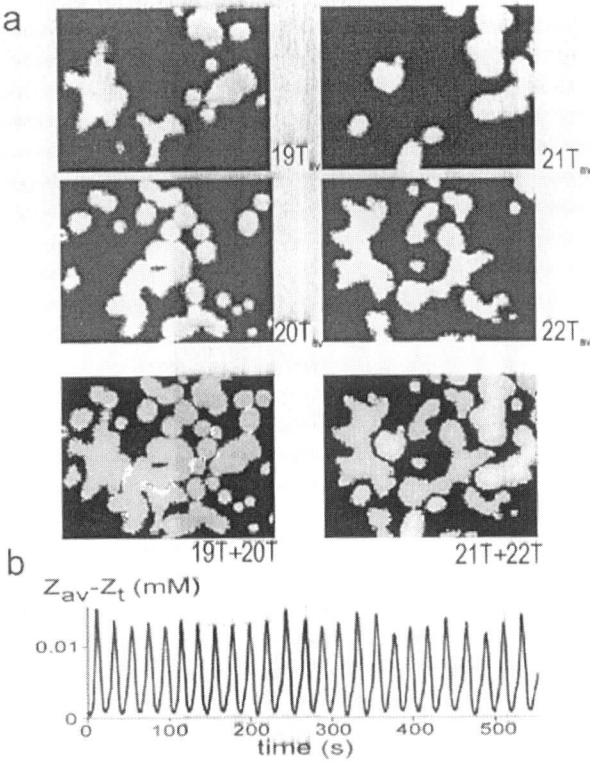

Figure 8. Irregular clusters. Bottom panel shows average concentration of oxidized form of catalyst. Peaks are separated by T_{av}. Upper two rows show pattern at intervals separated by one period, while third row shows superposition of upper two rows, demonstrating that domain boundaries are not stationary in time.

(Reproduced with permission from reference 11. Copyright 2000 Nature (London.))

Localized clusters arise if the initial reagent concentrations are close to the parameter space boundary between the oscillatory and the reduced steady state regions. Domains of antiphase oscillations in localized clusters occupy only part of the area, while no pattern can be seen in the remaining part of the system. Figure 9 shows two snapshots separated by half a period of oscillations. There are two adjacent large domains of antiphase oscillations separated by a nodal line. The two small domains at the right boundary of the first frame and one small domain near the left boundary are transients that subsequently die off.

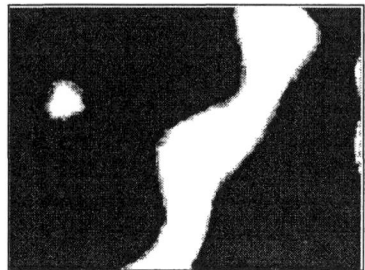 

Figure 9. Localized clusters.
(Reproduced with permission from reference 11. Copyright 2000 Nature (London.)

With increasing feedback coefficient, the portion of the medium occupied by the localized clusters shrinks. At higher g, clusters disappear, giving way to small amplitude bulk oscillations.

Computer Simulations

To simulate the pattern formation observed in our experiments, we employ a model of the BZ reaction (21). We add a global linear feedback term to account for the bromide ion production that results from the actinic illumination, $v_{GF} = g\varphi I_{max}(Z_{av} - Z_{ss})$, where φ is the quantum yield. The results of our simulations mimic those of the experiments. Bulk oscillations and travelling waves are observed in the model for smaller values of g. At higher g values, standing, irregular and localized clusters are observed in the same sequence and with the same patterns of hysteresis as in the experiments

Time-space plots in Figure 10 clarify the dynamics of the system. We see clearly the temporal periodicity of the standing and localized clusters as well as the aperiodicity in both time and space of the irregular clusters. Localized clusters may be of significance as a mechanism for distributed memory in natural systems. An enormous variety of patterns of this type can be created from different initial conditions, and their localized character makes them more convenient for information storage and retrieval than patterns that occupy the entire system. Cluster formation in models (20) of neural networks with negative global coupling has received considerable attention, but the experimental observation of such patterns is difficult. We hope that our findings will stimulate a search for analogous dynamic patterns in natural neural systems.

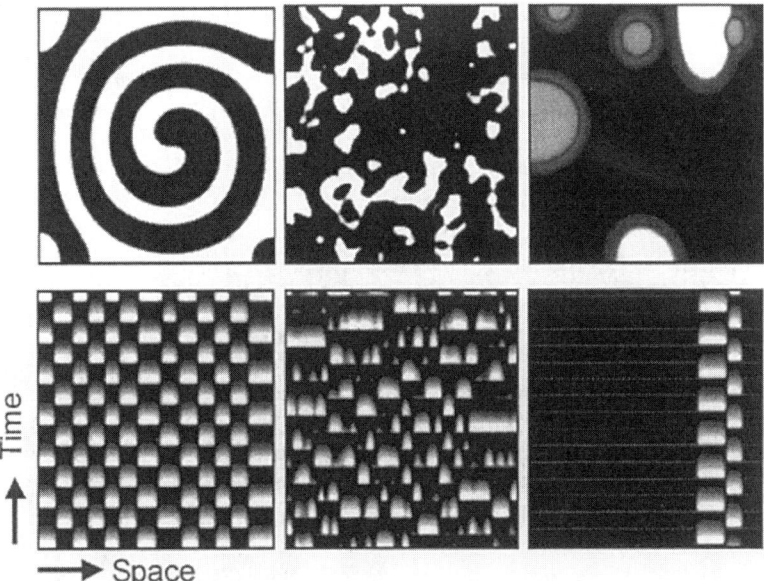

Figure 10. Computer simulations. (a) Standing clusters, (b) Irregular clusters, (c) Localized clusters. Top frames display snapshots of patterns; bottom frames show spatio-temporal behavior along bottom left–top right diagonal of corresponding squares during 380 s.

Acknowledgments. I am delighted to acknowledge the many contributions of Professor William N. Lipscomb to my scientific education. The work described here comes out of collaborations with many able scientific colleagues, the most recent of whom, Anatol Zhabotinsky, Milos Dolnik, Vladimir Vanag and Lingfa Yang, are responsible for the work described in the final section. I thank Dr. Dolnik in particular for his assistance in preparing this manuscript. This research has been supported by grants from the National Science Foundation Chemistry Division.

Literature Cited

1. Zhabotinsky, A. M. *Biofizika* **1964,** *9,* 306.

2. Belousov, B. P. *Sbornik Referatov po Radiatsionni Meditsine;* Medgiz: Moscow, 1958; p. 145.
3. Epstein, I. R.; Pojman, J. A. *Introduction to Nonlinear Chemical Dynamics. Oscillations, Waves, Patterns and Chaos*; Oxford University Press: New York, 1998.
4. Bray, W. C. *J. Am. Chem. Soc.* **1921,** *43,* 1262.
5. De Kepper, P.; Kustin, K.; Epstein, I. R. *J. Am. Chem. Soc.* **1981,** *103,* 2133.
6. Epstein, I. R.; Kustin, K.; De Kepper, P.; Orbán, M. *Sci. Amer.* **1983,** *248(3),* 112.
7. Citri, O.; Epstein, I. R. *J. Phys. Chem.* **1988,** *92,* 1865.
8. Luther, R.-L. *Z. Elekt. Angew. Phys. Chem.* **1906,** *12,* 506.
9. Murray, J. D. *Mathematical Biology, 2nd ed.*; Springer-Verlag: Berlin, 1993.
10. Turing, A. M. *Phil. Trans. Roy. Soc. B* **1952,** *237,* 37.
11. Castets, V.; Dulos, E.; Boissonade, J.; De Kepper, P. *Phys. Rev. Lett.* **1990,** *64,* 2953.
12. De Kepper, P.; Epstein, I. R.; Orbán, M.; Kustin, K. *J. Phys. Chem.* **1982,** *86,* 170.
13. Vanag, V. K.; Yang, L.; Dolnik, M.; Zhabotinsky, A. M.; Epstein, I. R. *Nature* **2000,** *406,* 389.
14. Yang, L.; Dolnik, M.; Zhabotinsky A. M.; Epstein, I. R. *Phys. Rev. E* **2000,** in press.
15. Kuhnert, L.; Agladze, K. I.; Krinsky, V. I *Nature* **1989,** *337,* 244.
16. Yamaguchi, T.; Kuhnert, L.; Nagy-Ungvarai, Zs.; Müller, S. C.; Hess, B. *J. Chem. Phys.* **1991,** *95,* 5831.
17. Grill, S.; Zykov, V. S;. Müller, S. C. *Phys. Rev. Lett.* **1995,** *75,* 3368.
18. Golomb, D.; Hansel, D.; Shraiman, B.; Sompolinsky, H *Phys. Rev. A* **1992,** *45,* 3516.
19. Hakim, V.; Rappel, W.-J.; *Phys. Rev. A* **1992,** *46,* 7347.
20. Golomb, D.; Rinzel, J. *Physica D* **1994,** *72,* 259.
21. Zhabotinsky, A. M.; Buchholtz, F.; Kiyatkin, A. B.; Epstein, I. R. *J. Phys. Chem.* **1993,** *97,* 7578.

Theory

Chapter 7

Electron Propagator Theory of Ionization Energies and Dyson Orbitals for μ-Hydrido, Bridge-Bonded Molecules: Diborane, Digallane, and Gallaborane

Gustavo Seabra, V.G. Zakrzewski, and J. V. Ortiz*

Department of Chemistry, Kansas State University,
Manhattan, KS 66506–3701
*ortiz@ksu.edu

Electron propagator calculations accurately account for the photoelectron spectra of diborane, digallane and gallaborane. Whereas electron correlation corrections to canonical, Hartree–Fock orbital energies are necessary for accurate results, large pole strengths confirm the qualitative validity of the Koopmans description of the first five cationic states. Only for the sixth final state is there evidence of significant multiconfigurational character. Dyson orbitals corresponding to each ionization energy are dominated by a single, Hartree–Fock orbital. The order of final states and the phase relationships between atoms in the Dyson orbitals are conserved for all three molecules. As the number of Ga atoms rises, ionization energies, splittings between cationic states and direct interactions between nonhydrogen atoms decrease.

A chemist learns to associate energetic quantities to orbitals at an early stage in his education. Aufbau principles for atomic structure are encountered typically in the first few weeks of an introductory course in chemistry. Hückel molecular orbital theory enables organic chemists to discern patterns in structure, spectra and reactivity without the need for complicated calculations. Model one–electron systems such as the particle in a box and H_2^+ are treated at length in typical physical chemistry courses.

Similar habits are reinforced by Hartree–Fock theory, where Koopmans's theorem [1] enables one to use canonical orbital energies as estimates of ionization energies and electron affinities. Here, orbitals that are variationally optimized for an N–electron state are used to describe final states with $N \pm 1$ electrons. Energetic consequences of orbital relaxation in the final states are ignored, as is electron correlation.

In the Kohn–Sham implementation of density functional theory [2], orbital energies have a distinct meaning. According to Janak's theorem, the eigenvalues of the Kohn–Sham equations are derivatives of the total energy, which may include exchange and correlation terms, with respect to orbital occupation numbers [3]. Orbital energies therefore are closely related to electronegativity concepts associated with density functional formalisms.

The most direct experimental tests that pertain to these models of electronic structure are measurements of electron binding energies. Photoelectron spectra, for example, provide ionization energies that may be compared with canonical, Hartree–Fock orbital energies. Discrepancies between theory and experiment are generally redressed by improved total energy calculations that consider final–state orbital relaxation and electron correlation in initial and final states. Often these corrections are necessary for correct assignment of the spectra.

While quantitative agreement between calculations and spectroscopic experiments solves the immediate problem of assigning peaks to states, the interpretation of the result is obscured by the complicated structure of many–electron wavefunctions and energies that account for orbital relaxation and electron correlation. Orbital concepts are apparently sacrificed in the pursuit of reliable energy differences that are experimentally observable.

Ab initio electron propagator theory provides a way to avoid this dilemma [4–6]. Exact ionization energies and electron affinities can, in principle, be calculated with this formalism. To each of these electron binding energies, electron propagator theory assigns an orbital that is rigorously related to the many–electron wavefunctions of the initial and final states. A formally exact association of energies and orbitals is realized.

Fortunately, the resulting picture of electronic structure also is associated with efficient, practical algorithms that can be executed routinely on molecules that are large by the contemporary standards of quantum chemistry. Two electron propagator approximations that have been derived for calculations on large, closed–shell, organic molecules are applied here to the assignment of photoelectron spectra of diborane, digallane and gallaborane [7].

Electron Propagator Theory

Electron binding energies and corresponding orbitals may be obtained by solving a pseudoeigenvalue problem

$$H^{eff}\phi_p = \varepsilon_p \phi_p. \tag{1}$$

The uncorrelated case is represented by the Hartree–Fock equations, where the effective, one–electron operator contains the usual kinetic (T), nuclear attraction (U), Coulomb (J) and exchange (K) components such that

$$H^{eff} = T + U + J - K = F. \tag{2}$$

Coulomb and exchange operators depend on the occupied ϕ orbitals and therefore self–consistent field iterations are performed until the occupied orbitals are eigenfunctions of the Fock operator, F.

Electron propagator formalism [4–6] allows for generalizations that include the effects of correlation. Here, the pseudoeigenvalue problem has the following structure

$$[F + \Sigma(E)]\phi_p = \varepsilon_p \phi_p \tag{3}$$

such that the Fock operator is supplemented by the self–energy operator, $\Sigma(E)$. The latter operator depends on an energy parameter, E, and is nonlocal. All orbital relaxation effects between initial and final states may be included in the self–energy operator, as well as all differences in the correlation energies of these states.

The energy dependence of the correlated effective operator, H^{eff}, where

$$H^{eff}(E) = F + \Sigma(E), \tag{4}$$

indicates that the correlated pseudoeigenvalue problem must also contain iterations with respect to E. A search for electron binding energies requires that a guess energy be inserted into $H^{eff}(E)$, leading to new eigenvalues which may be reinserted into $H^{eff}(E)$ in a cyclic manner until consistency is obtained between the operator and its eigenvalues. At convergence,

$$[F + \Sigma(\varepsilon_p)]\phi_p = \varepsilon_p \phi_p. \tag{5}$$

Approximations to $\Sigma(E)$ may be systematically extended until, in principle, exact ionization energies and electron affinities emerge as ε_p values.

Eigenfunctions that accompany these eigenvalues have a clear physical meaning that corresponds to electron attachment or detachment. These functions are known as Dyson orbitals, Feynman–Dyson amplitudes or generalized overlap amplitudes. For ionization energies, these orbitals are given by

$$\phi_p(x_1) = \sqrt{N} \int \Psi_N(x_1, x_2, x_3, \ldots, x_N)$$

$$\Psi^*_{N-1,p}(x_2,x_3,x_4,\ldots,x_N)dx_2dx_3dx_4\ldots dx_N, \tag{6}$$

where x_i is the space–spin coordinate of electron i. The Dyson orbital corresponding to the energy difference between the N–electron state Ψ_N and the pth electron–detached state $\Psi_{N-1,p}$ may be used to calculate cross sections for various types of photoionization and electron scattering processes. For example, photoionization intensities, I, may be determined via

$$I_p = \kappa |\langle \phi_p | \nabla \chi \rangle|^2 \tag{7}$$

where χ is a description of the ejected photoelectron and κ is a constant. For electron affinities, the formula for the Dyson orbital reads

$$\phi_p(x_1) = \sqrt{N+1} \int \Psi_{N+1,p}(x_1,x_2,x_3,\ldots,x_N,x_{N+1})$$

$$\Psi^*_N(x_2,x_3,x_4,\ldots,x_N,x_{N+1})dx_2dx_3dx_4\ldots dx_Ndx_{N+1}. \tag{8}$$

In the Hartree–Fock, frozen–orbital case, the reference state consists of a single determinant of spin–orbitals and the final states differ by the addition or subtraction of an electron in a canonical spin–orbital. The overlaps between states of unequal numbers of electrons represented by the Dyson orbital formulae reduce to occupied or virtual orbitals that are solutions of the canonical Hartree–Fock equations. Dyson orbitals also may be obtained from configuration interaction wavefunctions. Electron propagator calculations, however, avoid the evaluation of complicated, many–electron wavefunctions (and their energies) in favor of direct evaluation of electron binding energies and their associated Dyson orbitals. Note that for correlated calculations, the Dyson orbitals are not necessarily normalized. The pole strength, P, is given by

$$P_p = \int |\phi_p(x)|^2 dx. \tag{9}$$

In the Hartree–Fock, frozen–orbital case, P acquires its maximum value, unity. Correlation final states are characterized by low pole strengths. Transition intensities, such as those in equation 7, are proportional to P.

Canonical Hartree–Fock orbital energies are a convenient and powerful foundation for estimating the smallest vertical electron binding energies of closed–shell molecules. This approximation, which is based on Koopmans's theorem, is the most often used method for assigning the lowest peaks in photoelectron spectra. However, there are many classes of important molecules for which the Koopmans approximation fails to predict the correct order of final states. Average errors made by this frozen–orbital, uncorrelated method are between 1 and 2 eV for valence ionization energies [8]. More confident assignments require that these errors be reduced.

Perturbative expressions for the self–energy operator can achieve this goal for large, closed–shell molecules. The original derivation of the partial, third–order (P3) method was accompanied by test calculations on challenging, but small, closed–shell molecules with various basis sets [9]. The average absolute error was approximately 0.2 eV for vertical ionization energies below 20 eV. Since 1996, the P3 method has been applied chiefly to the ionization energies of organic molecules. For nitrogen–containing heterocyclics, P3 corrections to Koopmans results are essential in making assignments of photoelectron spectra [10]. Correlation corrections generally are much larger for hole states with large contributions from nonbonding, nitrogen–centered functions than for delocalized π levels. Therefore, P3 results often produce a different ordering of the cationic states. The accuracy of P3 predictions generally suffices to make reliable assignments. Several reviews on electron propagator theory have discussed relationships between P3 and other methods [4–6].

The P3 method is generally implemented in the diagonal self–energy approximation. Here, off–diagonal elements of the self–energy matrix in the canonical, Hartree–Fock orbital basis are set to zero. The pseudoeigenvalue problem therefore reduces to separate equations for each canonical, Hartree–Fock orbital:

$$\varepsilon_p^{HF} + \Sigma_{pp}(E) = E. \tag{10}$$

Only energy iterations are needed in the diagonal self–energy approximation. For example, $\Sigma_{pp}(E)$ may be evaluated at $E = \varepsilon_p^{HF}$ to obtain a new guess for E. The latter value is reinserted in $\Sigma_{pp}(E)$ and the process continues until consecutive energy guesses agree to within 0.01 millihartrees of each other. Neglect of off–diagonal elements of the self–energy matrix also implies that the corresponding Dyson orbital is given by

$$\phi_p = \sqrt{P} \phi_p^{HF}, \tag{11}$$

where P, the pole strength, is determined by

$$P = \left[1 - \frac{d\Sigma_{pp}(E)}{dE}\right]^{-1}. \tag{12}$$

In the latter expression, the derivative is evaluated at the converged energy. Diagonal self–energy approximations therefore subject a frozen Hartree–Fock orbital, ϕ_p^{HF}, to an energy–dependent, correlation potential, whose matrix elements are $\Sigma_{pp}(E)$.

Diagonal matrix elements of the P3 self–energy approximation may be expressed in terms of canonical Hartree–Fock orbital energies and electron repulsion integrals in this basis. For ionization energies, the bottleneck arithmetic operation has a scaling factor of O^2V^3, where O is the number of occupied orbitals and V is the number of virtual orbitals. Electron repulsion integrals in the Hartree–Fock basis with four virtual indices are not needed. In general, P3 calculations are as easily executed as second–order perturbation calculations.

An extension of the P3 method that has been useful in studying higher ionization energies is known as the nondiagonal, renormalized second–order (NR2) approximation [11]. Nondiagonal elements of the self–energy matrix are not neglected here. Terms of fourth and all higher orders are included in the renormalized self–energy expression as well. All self–energy terms that are present in the P3 method are retained. For valence ionization energies of closed–shell molecules, NR2 is somewhat more accurate than P3, but it is also applicable to correlation final states, where the Koopmans description of ionized states is qualitatively invalid [12, 13]. One pays for the enhanced versatility of the NR2 method with slightly increased arithmetic and storage requirements.

Methods

All calculations were done with the Gaussian 98 program [14]. Geometries were optimized at the MP2 level. 6–311G(2d,p) basis sets for hydrogen, boron and gallium [15, 16] were used. D_{2h} structures are found for diborane and digallane. A bridged, C_{2v} geometry obtains for gallaborane.

At each optimized geometry, we used the P3 [9] and NR2 [11] electron propagator approximations to calculate the vertical ionization potentials of the molecules. For these calculations, the 6–311G(2df,2pd) basis sets [15–17] were used. Gallium d-orbitals were included and all other core orbitals were dropped in the propagator calculations.

Plots of Dyson orbitals were generated with MOLDEN [18]. Contours in Fig.s 1–18 are set to ±0.05.

Results and Discussion

Diborane

Smith and Lipscomb established the crystal structure of the β phase of B_2H_6 [19] after earlier work by Stitt [20, 21], Price [22, 23] and Hedberg [24] on gas–phase diborane. Jones and Lipscomb [25, 26] used structure factors from calculations to explain apparent discrepancies between bond lengths determined from x–ray and electron diffraction experiments. Several recent theoretical studies have considered diborane with correlated, *ab initio* methods [27–30] and density functionals [31, 32]. Trinquier and Malrieu performed a general analysis of bonding in dibridged X_2H_6 compounds [33].

Table 1 compares our results, an estimate based on vibrational spectroscopy [34, 35], and two recent calculations [7, 36]. (Terminal and bridge hydrogens are represented by H_t and H_b, respectively.) The present geometries are in very good agreement with experiments and closely resemble the MP2/6–31G** results of Stanton, Bartlett and Lipscomb [37]. Table 2 shows the results of the elec-

Table 1: Diborane Structures

Parameter	Ref. [36]	Ref. [7]	This Work	Exp. [34]
r(B–B) /Å	1.777	1.793	1.760	1.743
r(B–H_t)/Å	1.182	1.188	1.186	1.184 ± 0.003
r(B–H_b)/Å	1.316	1.329	1.311	1.314 ± 0.003
MAD	.013	.023	.007	
H_t–B–H_t /deg	123.2	122.4	122.3	121.5 ± 0.5
H_b–B–H_b /deg		95.2	95.7	96.9 ± 0.5
MAD	1.7	1.3	1.0	

tron propagator calculations and compares them to the most recent photoelectron experiments [7]. Calculated values are listed as poles of the electron propagator [4–6]. Coresponding pole strengths are listed in parentheses. Previous experiments arrived at similar values for vertical ionization energies [38–41]. The Mean Absolute Deviations (MAD) show that the ionization potentials calculated with the P3 approximation deviate by 0.2 eV from experiment on the average. NR2 results are in even better agreement with experiment. The order of cationic states is the same as that predicted by Koopmans's theorem and is in agreement with previous photoelectron assignments, as well as configuration interaction calculations [7]. Hartree–Fock energy differences disagree on the order of the third and fourth states and density functional, transition–operator calculations disagree on the order of the fourth and fifth states [7]. For the first five final states, pole strengths are close to 0.9 in the P3 and NR2 calculations. These figures indicate that the Koopmans description of the final states is qualitatively valid. For the sixth state, there is a larger discrepancy between the P3 and NR2 ionization energies. In addition, the pole strength predicted at the NR2 level is lower. Lower pole strengths are characteristic of cationic states in carbon compounds where the Dyson orbitals are dominated by C 2s contributions [13]. Wavefunctions for these so-called inner valence states may exhibit strong mixings between determinants with one hole (that is, Koopmans determinants for final states) and determinants with two holes and one particle (a Koopmans determinant plus a single excitation). The greater the deviation of the pole strength from unity, the more important this kind of configuration mixing becomes. Low pole strengths are also an indication of nearby states with dominant two–hole, one–particle character. The breadth and the long, high–energy tail of the peak in the He II photoelectron spectrum at 21.4

Table 2: Diborane Ionization Energies (eV)

Final State	Expt Ref. [7]	P3 Pole (Str.)	NR2 Pole (Str.)	CI Ref. [7]	DFT Ref. [7]
$^2B_{1g}$	11.88 ± 0.01	12.13 (0.92)	12.06 (0.91)	12.19	11.44
^2A_g	13.35 ± 0.03	13.40 (0.91)	13.33 (0.90)	13.28	12.75
$^2B_{2u}$	13.93 ± 0.02	13.89 (0.91)	13.80 (0.89)	13.89	13.02
$^2B_{1u}$	14.76 ± 0.01	14.66 (0.91)	14.60 (0.91)	14.48	14.97
$^2B_{3u}$	16.08 ± 0.01	16.21 (0.90)	16.02 (0.87)	16.31	14.89
^2A_g	21.42 ± 0.09	21.98 (0.84)	21.42 (0.74)	a	20.93
MAD		.19	.09	.19	.64

^{a}Convergence problems encountered.

eV [7] may contain final electronic states of this type.

Dyson orbitals that accompany the propagator predictions consist chiefly of a single, canonical, Hartree–Fock orbital that is multiplied by a factor that is approximately equal to the square root of the pole strength. While correlation effects on ionization energies may be large, the corresponding Dyson orbitals are altered primarily by this multiplicative factor. B 2p contributions are more important than B 2s contributions in the Dyson orbitals for the first four ionization energies. (See Fig.s 1–4.) Bridge hydrogen functions interfere constructively with B functions in the a_g and b_{1u} molecular orbitals. The increased relative intensity of the fifth band when the radiation source is He II instead of He I has been ascribed to enhanced B 2s character in the b_{3u} Dyson orbital [7]. Figure 5 shows the corresponding Dyson orbital. No such comparison may be made for the sixth state, for it lies beyond the range of the He I spectrum. The chief contributors to the delocalized lobe of Fig. 6 are boron 2s functions.

Digallane

Digallane's synthesis was reported by Downs, Goode and Pulham in 1989 [42]. Infrared spectra and electron diffraction experiments established that digallane's structure resembles that of diborane [43]. Several theoretical studies have appeared as well [28, 29, 31, 44–46]. Table 3 shows the results of our geometry optimization, together with other recent calculations [7, 30, 32]. Our stucture is in very good agreement with experiment [43], with a MAD less then .02 Å in the bond distances. Results of the electron propagator calculations are presented in Table 4. P3 results are in better agreement with experiment [7] than NR2 results, presenting a MAD of 0.14 eV, against 0.24 eV for the NR2 calculations. The order of final states is the same as in diborane and is in agreement with Koopmans

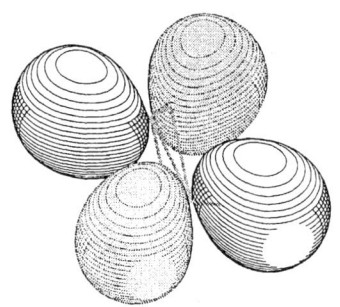

Figure 1: B_2H_6 X^2B_{1g} Dyson Orbital.

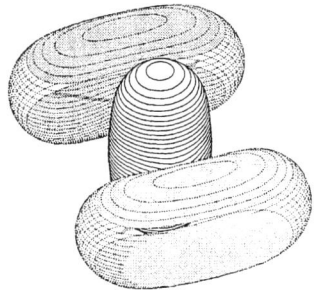

Figure 2: B_2H_6 A^2A_g Dyson Orbital.

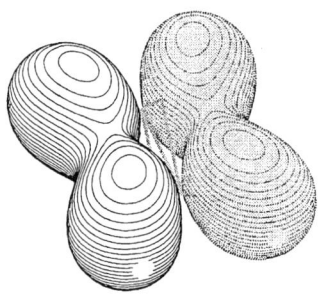

Figure 3: B_2H_6 B^2B_{2u} Dyson Orbital.

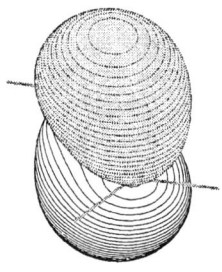

Figure 4: B_2H_6 C^2B_{1u} Dyson Orbital.

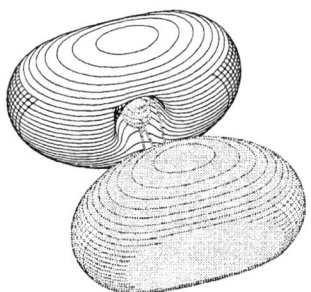

Figure 5: B_2H_6 D^2B_{3u} Dyson Orbital.

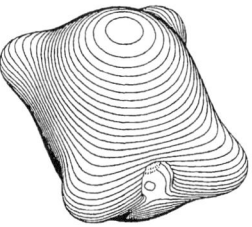

Figure 6: B_2H_6 E^2A_g Dyson Orbital.

Table 3: Digallane Structures

Parameter	Ref. [30]	Ref. [32]	Ref. [7]	This Work	Exp. [43]
r(Ga–Ga) /Å	2.608	2.618	2.650	2.572	2.580 ± 0.002
r(Ga–H_t) /Å	1.552	1.558	1.561	1.537	1.519 ± 0.035
r(Ga–H_b) /Å	1.753	1.761	1.775	1.736	1.710 ± 0.038
MAD	.035	.043	.059	.017	
H_t–Ga–H_t /deg	129.9	129.7	128.9	130.4	130.0 [a]
H_b–Ga–H_b /deg			83.4	84.4	82.1
MAD	.1	.3	1.2	1.4	

[a] Fixed at this value in the analysis of the electron diffraction results.

predictions. Hartree–Fock calculations reverse the order of the third and fourth states and configuration interaction calculations place the $^2B_{1u}$ state after the $^2B_{1g}$ state [7]. Density functional results with the transition operator method obtain the Koopmans ordering and the experimental assignment employed these predictions. An irreproducible, weak feature was reported in the vicinity of 18 eV [7]. P3 and NR2 predictions for the sixth state are at somewhat lower energies. Pole strengths

Table 4: Digallane Ionization Energies (eV)

Final State	Expt Ref. [7]	P3 Pole (Str.)	NR2 Pole (Str.)	CI Ref. [7]	DFT Ref. [7]
$^2B_{1g}$	10.88 ± 0.01	11.18 (0.91)	11.03 (0.89)	11.38	10.57
^2A_g	11.56 ± 0.03	11.57 (0.91)	11.43 (0.89)	11.76	11.07
$^2B_{2u}$	11.85 ± 0.03	11.80 (0.91)	11.65 (0.89)	11.94	11.12
$^2B_{1u}$	12.23 ± 0.03	11.89 (0.91)	11.77 (0.89)	11.68	11.96
$^2B_{3u}$	14.40 ± 0.01	14.38 (0.89)	14.13 (0.85)	14.77	13.74
^2A_g		17.87 (0.85)	17.35 (0.73)	[a]	17.34
MAD		.14	.24	.34	.49

[a] Convergence problems encountered.

are close to 0.9 for the first five final states. The Koopmans description of these states is qualitatively valid. NR2 calculations are needed to obtain a valid picture of the sixth final state, for the pole strength is below 0.8. A relatively large discrepancy between P3 and NR2 results is found for this case.

In general, the ionization energies are lower for digallane than for diborane. The calculated separation between the fourth and the first states is only 0.7 eV for digallane, whereas it is about 2.5 eV for diborane. In addition, the energy splitting between the fifth and sixth states is smaller for digallane.

Dyson orbitals for digallane (Fig.s 7–12) display nodal surfaces that resemble those of diborane. The radial nodes of Ga 4p functions can be discerned in Fig.s 7–10. Prominent positive and negative amplitudes are shifted toward hydrogen nuclei. Direct bonding or antibonding relationships between Ga atoms are not as strong as those between B atoms in the b_{2u} and b_{1g} orbitals.

Smaller energy separations between ionization energies and reduced direct interactions between nonhydrogen atoms suggest that bonding in digallane has more ionic character.

Gallaborane

A report on the synthesis, properties and structure of gallaborane was published in 1990 [47]. Since then, a number of articles appeared reporting calculations [7, 28, 36, 48, 49] and experimental studies [7]. Table 5 shows the results of our calculations, together with two of the most recent calculations found in the literature [7, 36]. Our geometries are in good agreement with experiment [47], with a precision about the same as the calculations on diborane and digallane. Electron

Table 5: Gallaborane Structures

Parameter	Ref. [36]	Ref. [7]	This Work	exp. [47]
r(Ga–B) /Å	2.209	2.229	2.190	2.179 ± 0.002
r(Ga–H_t) /Å	1.560	1.559	1.536	1.586 ± 0.008
r(Ga–H_b) /Å	1.769	1.784	1.762	1.826 ± 0.008
r(B–H_t) /Å	1.188	1.193	1.919	1.234 ± 0.008
r(B–H_b) /Å	1.290	1.305	1.285	1.334 ± 0.008
MAD	.041	.038	.043	
H_t–Ga–H_t /deg	129.2	128.8	130.4	145[a]
H_b–Ga–H_b /deg		71.7	71.9	75.3 ± 1.2
H_t–B–H_t /deg	121.8	121.2	120.7	120[a]
H_b–B–H_b /deg		106.3	107.1	113.4 ± 2.7
MAD	6.0	7.0	6.3	

[a]Fixed at this value in the analysis of the electron diffraction results.

propagator results are presented in Table 6. P3 and NR2 calculations are in comparable agreement with experimental ionization energies, with MADs of 0.11 and 0.15 eV, respectively. The order of the final states is in agreement with Koopmans predictions and density functional, transition–operator calculations [7]. Ionization

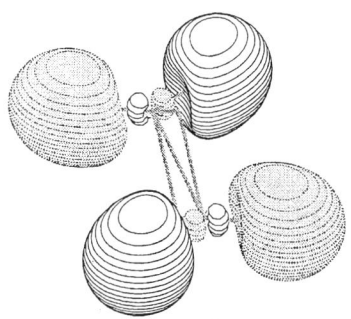

Figure 7: Ga_2H_6 X^2B_{1g} Dyson Orbital.

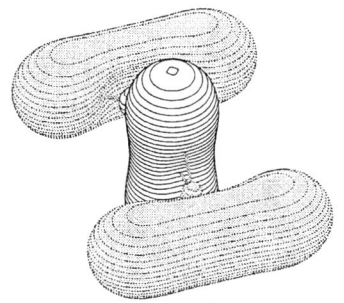

Figure 8: Ga_2H_6 A^2A_g Dyson Orbital.

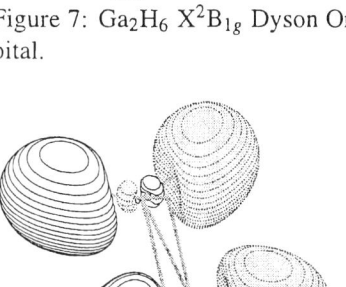

Figure 9: Ga_2H_6 B^2B_{2u} Dyson Orbital.

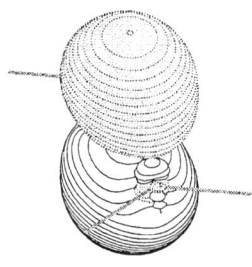

Figure 10: Ga_2H_6 C^2B_{1u} Dyson Orbital.

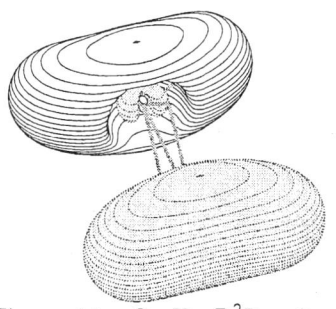

Figure 11: Ga_2H_6 D^2B_{3u} Dyson Orbital.

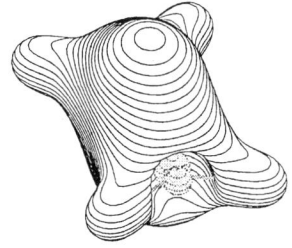

Figure 12: Ga_2H_6 E^2A_g Dyson Orbital.

energies for gallaborane are between their counterparts for diborane and digallane. 1.6 eV separates the first and fourth ionization energies in the P3 and NR2 calculations; this value lies between the corresponding values for diborane and digallane. For the fifth and sixth ionization energies, the NR2 splitting, 4.3 eV, is also between the results for diborane (5.4 eV) and digallane (3.2 eV). For gallaborane,

Table 6: Gallaborane Ionization Energies (eV)

Final State	Expt Ref. [7]	P3 Pole (Str.)	NR2 Pole (Str.)	CI Ref. [7]	DFT Ref. [7]
$^2B_2$	11.33 ± 0.01	11.58 (0.92)	11.44 (0.90)	11.68	10.84
$^2A_1$	12.15 ± 0.01	12.17 (0.91)	12.06 (0.90)	12.21	11.11
$^2B_2$	12.63 ± 0.01	12.55 (0.91)	12.46 (0.89)	a	12.15
$^2B_1$	13.31 ± 0.03	13.17 (0.91)	13.08 (0.90)	12.98	13.28
$^2A_1$	15.03 ± 0.01	15.10 (0.90)	14.87 (0.86)	a	13.91
$^2A_1$		19.68 (0.84)	19.18 (0.74)	a	18.21
MAD		.11	.15	.25	.63

^{a}Convergence problems encountered.

pole strengths remain close to 0.9 for the first five final states and each Dyson orbital is dominated a single, canonical, Hartree–Fock orbital. For the sixth final state, a relatively low pole strength obtains in the NR2 approximation. P3 and NR2 results differ by 0.5 eV.

Dyson orbitals for gallaborane (Fig.s 13–18) exhibit polarizations toward Ga for the first, second and fifth ionization energies. In the remaining cases, polarizations toward B occur. Nodal patterns resemble those of digallane.

Conclusions

Electron propagator calculations on the three bridged molecules considered here lead to a common order of cationic states: X^2B_{1g}, A^2A_g, B^2B_{2u}, C^2B_{1u}, D^2B_{3u}, E^2A_g for D_{2h} molecules and X^2B_2, A^2A_1, B^2B_2, C^2B_1, D^2A_1 and E^2A_1 for gallaborane. Pole strengths near 0.9 for the first five ionization energies in each molecule indicate that the Koopmans description of the final states is qualitatively valid. Only for the E states is there significant multiconfigurational character. Each of the Dyson orbitals for these ionization energies is dominated by a single, canonical, Hartree–Fock orbital.

Dyson orbitals for the first ionization energy of each molecule display σ bonding X–H_t and π antibonding X–Y relationships (X, Y = B, Ga). For the next final state, bonding σ X–H_t and three–center X–H_b–X interactions occur. In the Dyson

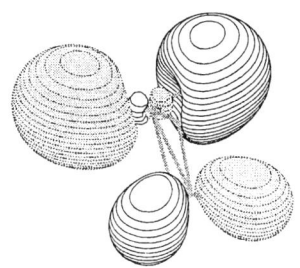

Figure 13: GaBH$_6$ X^{2}B$_2$ Dyson Orbital. (Ga atom on top.)

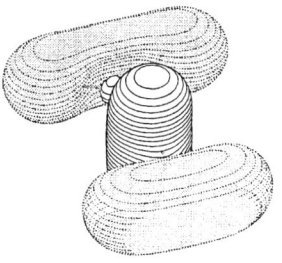

Figure 14: GaBH$_6$ A^{2}A$_1$ Dyson Orbital. (Ga atom on top.)

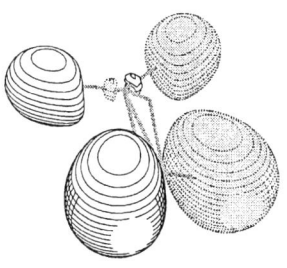

Figure 15: GaBH$_6$ B^{2}B$_2$ Dyson Orbital. (Ga atom on top.)

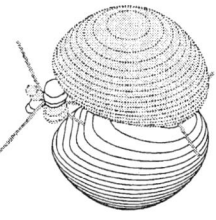

Figure 16: GaBH$_6$ C^{2}B$_1$ Dyson Orbital. (Ga atom on the left.)

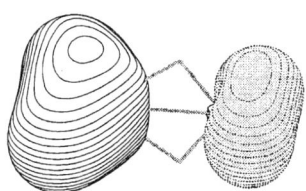

Figure 17: GaBH$_6$ D^{2}A$_1$ Dyson Orbital. (Ga atom on the left.)

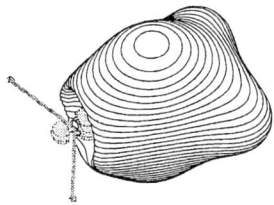

Figure 18: GaBH$_6$ E^{2}A$_1$ Dyson Orbital. (Ga atom on the left)

orbitals for the B final states, σ bonding X–H$_t$ and π bonding X–X relationships obtain. X–H$_b$–X bonding lobes dominate the Dyson orbitals for the C final states. Bonding X–H$_t$ and antibonding X–X interactions between valence s functions are seen for fifth ionization energies. A delocalized, bonding lobe is found for the E final states. Three–center bonding patterns involving the bridge hydrogens are evident in the a$_g$ and b$_{1u}$ Dyson orbitals.

Ionization energies decline as the number of Ga atoms increases. Splittings between the first and fourth final states, as well as separations between the D and E states, are greatest for diborane and smallest for digallane. Direct interactions between nonhydrogen atoms are most apparent in Dyson orbitals for diborane ionization energies and are weakest for digallane. These trends imply that more ionic bonding occurs when Ga is substituted for B.

Acknowledgments

William Lipscomb's career forever will be identified with the theory of the three–center bond in boron hydrides. His celebrated work in this field employed an incisive mixture of experimental and theoretical methods. In his laboratory, developers of conceptual and computational tools were given ample scope, for the Colonel has a knack for connecting new theoretical capabilities to significant chemical questions. We therefore offer this work, an application of the electron propagator picture of electronic structure, in tribute to his skills as a mentor of young scientists.

O. Dolgounitcheva provided essential technical assistance.

This work was supported by the National Science Foundation under grant CHE-9873897.

References

1. Koopmans, T. *Physica* **1933**, *1*, 104.
2. Parr, R. G.; Yang, W. *Density Functional Theory of Atoms and Molecules*, Oxford University Press: Oxford, 1989.
3. Janak, J. F. *Phys. Rev. B* **1978**, *18*, 7165.
4. Ortiz, J. V. In *Computational Chemistry: Reviews of Current Trends, Vol. 2*; Leszczynski, J., Ed.; World Scientific: Singapore, 1997; p. 1.
5. Ortiz, J. V.; Zakrzewski, V. G.; Dolgounitcheva, O. In *Conceptual Perspectives In Quantum Chemistry, Vol. 3*; Calais, J.-L., Kryachko, E., Eds.; Kluwer: Dordrecht, 1997; p. 465.
6. Ortiz, J. V. *Adv. Quantum Chem.* **1999**, *35*, 33.
7. Dyke, J. M.; Haggerston, D.; Warschkow, O.; Andrews, L.; Downs, A.; Souter, P. F. *J. Phys. Chem.* **1996**, *100*, 2998.

8. See, for example, Zakrzewski, V. G.; Ortiz, J. V.; Nichols, J. A.; Heryadi, D.; Yeager, D. L.; Golab, J. T. *Int. J. Quant. Chem.* **1996**, *60*, 29 and the following reference.
9. Ortiz, J. V. *J. Chem. Phys.* **1996**, *104*, 7599.
10. Ortiz, J. V.; Zakrzewski, V. G. *J. Chem. Phys.* **1996**, *105*, 2762.
11. Ortiz, J. V. *J. Chem. Phys.* **1998**, *108*, 1008.
12. Dolgounitcheva, O.; Zakrzewski, V. G.; Ortiz, J. V. *J. Phys. Chem. A* **2000**, *104*, 10032.
13. Dolgounitcheva, O.; Zakrzewski, V. G.; Ortiz, J. V. *J. Chem. Phys.* **2001**, *114*, 130.
14. Frisch, M. J.; Trucks, G. W.; Schlegel, H. B.; Scuseria, G. E.; Robb, M. A.; Cheeseman, J. R.; Zakrzewski, V. G.; Montgomery, J. A., Jr.; Stratmanni, R. E.; Burant, J. C.; Dapprich, S.; Millam, J. M.; Daniels, A. D.; Kudin, K. N.; Strain, M. C.; Farkas, O.; Tomasi, J.; Barone, V.; Cossi, M.; Cammi, R.; Mennucci, B.; Pomelli, C.; Adamo, C.; Clifford, S.; Ochterski, J.; Petersson, G. A.; Ayala, P. Y.; Cui, Q.; Morokuma, K.; Malick, D. K.; Rabuck, A. D.; Raghavachari, K.; Foresman, J. B.; Cioslowski, J.; Ortiz, J. V.; Stefanov, B. B.; Liu, G.; Liashenko, A.; Piskorz, P.; Komaromi, I.; Gomperts, R.; Martin, R. L.; Fox, D. J.; Keith, T.; Al-Laham, M. A.; Peng, C. Y.; Nanayakkara, A.; Gonzalez, C.; Challacombe, M.; Gill, P. M. W.; Johnson, B.; Chen, W.; Wong, M. W.; Andres, J. L.; Gonzalez, C.; Head-Gordon, M.; Replogle, E. S.; and Pople, J. A. *Gaussian 98 (Revision A.6)*; Gaussian, Inc.: Pittsburgh PA, 1998.
15. Krishnan, R.; Binkley, J. S.; Seeger, R.; Pople, J. A.; *J. Chem. Phys.* **1980**, *72*, 650.
16. Curtiss, L. A.; McGrath, M. P.; Blaudeau, J.–P.; Davis, N. E.; Binning, R. C.; Radom, L. *J. Chem. Phys.* **1995**, *103*, 6104.
17. Frisch, M. J.; Pople, J. A.; Binkley, J. S. *J. Chem. Phys.* **1984**, *80*, 3265.
18. Schaftenaar, G. MOLDEN 3.4, CAOS/CAMM Center, The Netherlands, 1998.
19. Smith, H. W.; Lipscomb, W. N. *J. Chem. Phys.* **1965**, *43*, 1060.
20. Stitt, F. *J. Chem. Phys.* **1940**, *8*, 981.
21. Stitt, F. *J. Chem. Phys.* **1941**, *9*, 780.
22. Price, W. C. *J. Chem. Phys.* **1947**, *15*, 614.
23. Price, W. C. *J. Chem. Phys.* **1948**, *16*, 894.
24. Hedberg, K.; Schomaker, V. *J. Am. Chem. Soc.* **1951**, *73*, 1482.
25. Jones, D. S.; Lipscomb, W. N. *J. Chem. Phys.* **1969**, *51*, 3133.
26. Jones, D. S.; Lipscomb, W. N. *Acta Cryst.* **1970** *A26*, 196.
27. Sana, M.; Leroy, G.; Henriet, C. *J. Mol. Struc.* **1989**, *187*, 233.
28. Bock, C. W.; Trachtman, M.; Murphy, C.; Muschert, B.; Mains, G. J. *J. Phys. Chem.* **1991**, *95*, 2339.
29. Duke, B. J.; Liang, C.; Schaefer, H. F. *J. Am. Chem. Soc.* **1991**, *113*, 2884.

30. Shen, M.; Schaefer, H. F. *J. Chem. Phys.* **1992**, *96*, 2868.
31. Barone, V.; Adamo, C.; Fliszár, S.; Russo, N. *Chem. Phys. Lett.* **1994**, *222*, 597.
32. Barone, V.; Orlandini, L.; Adamo, C. *J. Phys. Chem.* **1994**, *98*, 13185.
33. Trinquier, G.; Malrieu, J. P. *J. Am. Chem. Soc.* **1991**, *113*, 8634.
34. Duncan, J. L.; Harper, J. *Mol. Phys.* **1984**, *51*, 371.
35. Kuchitsu, K. *J. Chem. Phys.* **1968**, *49*, 4456.
36. Bennett, F. R.; Connelly, J. P. *J. Phys. Chem.* **1996**, *100*, 9308.
37. Stanton, J. F.; Bartlett, R. J.; Lipscomb, W. N. *Chem. Phys. Lett.* **1987**, *138*, 525.
38. Rose, T.; Frey, R.; Brehm, B. *J. Chem. Soc., Chem. Comm.* **1969**, 1518.
39. Lloyd, D. R.; Lynaugh, N. *Phil. Trans. Roy. Soc. London A* **1970**, *268*, 97.
40. Brundle, C. R.; Robin, M. B.; Basch, H.; Pinsky, M.; Bond, A. *J. Am. Chem. Soc.* **1970**, *92*, 3863.
41. Åsbrink, L.; Svensson, A.; Niessen, W. v.; Bieri, G. *J. Elec. Spectrosc. Relat. Phen.* **1981**, *24*, 293.
42. Downs, A. J.; Goode, M. J.; Pulham, C. R. *J. Am. Chem. Soc.* **1989**, *111*, 1936.
43. Pulham, C. R.; Downs, A. J.; Goode, M. J.; Rankin, D. W. H.; Robertson, H. E. *J. Am. Chem. Soc.* **1991**, *113*, 5149.
44. Liang, C.; Davy, R. D.; Schaefer, H. F. *Chem. Phys. Lett.* **1989**, *159*, 393.
45. Lammertsma, K.; Leszczyński, J. *J. Phys. Chem.* **1990**, *94*, 2806.
46. Souter, P. F.; Andrews, L.; Downs, A. J.; Greene, T. M.; Ma, B.; Schaefer, H. F. *J. Phys. Chem.* **1994**, *98*, 12824.
47. Pulham, C. R.; Brain, P. T.; Downs, A. J.; Rankin, D. W. H.; Robertson, H. E. *J. Chem. Soc., Chem. Comm.* **1990**, 177.
48. van der Woerd, M. J.; Lammertsma, K.; Duke, B. J.; Schaefer, H. F. *J. Chem. Phys.* **1991**, *95*, 1160.
49. Barone, V.; Minichino, C.; Lelj, F.; Russo, N. *J. Comp. Chem.* **1988**, *5*, 518.

Chapter 8

Application of Theoretical Methods to NMR Chemical Shifts and Coupling Constants

Michael L. McKee

Department of Chemistry, Auburn University, Auburn, AL 36849

Abstract: NMR has been called the third dimension of computational chemistry. The developing field of computational NMR has the potential of having a significant impact in all areas of chemistry by making available accurate and reliable NMR properties such as chemical shifts and spin-spin coupling constants. This chapter is a progress report outlining the status of the field.

Introduction

The interaction of a magnetic field with atoms and molecules has given scientists an invaluable method for probing matter through a variety of spectroscopies, most notably nuclear magnetic resonance spectroscopy, NMR.[1] A variation of the electron density around different nuclei in a molecule leads to changes in induced fields, which in turn can be measured. This leads to a molecular fingerprint which can be used in rationalizing the molecular structure. Relaxation processes limit the time resolution, which means that fluxional motions give rise to an average magnetic environment. Thus, the symmetry of a molecule on the NMR time scale may be different from that on an IR, UV, or X-ray time scale.

The intent of this book chapter is provide a reader with an overview of the different methods used to calculate NMR parameters. The mathematical details will be left to the experts,[2-36] but rather the variety of methods will be presented with some comments on the advantages of each.

Since this book commemorates the varied contributions of William N. Lipscomb, it is fitting to start with his contributions to the field of computational NMR.[37-41] In 1966, Lipscomb asked the question: "What does one need to know about a molecular wave function to compute the NMR shielding constants?" With co-workers Stevens and Pitzer, they carried out the first ab

© 2002 American Chemical Society

initio calculations for chemical shifts within the coupled-perturbed Hartree-Fock method. The diatomic chemical shifts were computed with a STO basis with a common gauge origin.

Theory

When a magnetic field $\vec{B}$ interacts with a molecule with a single nuclear moment $\vec{\mu}$ the effective spin Hamiltonian can be written as the sum of two parts, the classical moment-field term, $-\vec{\mu} \cdot \vec{B}$, and the interaction of the nuclear moment with the field induced by the electrons' motion, $\vec{\mu} \cdot \vec{\sigma} \cdot \vec{B}$ (eq 1).

$$\vec{H}_\mu^{spin} = -\vec{\mu} \cdot \vec{B} + \vec{\mu} \cdot \vec{\sigma} \cdot \vec{B} \tag{1}$$

The chemical shielding tensor, σ_{ij}, is an asymmetric tensor containing information about the coupling between the nuclear moments and the electrons. The shielding tensor can also be written as the second derivative of energy with respect to magnetic moment and field (eq 2).

$$\sigma_{ij} = \left(\frac{\partial^2 E}{\partial \mu_i \partial B_j} \right)_{\vec{B}=\vec{\mu}=0} \tag{2}$$

The shielding tensor σ_{ij} can be divided into two contributions, diamagnetic, and paramagnetic (eq 3). The diamagnetic part, associated with the ground state wave function, is usually aligned opposite to the field and

$$\vec{\sigma}^{total} = \vec{\sigma}^D + \vec{\sigma}^P \tag{3}$$

causes an upfield shift. The paramagnetic part, associated with the interaction between the ground state and excited states, is usually aligned with the field and causes a downfield shift. The contribution from the shielding due to the diamagnetic part can be calculated accurately; however, this part of the chemical shielding is not as sensitive to details of molecular environment. On the other hand, the paramagnetic part is difficult to calculate accurately, but it contains the most information about the electronic environment.

Early application of theoretical methods to the calculation of magnetic peroperties was hampered by the so called "gauge problem". The gauge problem arises from the fact that the Schrödinger equation contains the vector potential $\vec{A}$. The latter is, as shown by (eq 4), determined only to the gradient

$$\nabla \times \vec{A} = \vec{B} = \nabla \times \left[\vec{A} + \nabla f \right] \quad (4)$$

of an arbitrary scalar function f by the flux density $\vec{B}$. In approximate (but not in the exact) calculations, this indeterminancy can affect results. A poor choice of gauge leads to significant mathematical difficulties. Current methods use atomic orbitals or localized orbitals, each with its own gauge origin located on atomic centers.

Methods

The vast majority of calculations of NMR parameters use either GIAO[43-50] (gauge-including atomic orbitals) or IGLO[51-53] (individual gauge for localized orbitals). Evaluation of GIAO integrals was significantly improved by Pulay[43] who used recent advances in derivative theory. With the GIAO method, integrals are computed over atomic orbitals, while IGLO and LORG (localized orbitals, local origin)[54-55] make use of localized orbitals. In the limit of large basis sets all methods appear to give very similar results. However, the GIAO method shows more rapid convergence of chemical shifts as the basis set size is improved. Also, electron correlation is easier to introduce into the GIAO formalism. On the other hand, the IGLO method breaks down diamagnetic and paramagnetic contribution into a per orbital contribution which improves analysis. For example, the IGLO method allows a dissected analysis of NICS (see below) contributions from the σ and π orbitals. However, with the NCS[56-57] (natural chemical shielding) partitioning scheme developed by Bohmann, Weinhold, and Farrar and based on the HF-GIAO method, similar analysis can be made with the GIAO chemical shifts. CSGT[58] (continuous set of gauge transformations) and LORG are similar methods for computing chemical shielding.

The GIAO method has recently been implemented at the semiempirical MNDO/d level using analytical derivative theory.[59] With the standard parameters for H, C, N, and O, the variation of the paramagnetic contribution is overestimated. However, the one-center energies, orbital exponents, and resonance β parameters can be adjusted to significantly improve the agreement between calculated and experimental chemical shifts. It is also found that the NICS values (nucleus-independent chemical shifts; see below) can be calculated at the GIAO-MNDO/d level[60] such that aromatic or antiaromatic character is usually assigned correctly. Chemical shift calculations have also been reported using ZINDO and Fenske-Hall.[61-62]

While the computational methods yield the asymmetric absolute shielding tensor, the usually observable quantity is related to the symmetrized tensor (eq 5).

$$\sigma^s_{i,j} = \frac{1}{2}\left(\sigma_{i,j} + \sigma_{j,i}\right) \tag{5}$$

The isotropic chemical shielding is obtained by averaging the eigenvalues of the diagonalized symmetrized tensor. While the isotropic absolute shielding can be compared with experiment, values are more often reported relative to a standard (e.g. SiMe$_4$ for ^{13}C). The chemical shift δ is related to the isotropic chemical shielding through $\delta = \sigma_{ref} - \sigma$. Thus, the calculated chemical shifts can be compared to a standard value just like the experimental values. Finally, systematic errors in the chemical shifts/shielding for a particular method/basis set combination can be corrected if comparisons are made with experiment and a linear regression is performed. The chemical shifts are then, $\delta_{pred} = m\sigma_{calc} + i$, where m and i are constants of best fit.[63,64]

In principle (and using infinite basis sets), NMR chemical shifts can be computed using a single origin. However, practical (i.e. finite) basis sets require that the gauge problem be addressed. The several ways of doing this are listed below:

1. GIAO (gauge including atomic orbital): This method is coded in the popular Gaussian program and is therefore widely available. The GIAO chemical shifts seem to converge better than IGLO and other methods with respect to the size of the basis set. Correlated chemical shifts calculations including electron correlation are easier to implement within the GIAO formalism. Thus, it is possible to do MP2-GIAO, MP4-GIAO and CCSD-GIAO shift calculations.

2. IGLO (individual gauge for localized orbitals): This method is based on localized orbitals and has the advantage that the shielding contributions can be dissected into contributions from particular orbitals. This feature is particularly helpful when analyzing NICS results at the center of rings because π bond contributions to the NICS value can be separated from σ bond contributions. It is possible to include the effects of several electronic configurations with the MC-IGLO method.

3. LORG (localized orbitals, local origin) employs the random phase approximation to compute absolute NMR shielding using localized orbitals. Electron correlation can be included using the second-order polarization propagator approximation via SOLO (second-order LORG).

4. (QM:MM/QM:QM)-NMR Morokuma and co-workers have developed a layered approach to NMR shift calculations (ONIOM) where a higher level of theory (i.e. including electron correlation) is used for one part of the system which is combined with the shift calculation for the entire system at a lower level of theory.[65] Along similar lines, Qui and Karplus[66] have implemented a QM:MM method to predict shifts in proteins.

The method most frequently used for chemical shift calculations is the DFT approach where the formalism introduces electron correlation into the wavefunction.[21-32] While in principle the current dependency needs to be included for treating magnetic properties, it should be mentioned that almost all of the commonly used exchange/correlation functionals are current-independent. The innovation of density functionals dependent on the gradient of electron density was a breakthrough in developing a useful DFT method; however, it turns out that functionals dependent on the current density do not show an improvement in the prediction of magnetic properties. Fortunately, the present crop of exchange/correlational functionals are able to reproduce a number of magnetic properties quite well.

To quote from Schreckenback, Wolff, and Ziegler[67] "chemical shifts are known to be sensitive to everything". Their list of factors include the following: (1) relativity, (2) quantum mechanical approximation, (3) gauge problem, (4) basis set, (5) geometries, (6) reference compound, (7) condensed phase, temperature, and pressure. Relativistic effects, which are particularly important for heavy nuclei or light nucleus attached to a heavy one, will be discussed later.

Chemical shifts are not only dependent on the level of theory used to calculate the shift, but also dependent on the level of theory used to optimize geometries. In fact, the reliability of the molecular structure may be the most important factor in determining the quality of results. The recommended level of theory for geometry optimization includes electron correlation and a triple-ζ basis set with polarization functions.[68-69]

The calculated absolute shielding is converted to a chemical shift by computing the absolute shielding of a reference compound. A careful choice of reference compound will insure maximum cancellation of error in the shielding calculation. In fact, a regression fit of calculated shieldings against a set of known chemical shifts can produce linear correlation independent of reference compound. Since many of the errors in the shielding are systematic, these fits significantly increase the accuracy of the chemical shifts.

One known shortcoming of the current-independent DFT functionals is that they tend to overestimate the paramagnetic terms somewhat. Since the paramagnetic terms are dominated by excitation contributions from low-lying orbitals, Malkin et al.[33] introduced a correction term into the denominator of the expression for the paramagnetic part of the shielding tensor. This method is known as the sum-over-states (SOS-DFPT) method and improves agreement with experiment. A similar improvement is found if an adjusted 'exact-exchange' coefficient is used with the hybrid Kohn-Sham orbitals.[70]

Application

There is no question that the successful application of IGLO in the area of electron deficient compounds spurred the interest in computational NMR.

Even with rather modest basis sets, chemical shifts could be calculated for boranes (^{11}B), carboranes (^{11}B and ^{13}C) and carbocations (^{13}C) within about 5 ppm of experiment with a high degree of confidence. In this method, high-level theory is used to compute a number of potential candidate structures of a species for which the experimental NMR chemical shifts are known. A comparison of the calculated chemical shifts using the GIAO (or IGLO) method with the experimental chemical shifts often identifies the experimental structure with a high degree of confidence. This is true even though the calculations refer to the gas phase while the experimental measurements are in solution. Using this method (now called ab initio/(IGLO or GIAO)/NMR), the structures of several carboranes, determined on the basis of NMR evidence, were found to be incorrect. New structures were found with much better agreement between calculated and measured NMR chemical shifts.[71-84]

In addition, it is possible to determine whether the species in solution is static or a rapidly equilibrating pair of species by comparing the calculated NMR chemical shifts of the static structure, the averaged chemical shifts of the equilibrating structure, and the known NMR chemical shifts. In this way, it was determined that the experimental $C_2B_{10}H_{13}^-$ species had C_1 symmetry rather than C_s symmetry.[82]

An interesting application of computed NMR chemical shifts is to the fullerenes and the corresponding endohedral complexes.[85-89] The chemical shift of ^{3}He@C_{60} and ^{3}He@C_{70} are -6.3 and -8.8 ppm, respectively which can be compared to calculated values of -8.7 and -24.0 ppm, respectively. GIAO/B3LYP/6-31G(d) calculation have been reported for the fullerenes up to C_{84}.

Magnetic criteria are often used as a judge for aromaticity/antiaromaticity. Such effects include (1) anomalous proton chemical shifts, (2) large magnetic anisotropies, (3) diamagnetic/paramagnetic susceptibility exaltation, and (4) nucleus-independent chemical shifts (NICS).[90-100] NICS values, now widely used as a measure of aromaticity/antiaromaticity, are obtained by calculating absolute NMR shieldings at the ring or cage center. In addition, when NICS values are calculated above the ring center the notation NICS(n) is often used where n is the number of Ångstroms above the ring. The NICS value is the negative of the absolute shielding such that a negative NICS value indicates aromaticity and a positive value indicates antiaromaticity.

Besides the negative (paramagnetic, deshielding) contribution to the NICS value due to aromatic ring currents, a negative (diamagnetic, shielding) contribution is also made by the CC σ bonds. For example, the NICS(0) value is large and negative in the center of tetrahedrane. [92]

Calculation of nuclear spin-spin coupling constants have begun to appear recently.[101-118] There are four terms which contribute: diamagnetic spin-orbit, paramagnetic spin-orbital, spin-dipole, and Fermi contact terms. Often, the Fermi contact term dominates the coupling interaction and several workers have calculated $^1J(^{11}B^{-1}H)$ and $^1J(^{13}C^{-1}H)$ values at the B3LYP level using finite perturbation theory.[119] Recent work has shown that the ^{1}H NMR spectrum of organic molecules can be reasonably simulated using calculated chemical shifts and coupling constants.[120]

Although formally a second-order property (like NMR shielding constants), the spin-spin coupling constants are much more sensitive to the level of electron correlation and completeness of the basis set. Like NMR shielding constants, all-electron basis sets have been used to evaluate the spin-spin coupling constant at a nucleus of interest. Since the spin-spin coupling constant is sensitive to the quality of the wave function at the nucleus, very flexible basis sets are needed with tight s-functions. Relativistic effects have recently been included in the calculation of spin-spin coupling constants.[101]

Relativistic spin-free "scalar" effects on light nuclei in the vicinity of heavier atoms can be efficiently treated by replacing core electrons of heavy atom with quasirelativitic core potential (ECP).[67,121-124] It is possible to consider relativistic spin-orbit coupling effects by including the Fermi contact part of hyperfine interactions using finite perturbation theory (FPT).[125-130] In fact, there is an interesting parallel between the SO contribution to chemical shifts and the FC contribution to indirect spin-spin coupling.[129]

At first glance, NMR chemical shifts and spin-spin coupling constants of transition metal systems might seem out of reach to computational methods due to the high atomic number and importance of electron correlation and relativistic effects. Nevertheless, considerable progress has been made which shows that NMR chemical shifts (and coupling constants) can reliably be calculated.[131-138] All-electron basis sets are required for calculating spin-spin coupling constants at the nucleus of interest. However, if the nucleus of interest is attached to a heavier element, then a relativistic effective core potential (RECP) can be used to replace the core electrons of the heavier element. For different transition metals systems, either a pure DFT functional (such as BLYP) or a hybrid DFT functional (such as B3LYP) may give better results. For other transition metal systems spin-orbit corrections are important as are relativistic corrections via the zero-order regular approximation (ZORA) or via the Pauli approximation. Recently, Ziegler and co-workers have found that the ZORA method is more reliable than the Pauli method.[123] While relativistic effects can have a huge effect on the absolute shielding at heavy nuclei (transition metals and heavier), the effect on chemical shifts is much less because the relativistic effect is dominated by the nuclear charge and not by the chemical environment.

Since NMR is a relatively slow spectroscopic method, the observed chemical shifts are averaged over rotational and vibrational motion. The computational NMR results can also include robvibrational averaging. While not yet common, robvibrational averaging can make an important contribution.[138-140] Another effect which is beginning to be considered is solvation.[141-144] Solvation can affect the NMR chemical shift by causing a change in geometry or by electrostatic polarization. It is interesting to note that the chemical shifts of carbocations (where solvation effects are expected to be very large) are often predicted very well by calculating the gas-phase geometry at a reliable level of theory.[71].

The following is a partial list of current programs which calculate magnetic properties.

ACESII - correlated GIAO shifts and coupling constants
ADF - relativistic spin-orbital shifts
Dalton - coupling constants
deMon-NMR - sum over states (SOS-DFPT(IGLO))
Gaussian - GIAO and MP2-GIAO
Parallel Quantum Solutions (PQS) - Pulay's integrated software system
Turbomol -
Colgone99 -
DGauss -

Conclusion

The development of computational NMR is progressing very rapidly. This means that by the time this chapter is set into print, there will already be new developments. However, the progress to date is sufficient to guarantee that calculating magnetic properties will be a major focus in computational chemistry in the future. Thus, it can be said that WNL demonstrated yet again the ability to focus on an the important area of chemistry by performing one of the first ab initio calculations of an NMR shielding constant.

Acknowledgments: I would like to thank the Colonel for insight and encouragement given while I was a postdoc in his group.

References

1. "NMR Methods Blossom" *C&E News*, September 28, 1998.
2. Fleischer, U.; van Wüllen, C.; Kutzelnigg, W. "NMR Chemical Shift Computation: *Ab Initio*", in *Encyclopedia of Computational Chemistry*; Schleyer, P. v. R., Allinger, A., Clark, T., Gasteiger, J., Kollmann, P. A., Schaefer, H. F., Schreiner, P. R., Eds., Wiley: Chichester, U. K. 1998, pp 1827-1835.
3. Lazzeretti, P. "Ring currents", In *Progress in Nuclear Magnetc Resonance Spectroscopy* **2000**, *36*, 1-88.
4. (a) Fukui, H. "Theory and calculation of nuclear shielding constants", in *Progress in Nuclear Magnetc Resonance Spectroscopy* **1997**, *31*, 317-342.
   (b) Fukui, H. "Theory and calculation of nuclear spin-spin constants", in *Progress in Nuclear Magnetc Resonance Spectroscopy* **1999**, *51*, 267-294.
5. Webb, G. A. "Shielding: Overview of Theoretical Methods", In *Encyclopedia of Nuclear Magnetic Resonance*; Grant, D. M., Harris, R. K., Eds., Wiley: Chichester, U. K. 1996, pp 4316-4318.

6. Lazzeretti, P.; Malagoli, M.; Zanasi, R. "Shielding in Small Molecules", in *Encyclopedia of Nuclear Magnetic Resonance*; Grant, D. M., Harris, R. K., Eds., Wiley: Chichester, U. K. 1996, pp 4318-4327.
7. Jensen F. *Introduction to Computational Chemistry*, Wiley: New York, 1999.
8. Cremer, D.; Olsson, L.; Reichel, F.; Kraka, E. *Israel J. Chem.* **1993**, *33*, 369-385.
9. Gauss, J. *Ber. Bunsenges. Phys. Chem.* **1995**, *99*, 1001-1008.
10. Gauss, J. *J. Chem. Phys.* **1993**, *99*, 3629-3643.
11. Bühl, M. "Correlation between Transition-Metal NMR Chemical Shifts and Reactivities", in *Modeling NMR Chemical Shifts, Gaining Insights into Structure and Environment* (ACS Symposium Series, 732); Facelli, J. C., Ed., ACS: Washington DC, 1999, pp 240-250.
12. Kaupp, M.; Malkin, V. G.; Malkina, O. L. "NMR of Transitional Metal Compounds", in *Encyclopedia of Computational Chemistry*; Schleyer, P. v. R., Allinger, A., Clark, T., Gasteiger, J., Kollmann, P. A., Schaefer, H. F., Schreiner, P. R., Eds., Wiley: Chichester, U. K. 1998, pp 1857-1866.
13. Facelli, J. C. "Shielding Tensor Calculations", in *Encyclopedia of Nuclear Magnetic Resonance*; Grant, D. M., Harris, R. K., Eds., Wiley: Chichester, U. K. 1996, pp 4327-4334.
14. Facelli, J. C. "Shielding Calculations: Perturbation Methods", in *Encyclopedia of Nuclear Magnetic Resonance*; Grant, D. M., Harris, R. K., Eds., Wiley: Chichester, U. K. 1996, pp 4299-4307.
15. Jameson, C. J. "Recent Advances in Nuclear Magnetic Shielding Theory and Computational Methods", in *Modeling NMR Chemical Shifts, Gaining Insights into Structure and Environment* (ACS Symposium Series 732); Facelli, J. C., Ed., ACS: Washington DC, 1999, pp 1-23.
16. Jameson, C. J. "Chemical Shift Scales on an Absolute Basis", in *Encyclopedia of Nuclear Magnetic Resonance*; Grant, D. M., Harris, R. K., Eds., Wiley: Chichester, U. K. 1996, pp 1273-1281.
17. Cheeseman, J. R.; Trucks, G. W.; Keith, T. A.; Frisch, M. J. *J. Chem. Phys.* **1996**, *104*, 5497-5509.
18. de Dios, A. C.; Jameson, C. J. "The NMR Chemical Shift: Insight into Structure and Environment", in *Annual Reports on NMR Spectroscopy*, Vol. 29; Webb, G. A., Ed., Academic Press: London, 1994, pp 1-69.
19. Chesnut, D. B. "Ab Initio Calculation of NMR Chemical Shielding", in *Annual Reports on NMR Spectroscopy*, Vol. 29; Webb, G. A., Ed., Academic Press: London, 1994, pp 70-122.
20. Chesnut, D. B. "The Ab Initio Computation of Nuclear Magnetic Resonance Chemical Shielding", in *Reviews in Computational Chemistry*, Vol. 8; Lipkowitz, K. B.; Boyd, D. B., Eds., VCH: New York, 1996, pp 245-297.
21. Koch, W.; Holthausen, M. C. *A Chemist's Guide to Density Functional Theory*, Wiley-VCH: Weinheim, 2000, pp 195-209.

22. Alam, T. M. "Ab Initio Calculations of ^{31}P NMR Chemical Shielding Anisotropy Tensors in Phosphates: The Effect of Geometry on Shielding", In *Modeling NMR Chemical Shifts, Gaining Insights into Structure and Environment* (ACS Symposium Series 732); Facelli, J. C., Ed., ACS: Washington DC, 1999, pp 320-334.
23. Wilson, P. J.; Amos, R. D.; Handy, N. C. *Mol. Phys.* **1999**, *97*, 757-768.
24. Helgaker, T.; Wilson, P. J.; Amos, R. D.; Handy, N. C. *J. Chem. Phys.* **2000**, *113*, 2983-2989.
25. Schreckenbach, G.; Ziegler, T. *Theor. Chem. Acc.* **1998**, *99*, 71-82.
26. Wiberg, K. B. *J. Comput. Chem.* **1999**, *20*, 1299-1303.
27. Wiberg, K. B.; Hammer, J. D.; Keith, T. A.; Zilm, K. *J. Phys. Chem. A* **1999**, *103*, 21-27.
28. Wiberg, K. B.; Hammer, J. D.; Zilm, K. W.; Cheeseman, J. R. *J. Org. Chem.* **1999**, *64*, 6394-6400.
29. (a) Wiberg, K. B.; Hammer, J. D.; Zilm, K. W.; Cheeseman, J. R.; Keith, T. A. *J. Phys. Chem. A* **1998**, *102*, 8766-8773. (b) Wiberg, K. B.; Zilm, K. W. *J. Org. Chem.* **2001**, *66*, 2809-2817.
30. Bühl, M.; Kaupp, M.; Malkina, O. L.; Malkin, V. G. *J. Comput. Chem.* **1999**, *20*, 91-105.
31. Malkin, V. G.; Malkina, O. L.; Eriksson, L. A.; Salahub, D. R. "The Calculation of NMR and ESR Spectroscopic Parameters Using Density Functional Theory", in *Modern Density Functional Theory: A Tool for Chemistry*; Seminario, J. M., Politzer, P., Eds., Elsevier: Amsterdam, 1995, pp 273-347.
32. Schreckenbach, G.; Dickson, R. M.; Ruiz-Morales, Y.; Ziegler, T. "The Calculation of NMR Parameters by Density-Functional Theory", in *Chemical Applications of Density-Functional Theory*, (ACS Symposium Series 629); ACS: Washington DC, 1996, pp 328-341.
33. Malkin, V. G.; Malkina, O. L.; Casida, M. E.; Salahub, D. R. *J. Am. Chem. Soc.* **1994**, *116*, 5898-5908.
34. van Wüllen, C. *Phys. Chem. Chem. Phys.* **2000**, *2*, 2137-2144.
35. Olsson, L.; Cremer, D. *J. Phys. Chem.* **1996**, *100*, 16881-16891.
36. Adamo, C.; Barone, V. *Chem. Phys. Lett.* **1998**, *298*, 113-119.
37. Stevens, R. M.; Pitzer, R. M.; Lipscomb, W. N. *J. Chem. Phys.* **1963**, *8*, 550-560.
38. Lipscomb, W. N. "The Chemical Shift and Other Second-Order Magnetic and Electric Properties of Small Molecules", in *Advances in Magnetic Resonance*, Waugh, J. S., Ed., Academic Press: New York, 1966, pp 137-176.
39. Hegstrom, R. A.; Lipscomb, W. N. *J. Chem. Phys.* **1967**, *46*, 1594-1597.
40. Hegstrom, R. A.; Lipscomb, W. N. *Rev. Mod. Phys.* **1968**, *40*, 354-358.
41. Hegstrom, R. A.; Lipscomb, W. N. *J. Chem. Phys.* **1968**, *48*, 809-811.
42. Dichtfield, R. *J. Chem. Phys.* **1972**, *56*, 5688-5691.
43. Wolinski, K.; Hinton, J. F.; Pulay, P. *J. Am. Chem. Soc.* **1990**, *112*, 8251-8260.

44. Pulay, P.; Hinton, J. F.; Wolinski, K. "Efficient Imlementation of the GIAO Method for Magnetic Properties: Theory and Application", *Nuclear Magnetic Shieldings and Molecular Structure* (NATO ASI: Series C: Mathematical and Physical Sciences, Vol. 386); Tossell, J. A., Ed., Kluwer: the Netherlands, 1993, pp 243-262.
45. Pulay, P.; Hinton, J. F. "Shielding Theory: GIAO Method", in *Encyclopedia of Nuclear Magnetic Resonance*; Grant, D. M., Harris, R. K., Eds., Wiley: Chichester, U. K. 1996, pp 4334-4339.
46. Gauss, J.; Werner, H.-J. *Phys. Chem. Chem. Phys.* **2000**, *2*, 2083-2090.
47. Kollwitz, M.; Gauss, J. *Chem. Phys. Lett.* **1996**, *260*, 639-646.
48. Bühl, M.; Gauss, J.; Stanton, J. F. *Chem. Phys. Lett.* **1995**, *241*, 248-252.
49. Gauss, J. *Chem. Phys. Lett.* **1994**, *229*, 198-203.
50. Gauss, J. *Chem. Phys. Lett.* **1992**, *191*, 614-620.
51. Kutzelnigg, W.; Fleischer, U.; Schindler, M. "The IGLO-Method: Ab-initio Calculation and Interpretation of NMR Chemical Shifts and Magnetic Susceptibilities", in *NMR Basic Principles and Progress*, Vol. 23 Springer-Verlag: Berlin, 1990, pp 165-262.
52. Kutzelnigg, W.; van Wüllen, C.; Fleischer, U.; Franke, R.; Mourik, T. v. "The IGLO method. Recent developments", *Nuclear Magnetic Shieldings and Molecular Structure* (NATO ASI: Series C: Mathematical and Physical Sciences, Vol. 386); Tossell, J. A., Ed., Kluwer: the Netherlands, 1993, pp 141-161.
53. Kutzelnigg, W.; Fleischer, U.; van Wüllen, C. "Shielding Calculations: IGLO Method", in *Encyclopedia of Nuclear Magnetic Resonance*; Grant, D. M., Harris, R. K., Eds., Wiley: Chichester, U. K. 1996, pp 4284-4291.
54. Hansen, A. E.; Bilde, M. "Shielding Calculations: LORG & SOLO Approaches", in *Encyclopedia of Nuclear Magnetic Resonance*; Grant, D. M., Harris, R. K., Eds., Wiley: Chichester, U. K. 1996, pp 4292-4299.
55. Hansen, A. E.; Bouman, T. D. "Ab Initio Calculations and Analysis of Nuclear Magnetric Schielding Tensors: the LORG and SOLO Approaches", *Nuclear Magnetic Shieldings and Molecular Structure* (NATO ASI: Series C: Mathematical and Physical Sciences, Vol. 386); Tossell, J. A., Ed., Kluwer: the Netherlands, 1993, pp 117-140.
56. Bohmann, J. A.; Weinhold, F.; Farrar, T. C. *J. Chem. Phys.* **1997**, *107*, 1173-1184.
57. Weinhold, F. "Natural Bond Orbital Methods", in *Encyclopedia of Computational Chemistry*; Schleyer, P. v. R., Allinger, A., Clark, T., Gasteiger, J., Kollmann, P. A., Schaefer, H. F., Schreiner, P. R., Eds.; Wiley: Chichester, U. K., 1998, pp 1792-1811.
58. Galasso, V. *Chem. Phys.* **1999**, *241*, 247-255.
59. Patchkovskii, S.; Thiel, W. *J. Comput. Chem.* **1999**, *20*, 1220-1245.
60. Patchkovskii, S.; Thiel, W. *J. Mol. Model.* **2000**, *6*, 67-75.
61. Baker, J. D.; Zerner, M. C. *Int. J. Quantum Chem.* **1992**, *43*, 327-342.
62. Fehlner, T. P.; Czech, P. T.; Fenske, R. F. *Inorg. Chem.* **1990**, *29*, 3103-3109.

63. Baldridge, K. K.; Siegel, J. S. *J. Phys. Chem. A* **1999**, *103*, 4038-4042.
64. Forsyth, D. A.; Sebag, A. B. *J. Am. Chem. Soc.* **1997**, *119*, 9483-9494.
65. Karadakov, P. B.; Morokuma, K. *Chem. Phys. Lett.* **2000**, *317*, 589-596.
66. Cui, Q.; Karplus, M. *J. Phys. Chem. B* **2000**, *104*, 3721-3743.
67. Schreckenbach, G.; Wolff, S. K.; Ziegler, T., "Covering the Entire Periodic Table: Relativistic Density Functional Calculations of NMR Chemical Shifts in Diamagnetic Actinide Compounds', in *Modeling NMR Chemical Shifts, Gaining Insights into Structure and Environment* (ACS Symposium Series, 732); Facelli, J. C., Ed., ACS: Washington DC, 1999, pp 101-114.
68. Karadakov, P. B.; Webb, G. A.; England, J. A., "The Effect of Electron Correlation on the ^{19}F Chemical Shifts in Fluorobenzenes", in *Modeling NMR Chemical Shifts, Gaining Insights into Structure and Environment* (ACS Symposium Series, 732); Facelli, J. C., Ed., ACS: Washington DC, 1999, pp 115-125.
69. Robert, V.; Petit, S.; Dorshch, S. A.; Bigot, B. *J. Phys. Chem. A* **2000**, *104*, 4586-4591.
70. Wilson, P. J. *Mol. Phys.* **2001**, *99*, 363-367.
71. Bühl, M. "NMR Chemical Shift Computations: Structural Applications", in *Encyclopedia of Computational Chemistry*; Schleyer, P. v. R., Allinger, A., Clark, T., Gasteiger, J., Kollmann, P. A., Schaefer, H. F., Schreiner, P. R., Eds., Wiley: Chichester, U. K., 1998, pp 1835-1845.
72. Wolinski, K.; Hsu, C.-L.; Hinton, J. F.; Pulay, P. *J. Chem. Phys.* **1993**, *99*, 7819-7824.
73. Prakash, G. K. S.; Rasul, G.; Olah, G. A. *J. Phys. Chem. A* **1998**, *102*, 2579-2583.
74. Onak, T. in *The Borane, Carborane, Carbocation Continuum*; Casanova, J., Ed., Wiley: New York, 1998, pp 247-258.
75. Onak, T.; Diaz, M.; Barfield, M. *J. Am. Chem. Soc.* **1995**, *117*, 1403-1410.
76. DerHovanessian, A.; Rablen, P. R.; Jain, A. *J. Phys. Chem. A* **2000**, *104*, 6056-6061.
77. Ochsenfeld, C. *Phys. Chem. Chem. Phys.* **2000**, *2*, 2153-2159.
78. Hofmann, M.; Schleyer, P. v. R.; Williams, R. E. *Inorg. Chem.* **2000**, *39*, 1066-1070.
79. Beez, V.; Pritzkow, H.; Hofmann, M.; Schleyer, P. v. R.; Siebert, W. *Eur. J. Inorg. Chem.* **1998**, 1775-1779.
80. Plešek, J.; Štíbr, B.; Hnyk, D.; Jelínek, T.; Heřmánek, S.; Kennedy, J. D.; Hofmann, M.; Schleyer, P. v. R. *Inorg. Chem.* **1998**, *37*, 3902-3909.
81. Hofmann, M.; Schleyer, P. v. R. *Inorg. Chem.* **1998**, *37*, 5557-5565.
82. McKee, M. L.; Bühl. M.; Schleyer, P. v. R. *Inorg. Chem.* **1993**, *32*, 1712-1715.
83. McKee, M. L. *Inorg. Chem.* **1999**, *38*, 321-330.
84. Heřmánek, S. *Inorg. Chim. Acta* **1999**, *289*, 20-44.
85. Sun, G.; Kertesz, M. *J. Phys. Chem. A* **2000**, *104*, 7398-7403.
86. Bühl, M.; Thiel, W.; Jiao, H.; Schleyer, P. v. R.; Saunders, M.; Anet, F. A. L. *J. Am. Chem. Soc.* **1994**, *116*, 6005-6006.

87. Cioslowski, J. *Chem. Phys. Lett.* **1994**, *227*, 361-364.
88. Bühl, M.; Thiel, W. *Chem. Phys. Lett.* **1995**, *233*, 585-589.
89. Bühl, M. Z. *Anorg. Allg. Chem.* **2000**, *626*, 332-337.
90. Schleyer, P. v. R.; Maerker, C.; Hommes, N. J. R. v. E. *J. Am. Chem. Soc.* **1996**, *118*, 6317.
91. Using the calculated magnetic shielding at a non-nuclear position for use as a "virtual probe" has been proposed independently. Wolinski, K. *J. Chem. Phys.* **1997**, *106*, 6061-6067.
92. (a) McKee, M. L.; Wang, Z.-X.; Schleyer, P. v. R. *J. Am. Chem. Soc.* **2000**, *122*, 4781-4793. (b) Balci, M.; McKee, M. L.; Schleyer, P. v. R. *J. Phys. Chem. A* **2000**, *104*, 1246-1255.
93. Mauksch, M.; Gogonea, V.; Jiao, H.; Schleyer, P. v. R. *Angew. Chem. Int. Ed.* **1998**, *37*, 2395-2397.
94. Nendel, M.; Houk, K. N.; Tolbert, L. M.; Vogel, E.; Jiao, H.; Schleyer, P. v. R. *J. Phys. Chem. A* **1998**, *102*, 7191-7198.
95. Schulman, J. M.; Disch, R. L.; Jiao, H.; Schleyer, P. v. R. *J. Phys. Chem. A* **1998**, *102*, 8051-8055.
96. Gogonea, V.; Schleyer, P. v. R.; Schreiner, P. R. *Angew. Chem. Int. Ed.* **1998**, *37*, 1945-1948.
97. Xie, Y.; Schreiner, P. R.; Schaefer, III, H. F.; Li, X.-W.; Robinson, G. H. *J. Am. Chem. Soc.* **1996**, *118*, 10635-10639.
98. Baldridge, K. K.; Uzan, O.; Martin, J. M. L. *Organometallics* **2000**, *19*, 1477-1487.
99. Baldridge, K. K.; Siegel, J. S. *Angew. Chem. Int. Ed. Engl.* **1997**, *36*, 745-748.
100. Sawicka, D.; Houk, K. N. *J. Mol. Model.* **2000**, *6*, 158-165.
101. Helgaker, T.; Jaszuński, M.; Ruud, K. *Chem. Rev.* **1999**, *99*, 293-352.
102. Del Bene, J. E.; Bartlett, R. *J. Am. Chem. Soc.* **2000**, *122*, 10480-10481.
103. Del Bene, J. E.; Perera, S. A.; Bartlett, R. *J. Am. Chem. Soc.* **2000**, *122*, 3560-3561.
104. Perera, S. A.; Bartlett, R. *J. Am. Chem. Soc.* **2000**, *122*, 1231-1232.
105. Perera, S. A.; Bartlett, R. *J. Am. Chem. Soc.* **1996**, *118*, 7849-7850.
106. DelBene, J. E.; Perera, S. A.; Bartlett. R. J.; Alkorta, I.; Elguero, J. *J. Phys. Chem. A* **2000**, *104*, 7165-7166.
107. Perera, S. A.; Sekino, H.; Bartlett, R. J. *J. Chem. Phys.* **1994**, *101*, 2186-2191
108. Perera, S. A.; Bartlett, R. J.; Schleyer, P. v. R. *J. Am. Chem. Soc.* **1995**, *117*, 8476-8477.
109. Del Bene, J. E.; Perera, S. A.; Bartlett, R. J. *J. Phys. Chem. A* **1999**, *103*, 8121-8124.
110. Perera, S. A.; Noijen, M; Bartlett, R. J. *J. Chem. Phys.* **1996**, *104*, 3290-3305.
111. Sauer, S. P. A.; Raynes, W. T. *J. Chem. Phys.* **2000**, *113*, 3121-3129.
112. Pecul, M.; Leszczynski, J.; Sadlej, J. *J. Chem. Phys.* **2000**, *112*, 7930-7938.

113. Perera, S. A.; Bartlett, R.; Schleyer, P. v. R. *J. Am. Chem. Soc.* **1995**, *117*, 8486-8477.
114. Contreras, R. H.; Facelli, J. C. "Advances in Theoretical and Physical Aspects of Spin-Spin Coupling Constants", in *Annual Reports on NMR Spectroscopy*, Vol. 27; Webb, G. A., Ed., Academic Press: London, 1993, pp 255-356.
115. Sychrovský, Gräfenstein, J.; Cremer, D. *J. Chem. Phys.* **2000**, *113*, 3530-3547.
116. Autschbach, J.; Ziegler, T. *J. Chem. Phys.* **2000**, *113*, 936-947.
117. Autschbach, J.; Ziegler, T. *J. Chem. Phys.* **2000**, *113*, 9410-9418.
118. Helgaker, T.; Watson, M.; Handy, N. C. *J. Chem. Phys.* **2000**, *113*, 9502-9409.
119. (a) Onak, T.; Jaballas, J.; Barfield, M. *J. Am. Chem. Soc.* **1999**, *121*, 2850-2856. (b) Peralta, J. E.; Contreras, R. H.; Snyder, J. P. *Chem. Commun.* **2000**, 2025-2026.
120. Bagno, A. *Chem. Eur. J.* **2001**, *7*, 1652-1661.
121. Gilbert, T. M.; Ziegler, T. *J. Phys. Chem. A* **1999**, *103*, 7535-7543.
122. Bouten, R.; Baerends, E. J.; van Lenthe, E.; Visscher, L.; Schreckenbach, G.; Ziegler, T. *J. Phys. Chem. A* **2000**, *104*, 5600-5611.
123. Schreckenbach, G.; Wolff, S. K.; Ziegler, T. *J. Phys. Chem. A* **2000**, *104*, 8244-8255.
124. Kutzelnigg, W. *J. Comput. Chem.* **1999**, *20*, 1199-1219.
125. Visscher, L.; Enevoldsen, T.; Saue, T.; Jørgen, H.; Jensen, A.; Oddershede, J. *J. Comput. Chem.* **1999**, *20*, 1262-1273.
126. Vaara, J.; Ruud, K.; Vahtras, O. *J. Comput. Chem.* **1999**, *20*, 1314-1327.
127. Malkin, V. G.; Malkina, O. L.; Salahub, D. R. *Chem. Phys. Lett.* **1996**, *261*, 335-345.
128. (a) Kaupp, M.; Malkina, O. L. *J. Chem. Phys.* **1998**, *108*, 3648-3659 (b) Vaara, J.; Malkina, O. L.; Stoll, H.; Malkin, V. G.; Kaupp, M. *J. Chem. Phys.* **2001**, *114*, 61-71.
129. Kaupp, M.; Malkina, O. L.; Malkin, V. G.; Pyykkö, P. *Chem. Eur. J.* **1998**, *4*, 118-126.
130. Kaupp, M.; Malkina, O. L.; Malkin, V. G. *J. Comput. Chem.* **1999**, *20*, 1304-1313.
131. Kaupp, M.; Rovira, C.; Parrinello, M. *J. Phys. Chem. B* **2000**, *104*, 5200-5208.
132. Weller, A. S.; Fehlner, T. P. *Organometallics* **1999**, *18*, 447-450.
133. Bühl, M.; Baumann, W.; Kadyrov, R.; Börner, A. *Helv. Chim. Acta* **1999**, *82*, 811.
134. Bühl, M.; Thiel, W.; Fleischer, U.; Kutzelnigg, W. *J. Phys. Chem.* **1995**, *99*, 4000-4007.
135. Bühl, M. *J. Comput. Chem.* **1999**, *20*, 1254-1261.
136. Bühl, M. *Angew. Chem. Int. Ed.* **1998**, *37*, 142-144.
137. Jameson, C. J.; de Dios, A. C.; Jameson, A. K. *J. Chem. Phys.* **1991**, *95*, 9042.

138. Kozlowski, P. M.; Wolinski, K.; Pulay, P.; Ye, B.-H.; Li, X.-Y. *J. Phys. Chem. A* **1999**, *103*, 420-425.
139. de Dios, A. C.; Roach, J. L.; Walling, A. E., "The NMR Chemical Shift: Local Geometry Effects", in *Modeling NMR Chemical Shifts, Gaining Insights into Structure and Environment* (ACS Symposium Series 732); Facelli, J. C., Ed., ACS: Washington DC, 1999, pp 220-239.
140. Jameson, C. J.: de Dios, A. C. "The Nuclear Shielding Surface: The Shielding as a Function of Molecular Geometry and Intermolecular Separation", *Nuclear Magnetic Shieldings and Molecular Structure* (NATO ASI: Series C: Mathematical and Physical Sciences, Vol. 386); Tossell, J. A., Ed., Kluwer: the Netherlands, 1993, pp 95-116.
141. Yamazaki, T.; Sato, H.; Hirata, F. *Chem. Phys. Lett.* **2000**, *325*, 668-674.
142. Mikkelsen, K. V.; Ruud, K.; Helgaker, T. *J. Comput. Chem.* **1999**, *20*, 1281-1291.
143. Zhan, C.-G.; Chipman, D. M. *J. Chem. Phys.* **1999**, *110*, 1611-1622.
144. Manalo, M. N.; de Dios, A. C.; Cammi, R. *J. Phys. Chem. A* **2000**, *104*, 9600-9604.

Chapter 9

Does the Magnitude of NMR Coupling Constants Specify Bond Polarity?

Rodney J. Bartlett[1], Janet E. Del Bene[1,2], and S. Ajith Perera[1]

[1]Quantum Theory Project, University of Florida, Gainesville, FL 32611
[2]Department of Chemistry, Youngstown State University, Youngstown, OH 44555

In this study we have addressed the question of whether the existence of a spin-spin coupling constant between a pair of atoms can be related to bond polarity. We have applied external electric fields along the H-F, C-F, C-Cl, and C-Li bonds in the molecules HF, CH$_3$F, CH$_3$Cl, and CH$_3$Li, respectively, to reverse the dipole moment, and have computed one-bond EOM-CCSD spin-spin coupling constants $^1J_{H-F}$, $^1J_{C-F}$, $^1J_{C-Cl}$, and $^1J_{C-Li}$ as a function of field strength. The results of this investigation demonstrate that the coupling constant is not a maximum when the bonding between the atoms is very nonpolar (covalent) as the dipole moment approaches zero. In fact, we find no correlation between the polarity of a bond and its coupling. Hence, the experimental measurement of a coupling constant cannot be interpreted as proof of the existence of a covalent bond.

We have also computed EOM-CCSD two-bond coupling constants in hydrogen-bonded complexes, and addressed the question of the relationship between $^{2h}J_{X-Y}$ and the nature of the X-H-Y hydrogen bond. For this study we have examined the equilibrium structures of ClH:NH$_3$ and the value of $^{2h}J_{Cl-N}$ as a function of field strength and hydrogen bond type. In addition, we have computed the O-O spin-spin coupling constant $^{2h}J_{O-O}$ for O$_2$H$_5^+$, and for two H$_2$O molecules in the same orientation as in the O$_2$H$_5^+$ complex, but with the hydrogen-bonded proton removed. Our results again suggest that the experimental measurement of a coupling constant across a hydrogen bond does not prove that the hydrogen bond is covalent.

INTRODUCTION

Recently there has been a great deal of excitement regarding the observation of NMR spin-spin coupling constants $^{2h}J(^{15}N-^{15}N)$ across N-H-N hydrogen bonds(*1-8*). Since the H atoms cannot be seen in X-ray structures, verifying their presence and location via coupling constants provides experimental evidence of the existence of hydrogen bonds. This is of particular significance in biomolecules, where such bonds play an important role in structure determination. It is also possible with the aid of predictive theoretical calculations (*9-10*) to obtain N-N distances from experimental coupling constants, provided that the distance-dependence of the coupling constant has been established.

It has also been suggested that the measurement of coupling constants across hydrogen bonds is an indicator of covalent (nonpolar) character of such bonds. In a recent popular article it was stated that "the J couplings unambiguously establish the partial covalent character of H-bonds"(*1*). The controversy about ionic versus covalent and the size of the coupling constant is nothing new. Twenty years ago, Schleyer, et. al., argued similarly that CH_3Li is highly covalent because the coupling constant between Li and C appears to be quite high. They stated, "We conclude that $^1J(Li-C)$ in monomeric methyllithium should be very large, perhaps over 200 Hz. In our view, this indicates the predominantly covalent nature of C-Li bonding, unless a rationalization consistent with the 'wholly-ionic' interpretation can be found."(*11*). However, the opposite viewpoint was taken by Streitweiser, et al., who argued that by all theoretical measures they could apply, the C-Li bond was extremely ionic. They stated, "Methyllithium may be described as a charge transfer complex with substantial but not complete (0.8e) charge transfer." (*12*). These are two of many examples that attest to the fact that coupling constants as a measure of covalent or ionic character is a question of some significance. Unfortunately, like so many questions of this type in chemistry, much of the ambiguity pertains to the ill-defined nature of the terms "covalent" and "ionic". We prefer to address the question of how coupling constants depend upon bond polarity in terms of a quantity which can be given a more rigorous experimental and quantum mechanical meaning, namely, the dipole moment.

METHOD OF CALCULATION

In the last several years, we have developed coupled-cluster (CC) based quantum chemical methods (*13*), which are among the most accurate available in the field. Today, such methods are frequently those of choice when both high accuracy and wide applicability are required (*14*). In our CC efforts, we have generalized the ground state description of CC theory to that for excited states, via the equation-of-motion approach (extension for NMR spin-spin coupling constants is described in (*15-16*)). Because of the close correspondence between second-order properties like NMR coupling constants which require the first-order perturbed wavefunction and their description in terms of a sum over states, we may write

$$J = J_{DSO} + \langle\Psi_0|\hat{O}|\Psi^{(1)}\rangle$$
$$= \langle\Psi_0|\hat{O}_{DSO}|\Psi_0\rangle + \Sigma_k |\langle\Psi_0|\hat{O}|\Psi_k\rangle|^2/(E_0 - E_k) \quad (1)$$

where

$$\hat{O} = \hat{O}_{FC} + \hat{O}_{PSO} + \hat{O}_{SD} \quad (2)$$

and the Hamiltonian is

$$H = H_0 + \lambda\hat{O} + \lambda^2 \hat{O}_{DSO} \quad (3)$$

We have indicated the four components of the coupling constant, which in a non-relativistic treatment are the Fermi-contact (FC), paramagnetic spin orbit (PSO), spin dipole (SD), and diamagnetic spin orbit (DSO) terms. In equation (1), the DSO term is computed as an expectation value, and the remaining components are expressed in terms of a sum-over-states formula, where the denominator is the EOM-CCSD excitation energy. Rather than using the sum-over-states formula, which would almost certainly force an unwarranted truncation, we take advantage of the fact that all of the EOM-CCSD excited states are represented in terms of single and double excitations |h>, so we can write the first-order perturbed wavefunction as

$$\Psi^{(1)} = |h\rangle\langle h|E_0 - H_0|h\rangle^{-1} \langle h|\hat{O}|\Psi_0\rangle = |h\rangle a \quad (4)$$

$$\langle h|E_0 - H_0|h\rangle a = \langle h|\hat{O}|\Psi_0\rangle \quad (5)$$

and solve the linear equation for **a** to avoid inverting the resolvent matrix, which can easily have a rank exceeding 10^6. This avoids any truncation of the sum-over-states expression in the description of the NMR coupling constants. In calibrating the accuracy of our methods to experiment, for a large number of coupling constants we find an average error of ~3.5 Hz, which places the computed values within ~ 10% of the experimental value. To about this error we can argue that our treatment of coupling constants is predictive. Then, using this tool, we can directly address how coupling constants depend upon bond polarity for a few prototype systems, and answer the title question. This study also illustrates how predictive theory can be used to provide results when it would be difficult or impossible to control all of the parameters, such as geometries and polarity, to permit an experimental determination.

The structures of the molecules investigated in this study were optimized at the CCSD(T) level of theory (*13*) using the aug-cc-pVTZ basis set on C, F, Cl, and H atoms, and cc-pVTZ on Li (*17-19*). The coupling constants were computed at EOM-CCSD using the (qzp, qz2p) basis set of Ahlrichs, with (tzpl) on Li (*20*) The structures of hydrogen-bonded complexes discussed in this work had been optimized previously at second-order many-body perturbation theory [MBPT(2) = MP2] (*21-24*) either with the 6-31+G(d,p) basis set (*25-28*), or with Dunning's aug'-cc-pVDZ basis, where aug' indicates that diffuse functions are used only on nonhydrogen atoms. All calculations were performed with ACES II (*29*) on the SV1 computer at the Ohio Supercomputer Center.

RESULTS AND DISCUSSION

<u>The Hydrogen Fluoride Molecule.</u> The first molecule considered is HF. For any molecule containing F, the Fermi-contact term alone is insufficient to account for the value of the spin-spin coupling constant, and all terms must be included for quantitative agreement with experiment. Thus, the experimental value for the HF coupling constant is 529± 23 Hz (*30*). Our computed value is 515.0 Hz, with the PSO, DSO, Fermi-contact, and SD terms contributing 185.7, 0.5, 326.2, and 2.6 Hz, respectively.

To investigate the relationship between bond polarity and the value of J, we have employed an electric field applied along the F-H bond. We have varied the field strength, and thereby varied the polarity of the bond. That is, by modifying the Hamiltonian for the problem, we have allowed the molecule to adjust its polarity to oppose the imposed field, consistent with Lenz's Law. In using the finite-field approach, the field is added to the Hamiltonian for the molecule. It must be sufficiently small in magnitude not to invalidate the quantum chemical calculation, yet large enough to provide a significant change in energy as a function of field strength. We have checked to insure the validity of the numerical results, since by experimental standards, some of the electric fields used are large. When the field strength becomes too large, CC theory shows large amplitudes for configurations other than the reference configuration. This situation invalidates the basis of single-reference CC theory and leads to erroneous results. All such results have been eliminated from this work. (From a more formal viewpoint, the fact that the calculation is done in a Gaussian basis set imposes a "box-like" structure on the problem that maintains the usual square integrable nature of the bound solutions, even though technically, adding the electric field destroys the boundedness of the Hamiltonian. The latter could conceivably manifest itself in field ionization if plane waves were included in the basis.)

As the dipole moment for polyatomic molecules can vanish due to conflicting factors, we also monitor the bond polarity by observing the Mulliken atomic populations on F and H. We recognize that unlike the dipole moment, such net populations are not quantum mechanical observables, and that the

Mulliken definition is probably not even the best choice. However, for our purposes, they are adequate to show how the net charges change qualitatively as a function of field strength, and that these changes parallel changes observed in the dipole moment.

The values of the Mulliken populations on F and H, the dipole moment, the Fermi-contact term, and total J for HF as a function of field strength are reported in Table 1. At zero-field the dipole moment of HF is -0.744 au (-1.89 D), and the Mulliken net atomic charge on F is -0.34 (total population of 9.34 e). As a function of field strength the net Mulliken charge on F changes from -0.34 to 0.00, and then becomes positive. Correspondingly, the charge on H changes from +0.34 to negative values to maintain the neutral net charge on the molecule. At a field of 0.105 au the Mulliken charges on F and H are 0.00, and the dipole moment is quite small at -0.095 au. At a field of 0.118 au, the dipole moment is zero and the Mulliken charges are only about 0.05.

Figure 1 shows graphically the behavior of the dipole moment, the Fermi-contact term, and total J as a function of the strength of the imposed electric field. The value of the coupling constant is 515.0 Hz at zero field, decreases to its minimum value of 328.3 Hz at a field of 0.100 au, and then subsequently increases, with the largest value of J (692.6 Hz) found at a field of 0.175 au. At the same time, the dipole moment of HF is reduced from -0.744 au at zero field to approximately zero at a field between 0.115 and 0.120 au, and then increases to +0.50 au at a field of 0.175 au. Since we generally describe nonpolar molecules as those which are stabilized by "covalent" bonds, we would expect that if J is a measure of covalency, the largest J values should occur when

Table 1. Electron populations on F and H, dipole moment (au), and Fermi-contact term and total J (Hz) as a function of field strength (au) for HF.[a]

Field	F	H	Dipole	Fermi-contact	Total J
0.000	9.34	0.66	-0.744	326.2	515.0
0.010	9.31	0.69	-0.690	286.1	486.8
0.030	9.25	0.75	-0.580	205.3	433.3
0.050	9.19	0.81	-0.462	125.5	386.0
0.100	9.02	0.98	-0.131	-49.1	328.3
0.105	9.00	1.00	-0.095	-62.8	330.3
0.110	8.98	1.02	-0.057	-75.5	334.5
0.115	8.96	1.04	-0.019	-87.2	341.0
0.120	8.94	1.06	+0.021	-97.7	349.9
0.125	8.93	1.07	+0.061	-107.0	361.6
0.130	8.91	1.09	+0.102	-115.0	376.0
0.150	8.83	1.17	+0.277	-133.2	468.8
0.175	8.72	1.28	+0.520	-124.7	692.6

a) The F-H distance is fixed at the optimized CCSD(T)/aug'-cc-pVTZ value of 0.9191 Å.

the dipole moment is zero. As evident from Figure 1, this is obviously not the case.

CH$_3$F and CH$_3$Cl. Figures 2 and 3 show similar plots for methyl fluoride and methyl chloride. Using the molecular dipole moment to measure the polarity of the C-F and C-Cl bonds is not quite as straightforward as it is for HF, since the imposed electric fields affect the polarity of the C-H bonds as well. Nevertheless, we can still examine the change in the dipole moment of the molecule as a function of field strength, and also appeal to the Mulliken charges to assist in measuring the polarity of the C-F and C-Cl bonds. For the zero-field CH$_3$F molecule, the computed dipole moment is -0.777 au, with Mulliken charges of +0.07 on C and H, and -0.28 on F. The computed coupling constant for this molecule is -170.5 Hz. The molecule exhibits a dipole moment close to zero (0.025 au) at a field of 0.050 au, and has Mulliken charges of 0.00 on H, +0.08 on C, and -0.08 on F. These charges are much closer to neutral than the zero-field charges. The computed coupling constant at this field is -278.6 Hz. However, this value of the coupling constant is not a minimum (maximum absolute value) as would be anticipated if J were largest when the bond is highly nonpolar (covalent). Instead, J becomes increasingly more negative as the strength of the external field increases to 0.10 au. The increasing field induces an increase in the dipole moment and in the polarity of the C-F bond in the direction C$^{\delta-}$-F$^{\delta+}$. CH$_3$Cl exhibits similar behavior. Its zero-field dipole moment is -0.796 au, with Mulliken charges of +0.10, -0.05, and -0.24 on H, C, and Cl, respectively. The C-Cl coupling constant is -12.1 Hz. As the strength of the field increases, the value of the dipole moment approaches zero (+0.010 au) at a field of 0.025 au. At this field the charge on Cl is reduced to -0.07, but each H is still positively charged (+0.06) and C has a slightly larger negative charge (-0.09). Nevertheless, at this field strength J is -18.9 Hz, and it continues to increase as the field strength increases. Figure 3 shows that the coupling constant becomes increasingly more negative as the dipole moment changes from -0.8 au at zero field to +0.4 au at a field of 0.04 au. There is no extremum in the J curve to suggest that J has its maximum value when the C-Cl bond is nonpolar.

CH$_3$Li. As noted above, the polarity of CH$_3$Li was deduced incorrectly because a large value (~200 Hz) had been obtained for a coupling constant computed with semi-empirical theory. In addition to the theory being suspect (the EOM-CCSD value is 97.7 Hz), there is some question as to what is meant by "large" when a scale for coupling constants has not been established. Furthermore, any such scale should be based upon the reduced coupling constant K, rather than J. But, nevertheless, the fundamental premise that nonpolar bonds have larger coupling constants is contrary to the examples above. We have computed the dipole moment and C-Li coupling constant for CH$_3$Li as a function of field strength, and show these results in Figure 4. In the absence of a field, the

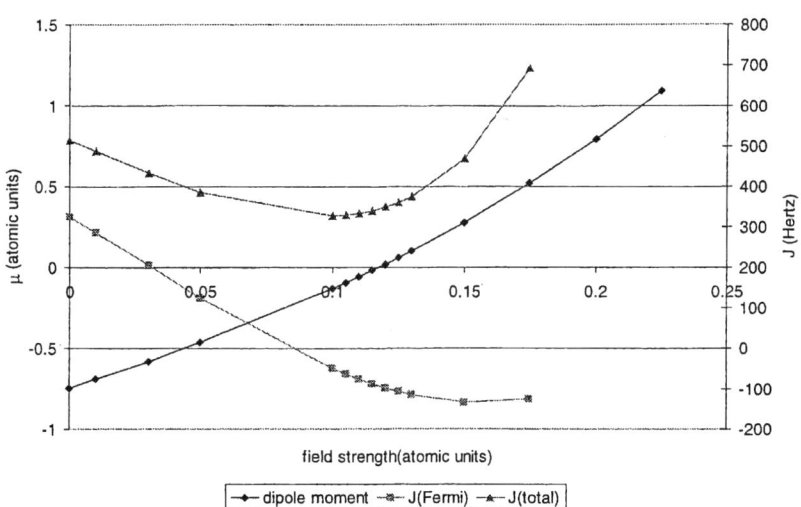

Figure 1. Dipole moment, Fermi-contact term, and total spin-spin coupling constant for HF as a function of field strength.

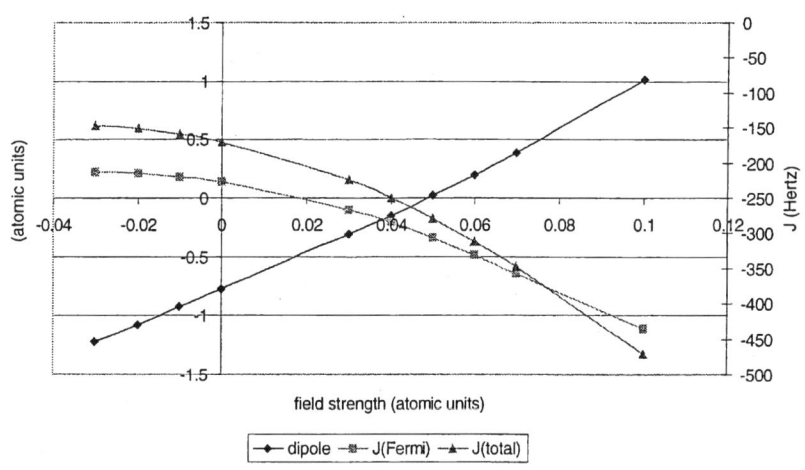

Figure 2. Dipole moment, Fermi-contact term, and total C-F spin-spin coupling constant for CH_3F as a function of field strength.

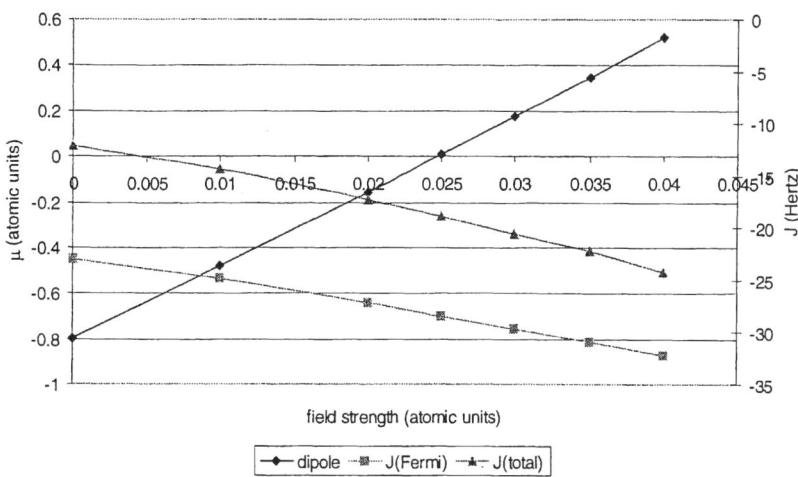

Figure 3. Dipole moment, Fermi-contact term, and total C-Cl spin-spin coupling constant for CH_3Cl as a function of field strength.

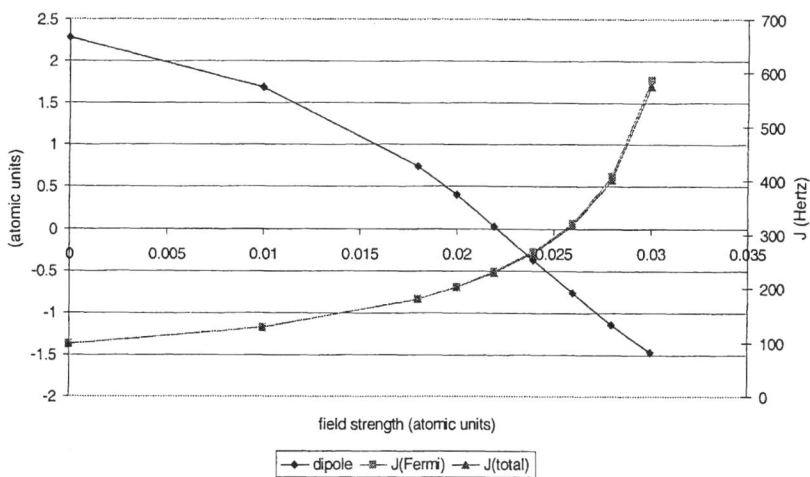

Figure 4. Dipole moment, Fermi-contact term, and total C-Li spin-spin coupling constant for CH_3Li as a function of field strength.

very large dipole moment of CH_3Li attests to its very high ionicity, which is substantially greater than that of HF, CH_3F, and CH_3Cl. Unlike the methyl halides, the dipole moment of CH_3Li has its positive end at Li. And, unlike the methyl halides, the only term of importance in determining J for CH_3Li is the Fermi-contact term, with the other terms being less than 0.1 Hz. At zero field, the charges on H, C, and Li are +0.07, -0.84, and +0.64, respectively. J has a value of 97.7 Hz. With increasing field strength, there is a migration of charge toward Li. At a field of 0.022 au, the dipole moment is close to zero (0.026 au), the charge on Li is only +0.05, while the charge on C has been reduced to -0.57. Each H bears a charge of +0.17. Though still polar, the C-Li bond is much less so than in the zero-field case. Yet, as Figure 4 demonstrates, J increases with decreasing polarity of the C-Li bond, and continues to increase when the polarity is reversed. This again does not support the supposition that bonds that are nonpolar (more covalent) have the largest J values.

Polarity in Hydrogen Bonds. We have investigated the two-bond spin-spin coupling constants ($^{2h}J_{X-Y}$) across X-H-Y hydrogen bonds in series of hydrogen-bonded complexes (*9,10, 31-34*). As noted in the Introduction, the measurement of N-N coupling constants across N-H-N hydrogen bonds has been used as evidence for the "covalent" character of hydrogen bonds. Support for this has also come from the theory community, who show orbital plots that suggest that there are electrons on H (*1,6*). Obviously, any proton would have an enormous electric field associated with it, so it is unreasonable to think that there would not be electrons attracted to that field, if that is all that is meant by "covalent" character. More than 60 years ago in referring to the water dimer, Pauling said, "The amount of partial ionic character expected for the O-H bond from the electronegativity difference of the atoms is 39 percent. Hence the 1s orbital of the hydrogen atom is liberated from use in covalent-bond formation with the adjacent oxygen atom to the extent of 39 percent, and hence available for formation of a fractional covalent bond with the more distant oxygen atom of the hydrogen-bonded group O-H...O."(*35*). This is still an accurate statement.
Unfortunately, we cannot use an electric field to bring the net electron populations on X, H, and Y in an X-H-Y hydrogen bond to zero. Because atoms X and Y normally bear a negative charge while H is positively charged, imposing an electric field increases the electron density on either X or Y (depending on the direction of the field) while H remains positively charged. Hence, the approach that was used successfully for other molecules is not helpful here.

Are there other ways in which the question of the covalency of the hydrogen bond, and the relationship between covalency and J, can be addressed? Oldfield (*37*) would answer this question in the affirmative. Based on his analysis of hydrogen bonding in proteins using three-bond spin-spin couplings ($^{3h}J_{N-C}$), chemical shifts, and AIM theory (*38*), Oldfield suggested that the trans hydrogen bond $^{3h}J_{N-C}$ coupling observed between peptide groups in proteins is mediated by a closed-shell noncovalent interaction between the donor hydrogen atom and the acceptor oxygen atom. He states that, "These results support the idea that the existence of electron-coupled nuclear spin-spin coupling requires neither a covalent bond nor an attractive electrostatic bond between the coupled nuclei." This implies that measurements of such couplings across hydrogen bonds are not a proof of covalency. However, Oldfield also characterized "low-barrier hydrogen bonds (LBHBs) as genuine electron-shared, covalent hydrogen bonds, and identified the symmetric hydrogen bond as the limiting case of an LBHB." He noted that such hydrogen-bonded complexes would have maximum chemical shifts of the hydrogen-bonded proton. Thus, the existence of a coupling constant across a hydrogen bond requires neither a covalent nor an attractive electrostatic bond between the coupled nuclei, although the largest coupling constants are found when the hydrogen bond is covalent, and we might add, when it is symmetric (*9,31*).

Another approach to the question of covalency and its relationship to J is to employ external electric fields to change the structure of hydrogen-bonded complexes, the two-bond coupling constants across the hydrogen bond, and the chemical shift of the hydrogen-bonded proton. This has been done for the ClH:NH$_3$ complex (*31,34,36*). Table 2 reports equilibrium Cl-N and Cl-H distances, coupling constants, and the chemical shift of the hydrogen-bonded proton for ClH:NH$_3$ as a function of field strength. The changes in the Cl-N and Cl-H distances as a function of field strength are indicative of a change in hydrogen-bond type from traditional at lower fields, to proton-shared at intermediate fields, to ion-pair at the highest fields. The structure at a field of 0.0055 au approaches that of a quasi-symmetric hydrogen-bonded complex, which is characterized by a very short Cl-N distance, a low proton-stretching frequency in the infrared spectrum (*31,36*) and extremum values of the proton chemical shift and the Cl-N coupling constant. Does this mean that the hydrogen bond in this structure is most covalent?

It is apparent from the data of Table 2 that as a function of field strength, both the Cl-N coupling constant and the chemical shift of the hydrogen-bonded proton in ClH:NH$_3$ complexes exhibit maximum values for the proton-shared hydrogen-bonded complex which is closest to quasi-symmetric. And indeed, we had previously suggested that these NMR observables are fingerprints of hydrogen bond type (*9,31*). We also noted that such structures have the shortest intermolecular distance, and the lowest electron density on the hydrogen-bonded proton. Symmetric X-H-Y hydrogen bonds do have the largest X-Y spin-spin coupling constants, and this has been correlated with the very short X-Y distances in these complexes. Whether a short X-Y distance equates with increased covalent character is an open question. However, the smooth variation in both structural and spectroscopic properties of ClH:NH$_3$ as a

Table 2. Equilibrium Cl-N and Cl-H distances {R_e(Cl-N) and R_e(Cl-H), Å}, Cl-N spin-spin coupling constants ($^{2h}J_{Cl-N}$, Hz), and proton chemical shifts [δ (ppm)] for ClH:NH$_3$ as a function of field strength (au).

Field	R_e(Cl-N)	R_e(Cl-H)	$^{2h}J_{Cl-N}$ [a]	δ(ppm)[b]
0.0000	3.080	1.341	-5.7	10.0
0.0010	3.056	1.349	-6.1	10.8
0.0020	3.019	1.363	-6.9	12.0
0.0040	2.975	1.383	-8.0	13.6
0.0055	2.832	1.575	-11.8	20.9
0.0100	2.896	1.766	-9.1	19.1
0.0150	3.004	1.857	-6.7	16.4

a) The coupling constants reported here are implicit functions of field strength, that is, they were computed at the equilibrium geometries, but with the field turned off, and are taken from ref. 31. Coupling constants as explicit functions of field strength are reported in ref. 34. At a given field strength, the explicit coupling constant is greater than the implicit value, but the trend with respect to increasing field strength is the same.
b) Relative to HCl.

function of field strength suggests that the nature of the bonding in these complexes also changes smoothly as a function of field strength.

In another experiment, we computed the O-O coupling constant for the equilibrium structure of $O_2H_5^+$ which has C_2 symmetry with a symmetric O..H...O hydrogen bond and a very short O-O distance of 2.385 Å (9). The computed value of $^{2h}J_{O-O}$, as estimated from the Fermi-contact term, is 39.9 Hz. (The remaining terms contribute 0.2 Hz or less to total J.) We then removed the hydrogen-bonded proton and its basis functions, leaving two neutral water molecules which are not properly oriented for hydrogen bonding, and are certainly not covalently bonded. Indeed, in this orientation, the interaction between the two is repulsive. Nevertheless, $^{2h}J_{O-O}$ estimated from the Fermi-contact term for this arrangement of the two H_2O molecules is 25.0 Hz. Therefore, large J does not even require a bonding interaction.

Finally, we note that there is another experimental observable that can be used to argue that hydrogen-bonded complexes have some covalent character, although this observable does not address the relationship of covalency to J. The observable is the structures of small neutral complexes such as $(H_2O)_2$ (39) and $(HF)_2$ (40). In both of these complexes the direction of hydrogen bond formation relative to the proton-acceptor atom coincides with the direction in which a proton is added to form covalent O-H or F-H bonds in H_3O^+ and H_2F^+. That is, both protonation and hydrogen bond formation occur in the direction of one of the lone pairs of electrons on O or F. In contrast, a cation such as Li^+ adds to H_2O and HF along the dipole moment vector of these molecules, giving structures of C_{2v} and $C_{\infty v}$ symmetry, respectively, which are stabilized by strong electrostatic ion-dipole interactions. The structures of hydrogen-bonded complexes argue that hydrogen bonds have 'directional' character, whether that be termed 'covalent' or not.

CONCLUSIONS

In this study we have employed external electric fields to reverse the dipole moment vectors of HF, CH_3F, CH_3Cl, and CH_3Li, and have computed one-bond EOM-CCSD spin-spin coupling constants $^1J_{H-F}$, $^1J_{C-F}$, $^1J_{C-Cl}$, and $^1J_{C-Li}$, respectively, as a function of field strength. The results of this investigation demonstrate that the coupling constant is not a maximum when the bonding between the atoms is nonpolar (covalent) as the dipole moment approaches zero. Hence, the experimental measurement of a coupling constant cannot be interpreted as proof of the existence of a covalent bond.

We have also addressed the question of whether the measurement of an X-Y coupling constants across a hydrogen bond proves that the hydrogen bond is covalent. Again, our EOM-CCSD results suggest that this is not the case.

However, coupling constants are fingerprints of hydrogen bond type. They are greatest in complexes with proton-shared hydrogen bonds, which have the shortest X-Y distances. We do not propose that hydrogen bonds do not have 'covalent' character, but rather that the existence of coupling constants across hydrogen bonds does not prove covalency.

ACKNOWLEDGMENTS

This work, dedicated to Professor W. N. Lipscomb, was presented on the occasion of the Colonel's 80th birthday celebration. With the help of the Guggenheim Foundation, one of us (RJB) had the pleasure of spending an unusually profitable semester with the Colonel in 1986, which initiated long-term collaborations between our groups. Colonel, congratulations on your 80th, and we look forward to reconvening for your 90th birthday. We appreciate Professor Paul Schleyer calling our attention to the problem of polarity and J for CH_3Li. This work was supported by the Air Force Office of Scientific Research through AFOSR F49620-98-0116 (RJB and SAP) and by the National Science Foundation through grant CHE-9873815 (JEDB). The authors are grateful to the Ohio Supercomputer Center for continuing computational support.

REFERENCES

1. Borman, S. Hydrogen Bonds Revealed by NMR, *Chem. And Eng. News*, American Chemical Society, Washington, D.C. **1999**, 36.
2. Dingley, A. J.; Grzesiek, S. *J. Am. Chem. Soc.* **1998**, 120, 8293.
3. Shenderovich, I. G.; Smirnov, S. N.; Denisov, G. S.; Gindin, V. A.; Golubev, N. S.; Dunger, A.; Reibke, R.; Kirpekar, S.; Malkina, O. L.; Limbach, H.-H, *Ber. Gunsen-Ges. Phys. Chem.* **1998**, 102, 422.
4. Dingley, A. J.; Masse, J. E.; Peterson, R. D.; Barfield, M.; Feigon, J.; Grzesiek, S. *J. Am. Chem. Soc.* **1999**, 121, 6019.
5. Scheurer, C.; Brüschweiler, R. *J. Am. Chem. Soc.* **1999**, 121, 8661.
6. Benedict, H.; Shenderovich, I. G.; Malkina, O. L.; Malkin, V. G.; Denisov, G. S.; Golubev, N. S.; Limbach, H.-H. *J. Am. Chem. Soc.* **2000**, 122, 1979.
7. Pecul, M.; Leszczynski, J.; Sadlej, J. *J. Phys. Chem. A.* **2000**, 104, 8105.
8. Barfield, M.; Dingley, A. J.; Feigon, J.; Grzesiek, S. *J. Am. Chem. Soc.* **2001**, 123, 4014.
9. Del Bene, J. E.; Perera, S. A.; Bartlett, R. J. *J. Am. Chem. Soc.* **2000**, 122, 3560.
10. Del Bene, J. E.; Bartlett, R. J. *J. Am. Chem. Soc.* **2000**, 122, 10480.

11. Clark, T.; Chandrasekhar, J.; Schleyer, P. v. R. *J. Chem. Soc. Chem. Comm.* **1980**, 14, 672. See also Bauer, W.; Winchester, W. R.; Schleyer, P. v. R.*Organometallics,* **1987**, 6, 2371.
12. Streitwieser, A.; Williams, J. E., Jr.; Alexandratos, S.; McKelvey, J. M. *J. Am. Chem. Soc.* **1976**, 98, 4778.
13. Bartlett, R. J., Coupled-Cluster Theory: An Overview of Recent Developments, *Modern Electronic Structure Theory, Part I,* Yarkony, D. R., Eds, World Scientific Publishing, **1995**, 1047.
14. Dunning, T. H., Jr. *J. Phys Chem. A.* **2000**, 104, 9062.
15. Perera, S. A.; Sekino, H.; Bartlett, R. J. *J. Chem. Phys.* **1994**, 101, 2186.
16. Perera, S. A.; Nooijen, M.; Bartlett, R. J. *J. Chem. Phys.* **1996**, 104, 3290.
17. Dunning, T. H., Jr. *J. Chem. Phys.* **1989**, 90, 1007.
18. Kendall, R. A.; Dunning, T. H., Jr.; Harrison, R. J., *J. Chem. Phys.* **1992**, 96, 6796.
19. Woon, D. E.; Dunning, T. H., Jr. *J. Chem. Phys.* **1993**, 98, 1358.
20. Schäfer, A.; Horn, H.; Ahlrichs, R. *J. Chem. Phys.* **1992**, 97, 2571.
21. Bartlett, R. J.; Silver, D. M. *J. Chem. Phys.* **1975**, 62, 3258.
22. Bartlett, R. J.; Purvis, G. D. *Int. J. Quantum Chem.* **1978**, 14, 561.
23. Pople, J. A.; Binkley, J. S.; Seeger, R. *Int. J. Quantum Chem. Quantum Chem. Symp.* **1976,** 10, 1.
24. Krishnan, R.; Pople, J. A. *Int. J. Quantum Chem.* **1978**, 14, 91.
25. Hehre, W. J.; Ditchfield, R.; Pople, J. A. *J. Chem. Phys.* **1972**, 56, 2257.
26. Hariharan, P. C.; Pople, J. A. *Theor. Chim. Acta* **1973**, 28, 213
27. Spitznagel, G. W.; Clark, T.; Chandrasekhar, J.; Schleyer, P. v. R. *J. Comput. Chem.* **1982**, 3, 363.
28. Clark, T.; Chandrasekhar, J.; Spitznagel, G. W.; Schleyer, P. v. R. *J. Comput. Chem.* **1983**, 4, 294.
29. ACES II is a program product of the Quantum Theory Project, University of Florida. Authors: Stanton, J. F.; Gauss, J.; Watts, J. D.; Nooijen, M.; Oliphant, N.;Perera, S. A.; Szalay, P. G.; Lauderdale, W. J.; Gwaltney, S. R.; Beck, S.; Balkova, A.; Bernholdt, D. E.; Baeck, K.-K.; Tozyczko, P.; Sekino, H.; Huber, C.; Pittner J.; Bartlett, R. J. Ingetral packages included are VMOL (Almlof, J.; Taylor, P. R.); VPROPS (Taylor, P. R.); ABACUS (Helgaker, T.; Jensen, H. J. Aa.; Jorgensen, P.; Olsen, J.; Taylor, P. R.).
30. Muenter J. S.; Klemperer. W. *J. Chem. Phys.* **1970**, 52, 6033.
31. Del Bene, J. E.; Jordan, M. J. T. *J. Am. Chem. Soc.* **2000**, 122, 4794.
32. Del Bene, J. E.; Perera, S. A.; Bartlett, R. J. *J. Phys. Chem. A.* **2001**, 105, 930.
33. Del Bene, J. E.; Perera, S. A.; Bartlett, R. J. *Magn. Reson. Chem.*, in press.
34. Chapman, K.; Crittenden, D.; Bevitt, J.; Jordan, M. J. T.; Del Bene, J. E. *J. Phys. Chem. A.* **2001**, 105, 5442.

35. Pauling, L. *The Nature of the Hydrogen Bond*, 3rd edition, Cornell University Press, New York, **1960**.
36. Jordan, M. J. T.; Del Bene, J. E. *J. Am. Chem. Soc.* **2000**, 122, 2101.
37. Arnold, W. D.; Oldfield, E. *J. Am. Chem. Soc.* **2000**, 122, 12835.
38. Bader, R. F. W. *Atoms in Molecules - A Quantum Theory*, Clarendon Press, Oxford, **1994**.
39. Dyke, T. R.; Muenter, J. S. J Chem. Phys.. **1974**, 60, 2929.
40. Howard, B. J.; Dyke, T. R.; Klemperer, W. *J. Chem. Phys.* **1984**, 81, 5417.

Chapter 10

Aluminosilicate Inorganic Compounds, Minerals, and Mineral Glasses: Connections Forged by Quantum Chemistry and NMR Spectroscopy

John A. Tossell

Department of Chemistry and Biochemistry, University of Maryland, College Park, MD 20742

Aluminosilicates are important as both inorganic compounds and as minerals and mineral glasses. Quantum mechanical calculations on cluster models for aluminosilicates can now accurately reproduce their energetic, structural and spectral properties. Species in glasses can be identified by matching their calculated spectral properties with experiment. More important, such calculations provide a framework for relating the structures and properties of inorganic aluminosilicates and their mineral counterparts.

As a consequence of recent enormous advances in computer technology and in quantum mechanical methodology, quantum mechanics can now be used in new ways within chemistry. For example, reports on the synthesis of new materials now commonly include ab initio calculations of properties in addition to experimental determinations (1). Good comparison of calculation and experiment is used to support the characterization of the compound.

Computation can also be used to decide between various alternative structures for new materials (2). We describe here some studies on aluminosilicates in which quantum chemistry is used as a tool both to explore the possible synthesis of new compounds with new properties and to characterize species occuring in glasses.

Aluminosilicates occur both as inorganic compounds and as minerals and their analog glasses. They are of great interest to mineralogists and geochemists because O, Si and Al rank 1, 2 and 3 in elemental abundance in the Earth's crust. They also form the basis of many important ceramics. Traditionally, the strucural theories used to describe aluminosilicates in the inorganic chemistry literature and the mineralogical literature have been substantially different. Recently inorganic chemists have begun to recognize the links between these different areas, particularly links through the chemistry of materials to the chemistry of inorganic natural products, i.e. minerals (3). Mineralogists are still rather unfamiliar with the inorganic chemistry literature focused on aluminosilicates. Quantum mechanical calculations on small molecular cluster models for such species provide a natural connection between the two fields, helping to identify and explain common strucural features of aluminosilicates in the two domains. A very important structural probe for both aluminosilicate mienrals and inorganic materials is NMR spectrsocopy, particualarly in its solid-state magic-angle-spinning form. Quantum mechanical calculation of NMR parameters can help to elucidate local structure in both minerals and materials

To illustrate the relationship between the two areas we will consider in detail the effect of coordinative unsaturation or underbonding at O in aluminosilicates on their structures, stabilities and properties, using examples from both inorganic chemistry and mineralogy. We use traditional methods of quantum chemistry in order to calculate the properties of such compounds. Similar relationships could be drawm between the bulk and surface structural properties of Fe (oxy)hydroxide) minerals and analog inorganic materials, but we will focus only upon the aluminosilicates, for which the structural and NMR data is most definitive.

Computational methods

Calculations were performed with the ab initio Hartree-Fock method (4) and in some cases with methods incorporating electron correlation such as MP2 (5) or BLYP (6). Geometry optimizations were initally done with relativistic effective core potential polarized valence double-zeta bases (7) and then refined with standard 6-31G* bases (8). NMR shieldings were evaluated using the GIAO-SCF method (9) and 6-31G* bases. We used the quantum chemical software GAMESS (10) and GAUSSIAN (11).

Results

In the 1920's Pauling developed a theory for predicting the structures of solids, culminating in a set of rules (12). Structures which violated such rules were predicted to be unstable or at least of lesser stability than alternative structures. Pauling's 2nd rule, or the electrostatic valence sum rule, predicts that the sum of the bond strengths received at an anion is equal to its valence, with the bond strength evaluated as the formal charge of the cation divided by its coordination number. This rule has since been refined to incorporate the effect of variations in bond length (13, 14), but it has generally stood the test of time. The rule has imporant applications in aluminosilicate mineralogy, where it suggests for a feldspar mineral such as $NaAlSi_3O_8$ that there should be an ordered distribution of Al and Si, so that O exists in the bridging bond linkages Si-O-Si and Si-O-Al, but not in Al-O-Al linkages. The absence of such Al-O-Al linkages is often called Loewenstein's Rule (15). The sums of Pauling bond strengths received at the O's are 2, 1.75 and 1.5 for the three types of linkages Si-O-Si, Si-O-Al and Al-O-Al, respectively, ignoring coordination with other atoms such as Na, which have low charge and large (and often somewhat uncertain) coordination number. Such an ordered distribution strongly influences the entropy of the crystalline material. In glasses and in crystalline materials at high temperature, however, there is evidence for Si, Al disorder resulting in some Al-O-Al linkages. At first, the evidence for such linkages was indirect, involving Si NMR (16) and some thermochemical studies (17).

In 1993 I developed a simple model system to calculate the energetics and properties expected for such a Al-O-Al bridging species (18). Previous quantum mechanical calculations addressing this question have given contradictory results since the systems studied were not charge balanced. For example one gas-phase reaction studied was:

$$2\ (OH)_3SiOAl(OH)_3^- \Rightarrow (OH)_3SiOSi(OH)_3 + (OH)_3AlOAl(OH)_3^{-2}$$

Although this at first seems like a reasonable reaction for studying the stabilities of different types of bridging O's, closer inspection shows that it simply involves the approach of two negative ions, which must be unfavorable.

I instead looked at the isomerization reaction of a neutral molecule, $Si_2Al_2O_4H_8^{-2}$, a molecule with four tetrahedrally coordinated atoms, two Al's and two Si's in a ring structure. Alternation of the Al's and Si's produced four Al-O-Si linkages while pairing of the Si's and Al's produced a mixture of two Si-O-Al, one Si-O-Si and one Al-O-Al linkages. The two isomers of this molecule are shown in Fig. 1.

From the difference in energy of the two isomers the relative stabilities of Si-O-Si, Si-O-Al and Al-O-Al bonds could be obtained. The results matched reasonably well with experiment, using the 3-21G* basis and the SCF method. This provided a simple molecular interpretation of what was essentially a solid-state structural rule. Note that Pauling's second rule has never been given a general quantum mechanical proof, although Burdettt and McLarnan (19) did establish that it was obeyed equally well by several different types of calculations using ionic model or qualitative MO approaches.

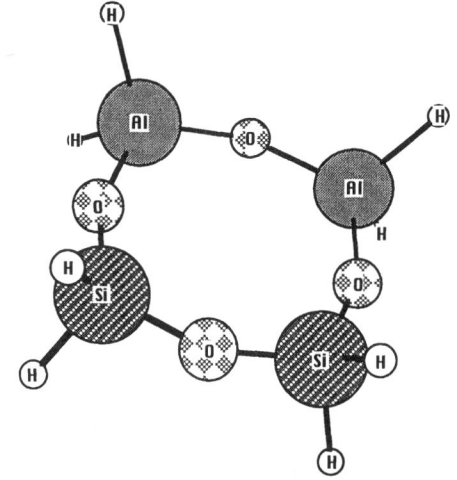

**Si $_2$ Al $_2$ O$_4$ H$_8$
paired**

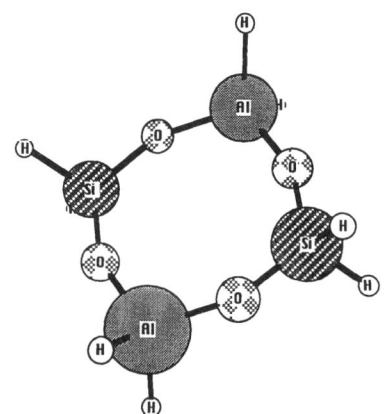

**Si $_2$ Al $_2$ O$_4$ H$_8$
alternating**

Fig. 1. Calculated structures for two isomers of $Si_2Al_2O_4H_8$

I was also able to evaluate NMR shieldings, electric field gradients at O and vibrational spectra, with the hope that at least one such property would show a unique value for a Al-O-Al linkage and could therefore be used to identify it spectroscopically. Indeed, the EFG at O in the Al-O-Al linkage was found to be considerably smaller than that in Si-O-Al or Si-O-Si. Typical electric field gradients calculated at the 3-21G* SCF level were about 0.95, 0.75 and 0.50 atomic units for Si-O-Si, Si-O-Al and Al-O-Al linkages, respectively. Using the atomic SCF results of Schaeffer, et al. (20) to determine an ^{17}O nuclear quadrupole moment appropriate for this level of treatment, we obtain nuclear quadrupole coupling constants of 5.0, 3.9 and 2.6 MHz for Si-O-Si, Si-O-Al and Al-O-Al bonds. In 1999 Stebbins, et al. (21) obtained the ^{17}O NMR spectra of $NaAlO_2$ and $CaAl_2O_4$ and identified peaks with small NQCC's of $1.5 - 2.4$, assigning them to Al-O-Al, basically confirming the calculated result. This was one of the very few instances in which the properties of a mineral species were correctly predicted before the species was characterized.

Of course, there are a number of problems with the calculations described in the paragraph above, some of them inevitable at the time with the limited computer resources and methodology available and some of them arising from the fact that this work was basically bootlegged off grants devoted to other topics. Some of the limitations are: (1) the use of –H rather than –OH or more complicated groups terminating the Si's, (2) the use of "four-ring" geometries (i.e. there were four tetrahedral (T) atoms in the ring), when aluminosilicates typically have rings with $6 - 8$ T atoms, (3) the neglect of counterions such as Na^+, which would neutralize the molecule, (4) the use of a small basis and (5) the neglect of correlation. We have since repeated these calculations replacing –H by –OH, expanding the basis to 6-311(2d,p) and incorporating correlation at the BLYP level, obtaining comparable results, as shown in Table 1.

Table 1. Difference in energy between alternating and paired isomers of $Si_2Al_2O_4R_8$

Molecule and method	Energy difference (kcal/mol)
R=H, 3-21G* SCF	17.9
R=H, 6-31G* SCF	17.0
R=OH, 6-31G* SCF	17.5
R=OH, 6-311(2d,p) SCF	18.2
R=OH, 6-311(2d,p) BLYP	15.4
Exp. (ref. 17)	8.8 – 21.6 depending on counterion

Once the basic energetic relationships above were established it was possible to do some structural chemistry on the computer, i.e. to play games in which ions or groups of atoms were added to the $Si_2Al_2O_4H_8^{-2}$ molecule to influence the relative stability of its isomers. For example when two protons or a single Ca^{+2} were added the lowest enegy geometry became the paired one with the Al-O-Al linkage, and H^+ or Ca^{+2} coordinated to this O. For the compound $Si_2Al_2O_4H_8(CH_3)_2$ the two isomers were of almost equal energy (22). Our goal in studying such neutral compounds was to suggest new materials for synthesis. The results indicated that Al-O-Al linkages could in fact be stabilized if additional atoms or groups were coordinated to the O. This idea was later vindicated by the synthesis of the molecule $[tBuOAlH_2]_2$ which has a central Al_2O_2 ring with each O also coordinated to a tBu group.(23). Later compounds with Al_4O_4 rings were characterized (24). As expected from the underbonded nature of O in a Al-O-Al linkage, each O was also coordinated to a H^+ or a Li^+.

We also studied species in which the H's terminating the Si's were replaced by –OH's, to obtain a better model for species occuring in aluminosilicate mineral glasses (22). Our goal was to better understand the mechanism by which H_2O interacts with an aluminosilicate melt. This topic is still being hottly debated (23). The topic is of great importance in geochemistry and petrology since the presence of small amounts of water (and other volatiles like CO_2) has a strong influence on the properties of aluminosilicate melts. The experimental data is not particularly illuminating, with O, Al and Si NMR showing little effect of hydration. There are significant changes in the Na NMR but these are hard to interpret. A model has recently been developed to calculate Na NMR, based on GIAO calculations on molecular model systems (26), which may eventually be able to explain the Na NMR of the hydrous glasses. However, to properly describe the interaction of water with the Na^+ in aluminosilicate glasses we must include a large number of atoms in our molecular cluster, while at the same time using a robust basis set and incorporating electron correlation. Krossner and Sauer and (27) have discussed how correlation effects can subtly but significantly change the preferred path for water interaction with zeolites. We have found that interaction of water with the dimer $SiAlO_7H_6NaH_3O_2$ yields significantly different geometries at the SCF and MP2 level.s (with the SBK bases, ref. 7, as shown in Fig. 2 below.
These two different structures will give significantly different ^1H and ^{23}Na NMR, as well as energetics for the hydrolysis reaction. The results are probably not yet converged with respect to basis set or correlation level.

At the same time that Al-O-Al linkages are being studied in mineralogy, other Al-O-Al linkages have been identified in inorganic systems. Alumoxanes, obtained as products of the hydrolysis and condensation in solution of starting materials like $Al(CH_3)_3$, have cage or double ring structures such as that shown in Fig. 3 for $Al_6O_6(CH_3)_6$, often called a double 3-ring (D3R) (28).

It is interesting to examine the structure of such compounds from the perspective of Loewenstein's rule. A single ring such as $Al_3O_3(CH_3)_3$ would have each O coordinated to only two Al's, so that it would be seriously underbonded. By combining the two 3-rings, we produce a three-coordinate O which is coordinated to three four-coordinated Al's for a Pauling bond strength sum of 3(3/4)=2.25, so that it is now saturated. The formation of the D3R also changes the Al from three to four-coordination. We have calculated structural,

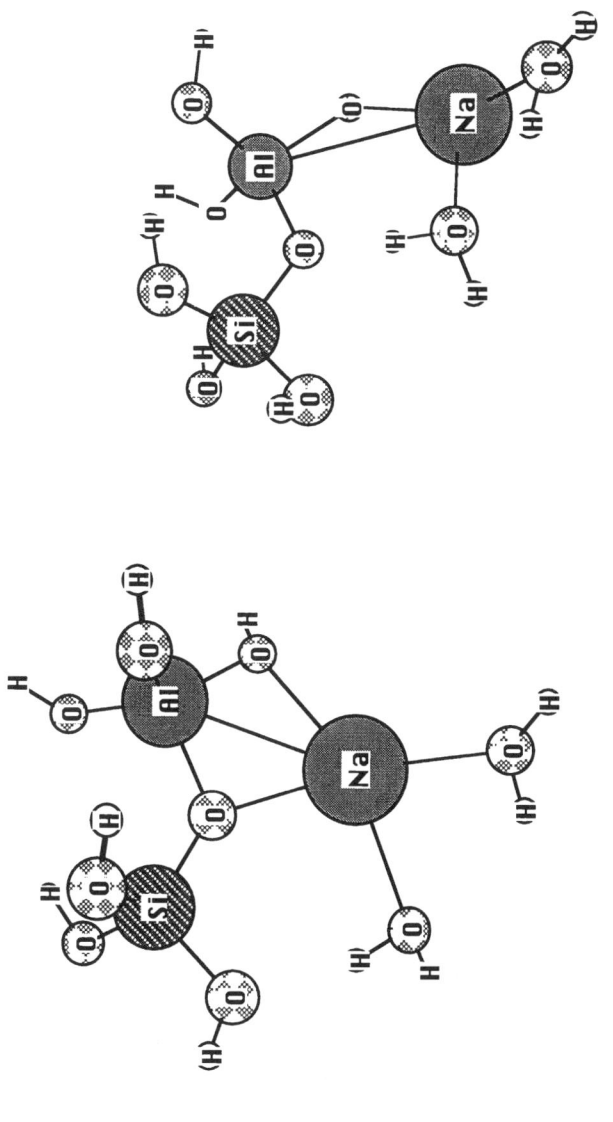

Fig. 2. Calculated structures for SiAlO$_7$H$_6$NaH$_3$O$_2$ at different levels

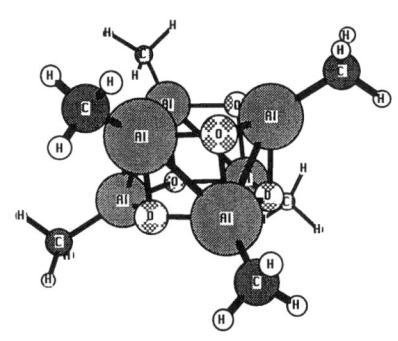

Al₆O₆(CH₃)₆

Fig. 3. Calculated geometry of Al$_6$O$_6$(CH$_3$)

energetic and spectral properties for a number of such alumoxane and aluminosiloxone "drum " molecules (29). In such "drum" shape molecules calculated NMR shieldings indicate that the Si's were deshielded compared to their component monomeric units. This is unusual since the Si's in normal silicate oligomers are invariably shielded compared to the monomer, with the degree of shielding increasing with the number of shared O's.

Such "drum" aluminosilicates do not appear to exists in minerals since their starting materials must contain three-cordinate Al. There are however another class of aluminosiloxanes which have central Al_4O_2 ring systems (30) and recent calculations indicate that they may represent the species which have been assigned to "tricluster" O's (O coordinated to three four-coordinate Al and/or Si) in Ca aluminosilicate glasses (31) . Such species were first seen as reaction products between $Al(CH_3)_3$ and silicone stopcock grease but have now been prepared in more conventional ways. The "tricluster" species seen in Ca aluminosilicates show nuclear quadrupole coupling constants at the O on the order of 2.3 MHz, while conventional planar triclusters like $Al_3O(OH)_9^{-3}$ have values around 3.9 MHz and the shared edge species like
$Al_2O_2(OH)_4[Al(OH)_3]_2^{-2}$ have values around 2.8 MHz (32), as well as ^{17}O NMR shifts consistent with experiment. Geometries for $Al_3O(OH)_9^-$ and $Al_2O_2(OH)_4[Al(OH)_3]_2^-$ are shown in Fig. 4 below. The energetics for the formation of such species are also more favorable than for conventional, planar triclusters.

Conclusion

Quantum mechanical calculations have now reached a level of applicability and reliability which allows them to be used to explore the possible synthesis of new materials with new properties. They can also help to bridge the gaps which separate disparate groups of experimentalists, such as silicate mineralogists and aluminosilicate synthetic inorganic chemists. By focusing on and identifiying crucial local groupings of atoms computational studies accentuate and elucidate the relations between these different classes of compounds. The calculations help to establish the local, molecular character of the structural rules which govern the materials, as well as identifying possible strategies which will allow such rules to be violated.

Acknowledgments

This work was supported by DOE Office of Basic Energy Sciences, Geosciences Program, Grant DE-FG02-94ER14467

Literature Cited

(1) Kornath, A. and Kadzimirsz, D., *Inorg. Chem.*, **1999**, *38*, 3066.
(2) Schleyer, P. v. R. and Maerker, A., *Pure and Appl. Chem.*, **1995**, *67*, 755.
(3) Montero, M. L., Voigt, A., Teichert, M., Uson, I. And Roesky, H. W., *Angew. Chem. Int. Ed. Engl.*, **1995**, 34, 2504.

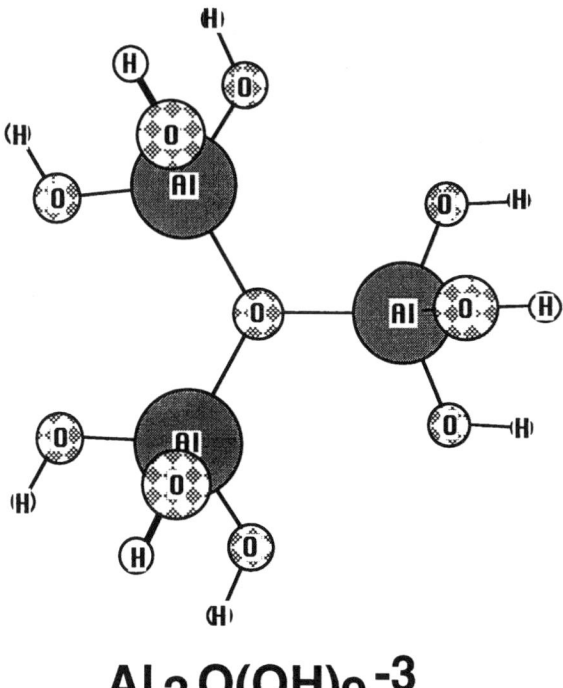

Al$_3$O(OH)$_9$$^{-3}$

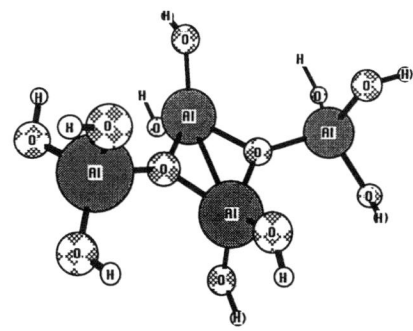

Al$_2$O$_2$(OH)$_4$[Al(OH)$_3$]$_2$$^{-2}$

Fig. 4. Structures of two different species with three-coordinate O, as models for tricluster O's in aluminosilicates

(4) Jensen, F., *Introduction to Computational Chemistry*, Wiley, **1999**
(5) Pople, J. A., Binkely, J. S. and Seeger, R., *Int. J. Quant. Chem. Symp.*, **1976**, *10*,1
(6) (a) Necke, A. D., *J. Cnem. Phys.*, **1992**, *96*, 2155; (b) Lee, C., Yang, W. and Parr, R. G., *Phys. Rev. B*, **1988**, *37*, 785
(7) Stevens, W. J., Krauss, M., Basch, H. and Jansen, P. G., *Canad. J. Chem.*, **1992**, *70*, 612.
(8) Hehre, W. J., Radom, L., Schleyer, P. v. R. and Pople, J. A., *Ab Initio Molecular Orbital Theory*, Wiley, **1986**
(9) Wolinski, K., Hinton, J. F. and Pulay, P, *J. Am. Chem. Soc.*, **1990**, *112*, 8251.
(10) Schmidt, et al., *J. Comp. Chem.*, **1993**, *14*, 1347.
(11) Frisch, M. J., et al., *GAUSSIAN94*, Rev. B.3, Gaussian, Inc., Pittsburgh, PA.
(12) Pauling, L., *The Nature of the Chemical Bond and the Structure of Molecules and Crystals*, 2nd Ed., Cornell Univ. Press, **1940**.
(13) Baur, W. H., *Amer. Mineral.*, **1971**, *56*, 157.
(14) Brown, I. D. and Altermatt, D., *Acta Cryst B*, 1985, *41*, 244.
(15) Loewenstein, W., *Amer. Mineral.*, **1954**, *39*, 92.
(16) Phillips, B. L., Kirkpatrick, R. J. and Carpenter, M. A., *Amer. Mineral.*, 1992, *77*, 484.
(17) Navrotsky, A., Geisinger, K. L., McMillan, P. and Gibbs, G. V., *Geochim. Cosmochim. Acta*, **1985**, *46*, 2039.
(18) Tossell, J. A., *Amer. Mineral.*, **1993**, *78*, 911.
(19) Burdett, J. K. and McLarnan, T. J., *Amer. Mineral.*, **1984**, *69*, 601.
(20) Schaefer, H. F. III, Klemm, R. A. and Harris, F. E., *Phys. Rev.*, **1968**, *176*, 49.
(21) Stebbins, J. F., Lewe, S. K. and Oglesby, J. V., *Amer. Mineral.*, **1999**, *84*, 983.
(22) Tossell, J. A. and Saghi-Szabo, G., *Geochim. Cosmochim. Acta*, **1997**, *61*, 1171.
(23) Veith, M., Faber, S,. Wolfanger, H and Huch, V, *Chem. Ber.*, **1996**, *129*, 381.
(24) (a) Veith, M., Jarczyk, M. and Huch, V., *Angew. Chem. Int. Ed. Engl.*, **1997**, *36*, 117; (b) Veith, M., Jarczyk, M. and Huch, V., *Angew. Chem. Int. Ed. Engl.*, **1998**, *37*, 105.
(25) (a) Sykes, D., Kubicki, J. D. and Farrar, T. C., *J. Phys. Chem. A*, **1997**, *101*, 2715; (b) Koh, S. C., Smith, M. E., Dirken, P. J., van Eck, E. R. H., Kentgens, A. P. M. and DuPree, R., *Geochim. Cosmochim. Acta*, **1998**, *62*, 79; (c) Zeng, Q., Nakvasil, H. and Grey, C. P., *Geochim. Cosmochim. Acta*, **2000**, *64*, 883.
(26) Tossell, J. A, . *Phys. Chem. Minerals*, **1999**, *27*, 70.
(27) Krossner, M. and Sauer, J., *J. Phys. Chem.*, **1996**, *100*, 6199.
(28) Mason, M. R., Smith, J. M., Bott, S. G. and Barron, A. R., *J. Am. Chem. Soc.*, **1993**, *115*, 4971.

(29) Tossell, J. A., *Inorg. Chem.*, **1998**, *37*, 2223.
(30) (a) Atwood, J. L. and Zaworotko, M.J., *J. Chem. Soc., Chem. Comm.*, **1983**, 302; (b) Apblett, A. W. and Barron, A. R., *Organometallics*, **1990**, *9*, 2137.
(31) Stebbins, J. F. and Xu, Z., *Nature*, **1997**, *390*, 60.
(32) Tossell, J. A. and Cohen, R. E., *J. Non-cryst. Solids*, **2001**, *286*, 187

Chapter 11

Bragg Diffraction and the Interference of Two Atom Lasers: An Analogy

Roger A. Hegstrom

Department of Chemistry, Wake Forest University, Winston-Salem, NC 27109

The interference pattern produced by the beams from two atom lasers can be understood in a relatively easy way by exploiting a formal equivalence between the expression for the quantum mechanical probability density for the atoms in the laser pair and the expression for the intensity of x-rays diffracted from a crystal lattice.

Introduction

Within the past decade, an exciting new form of matter, the Bose-Einstein (BE) condensate, has been produced (*1*). In a typical experiment, a gaseous sample of pure sodium containing several million atoms is cooled to about a microkelvin and trapped in the $F = 1$, $m_F = -1$ hyperfine state by a combination of stationary magnetic fields and laser fields. Under these conditions, the sodium atoms are bosons and essentially all of them enter the lowest energy state of the trapping potential, which is similar to the familiar three-dimensional particle-in-a-box system. When two such "boxes" of sodium atoms are prepared, and the atoms are subsequently released from these traps, the condensates expand, overlap, and are found to form beautiful interference fringes reminiscent of a two-slit interference pattern. Indeed, this interference is a direct confirmation of coherence, so it has now become common to describe the BE condensate (plus the associated apparatus) as an "atom laser".

The usual theoretical description of a BE condensate is given in terms of a macroscopic wavefunction with a definite phase, the latter supposedly being produced by a process known as "spontaneous symmetry breaking". A definite phase for each member of a pair of condensate wavefunctions is necessary for interference to occur when the condensates overlap. Justification of this description, which is approximate for a finite number of atoms, has been the subject of much theoretical work (1).

In this paper I begin with a simple, direct yet novel theoretical description of a pair of BE condensates which is exact for any number of atoms in the limit of an ideal gas (2). I emphasize a formal equivalence to the theory of Bragg diffraction of a beam of x-rays from a crystal. Once this equivalence is recognized, the location of the regions of greatest atom density in the interference pattern follows directly from the analog of the Bragg diffraction law.

The Exact Wavefunction for a Pair of Ideal Gas Bose-Einstein Condensates

Consider two electromagnetic traps a and b, one centered at coordinate x_a and the other at x_b, respectively. Let the wavefunction for a single atom in the ground state of the electromagnetic trap at a or b be denoted $a(x)$ or $b(x)$ respectively, where x is the coordinate of the center of mass of the atom. For N noninteracting identical atoms with integer total spin, and assuming for simplicity that half of the atoms are in each trap, the exact many-atom wave function is given by a symmetrized product of the orbitals $a(x)$ and $b(x)$:

$$\Psi(1,2,...,N) = \frac{1}{\sqrt{W_N}} \sum_{j=1}^{W_N} \hat{P}_j \, a(1)a(2)...a(\tfrac{N}{2})b(\tfrac{N}{2}+1)...b(N) \qquad (1)$$

where $W_N = N! / (N/2)!(N/2)!$ is the number of distinct permutations of the a's and b's in the product $a(1)a(2)...a(N/2)b(N/2+1)b(N/2+2)...b(N)$, and $\hat{P}_j$ denotes a permutation operator. In this notation $a(j)$ is an abbreviation for either $a(x_j)$ or $a(p_j)$ where x_j and p_j denote the spatial coordinate or the momentum, respectively, for the jth atom. The orbitals a and b and hence also the total wavefunction Ψ have a time dependence which is not denoted explicitly. In obtaining Eq. (1) overlap integrals between the orbitals a and b have been set equal to zero, which does not introduce any noticeable error because the trapped condensates are typically each about 10 μm wide and their centers separated by about 40 μm and hence their overlap is effectively zero. The assumption of noninteracting atoms is made here for simplicity and because it is a reasonably good approximation; more accurate results taking into account the atomic interactions can be made using the Gross-Pitaevskii equation (which can be considered "the Hartree-Fock equation for bosons") but the results obtained are essentially the same as in the present treatment.

It is important to realize that, although there are no dynamic interactions between them, the atoms are correlated, even when they are spatially separated by a large distance, due to the symmetrization of the many-atom wavefunction In fact it is this correlation which leads to the interference pattern which the theory predicts and which has been observed experimentally.

Analogy with Bragg Diffraction

It is now possible to see how interference between the two condensates occurs by drawing an analogy with Bragg diffraction. In order to do this most simply, we consider the wavefunction in momentum space, and assume that the condensate orbitals a and b are identical except for their spatial location. It then follows from translational symmetry that both condensates have the same momentum distribution, and more specifically that the condensate orbitals must have the form

$$a(\mathbf{p}) = e^{\frac{-i\mathbf{p}\cdot\mathbf{x}_{ab}}{2\hbar}} f(\mathbf{p}), \qquad b(\mathbf{p}) = e^{\frac{+i\mathbf{p}\cdot\mathbf{x}_{ab}}{2\hbar}} f(\mathbf{p}) \qquad (2)$$

when the origin of the coordinate system is chosen to be at the midpoint between the condensate centers, where $\mathbf{x}_{ab} = \mathbf{x}_a - \mathbf{x}_b$, and where $f(\mathbf{p})$ is the *same* function of the momentum and time for each of the two condensates. The absolute square of the function $f(\mathbf{p})$ is independent of time and gives the probability distribution of the momentum for a single isolated condensate, both during the time the condensate is in its trap and also after the condensate has been released from the trap. (After their release, the atoms are free particles.) Typically $|f(\mathbf{p})|^2$ resembles a Gaussian function peaked at p=0 with a width of magnitude $\Delta p = \hbar/s$ where $s \approx 10$ μm is the width of the trapped condensate.

Note that alternatively we could work with the wavefunction in coordinate space and obtain the same final result, but the mathematical expression for the total probability density $|\Psi|^2$, which we analyze below, is more complicated in coordinate space for two reasons. First, the orbital probability distributions in coordinate space are different for each condensate simply because the two condensates are, in general, located in different regions of space. Second, these spatial orbital probability distributions are, unlike their momentum space counterparts, time dependent after the trapping potentials are turned off because, according to the uncertainty principle, each condensate will then expand. Hence the treatment is much simpler and more transparent in momentum space. If desired, the total wavefunction in coordinate space can be obtained at any time from its momentum space counterpart by performing a Fourier transform.

Substitution of the orbital expressions in Eq. (2) into the expression for the total wavefunction in Eq. (1) gives for the total wavefunction in momentum space

$$\Psi(\mathbf{p}_1,\mathbf{p}_2,...,\mathbf{p}_N) = \frac{1}{\sqrt{W_N}} f(\mathbf{p}_1)f(\mathbf{p}_2)...f(\mathbf{p}_N)F(\mathbf{x}_{ab}) \qquad (3)$$

where

$$F(\mathbf{x}_{ab}) = \sum_{j=1}^{W_N} e^{-i\mathbf{x}_{ab}\cdot\mathbf{K}_j} \qquad (4)$$

with

$$\mathbf{K}_j = \frac{1}{2\hbar}\hat{P}_j(\mathbf{p}_1 + \mathbf{p}_2 + ... + \mathbf{p}_{\frac{N}{2}} - \mathbf{p}_{\frac{N}{2}+1} - \mathbf{p}_{\frac{N}{2}+2} - ... - \mathbf{p}_N). \qquad (5)$$

The formal resemblance to Bragg diffraction can now begin to be seen. The function $F(\mathbf{x}_{ab})$ defined in Eq. (4) is the analog of the x-ray scattering amplitude (*3*)

$$F(\mathbf{q}) = \sum_{j=1}^{N_L} e^{-i\mathbf{q}\cdot\mathbf{x}_j} \qquad (6)$$

for a crystal composed of identical atoms, where $\mathbf{q} = \Delta\mathbf{k} = \mathbf{k}' - \mathbf{k}$ is the scattering vector, where N_L is the number of atoms in the crystal lattice, and where x_j are the spatial coordinates of the atoms. It is well known that, for N_L sufficiently large, the intensities of the scattered x-rays, proportional to $|F(\mathbf{q})|^2$, are large only if the Bragg diffraction law $qd = 2\pi n$ is satisfied, where d is the distance between adjacent scattering planes, q is the component of the scattering vector in a direction perpendicular to these planes, and n is an integer. (Equating $q = (4\pi/\lambda)\sin\Theta$ gives the more familiar form of the Bragg law $2d\sin\Theta = n\lambda$). Equivalent to the Bragg diffraction law is the condition

$$\mathbf{q} \cdot \mathbf{x}_j = 2\pi N_j + const \tag{7}$$

where N_j is an integer and *const* is an arbitrary real constant. (The arbitrariness of *const* is a consequence of the invariance of the x-ray intensity with respect to arbitrary spatial translations of the crystal, and has an interesting analog in the BE case as we will see below).

Now, just as the intensity of the scattered x-ray beam is proportional to $|F(\mathbf{q})|^2$, the probability of finding the atoms with the momentum values $\mathbf{p}_1$, $\mathbf{p}_2$, ...$\mathbf{p}_N$ is proportional to $|\Psi(\mathbf{p}_1,\mathbf{p}_2,...,\mathbf{p}_N)|^2$ and therefore proportional to $|F(\mathbf{x}_{ab})|^2$ according to Eq. (3). Hence there is a analogous Bragg law in the BE case: for N sufficiently large, the probability of finding momentum values $\mathbf{p}_1$, $\mathbf{p}_2$, ...$\mathbf{p}_N$ is negligible unless the condition $|\mathbf{x}_{ab}|d_K = 2\pi n$ is satisfied, where d_K is the spacing between adjacent planes in a fictitious wave vector space, or equivalently, unless the condition

$$\mathbf{x}_{ab} \cdot \mathbf{K}_j = 2\pi N_j + const \tag{8}$$

is satisfied, where N_j is an integer. Note that, in this analogy, the vector $\mathbf{x}_{ab}$ giving the spatial separation of the condensate centers is the analog of the scattering vector **q**, and the wave vector $\mathbf{K}_j$ is the analog of the coordinate x_j of an atom in the crystal, so that the roles of coordinate space and momentum space have been interchanged, so to speak.

To see the physical meaning of Eq. (8), we note that according to the definition of $\mathbf{K}_j$ given in Eq. (5), the constraint on the $\mathbf{K}_j$ expressed in Eq. (8) implies a corresponding constraint on the momenta $\mathbf{p}_j$ of the atoms in the BE condensate, which is expressible as

$$\mathbf{x}_{ab} \cdot \mathbf{p}_j = 2\pi \left(n_j + \frac{\theta}{\pi} \right) \hbar \tag{9}$$

where n_j is an integer, and the arbitrary additive constant has been written as $2\theta\hbar$. Hence, physically, the analog of the Bragg diffraction law requires that the probability is large only for momentum values $\mathbf{p}_1$, $\mathbf{p}_2$, ...$\mathbf{p}_N$ which form a lattice in momentum space such that the lattice planes are perpendicular to the vector $\mathbf{x}_{ab}$ and are separated by the distance $h/|\mathbf{x}_{ab}|$ where h is Planck's constant. The condensate atoms have acquired preferred speeds and directions of motion. This prediction could be tested experimentally by measuring directly the momentum distribution of the atoms in a pair of trapped BE condensates.

When the atoms are released from the trap, they move as free particles with the momentum values they possessed at the time of release. Hence, at a time t after release, with t sufficiently large so that the linear dimension of each condensate has become much greater than the distance between their centers $|\mathbf{x}_{ab}|$, the positions of the atoms in the region of condensate overlap will be, with exceedingly high probability, close to values given by $\mathbf{x}_j = \mathbf{p}_j \, t / m$ where m is the mass of an atom. Then, if the z-axis is chosen to be in the direction of $\mathbf{x}_{ab}$, it follows from Eq. (9) that the regions of maximum atomic density have the spatial z-coordinates

$$z_j = \left(n_j + \frac{\theta}{\pi}\right) \frac{ht}{m|\mathbf{x}_{ab}|}. \tag{10}$$

Eq. (10) can also be obtained somewhat more directly but much less simply by calculating the probability density in coordinate space and then identifying the values of the spatial coordinates for which $|\Psi|^2$ is a maximum. According to Eq. (10), the condensate atoms form an interference pattern with a fringe spacing $ht/m|\mathbf{x}_{ab}|$. This agrees with the experimental observations (1).

The theory presented here can be developed so that a detailed expression for the atom density is obtained, and a justification for the usual approximation using a macroscopic wavefunction can be given (2). Two interfering macroscopic wave functions, one for each condensate, are found to have a phase difference 2θ, where θ is the constant appearing in Eqs. (9) and (10). This theory completely solves the problem of the mysterious origin of the phase by showing in detail its necessary presence in the probability density maxima calculated from the symmetrized many-particle wavefunction. In essence, the phase arises from "the Pauli Principle for bosons" (symmetric wavefunction) and nothing more than standard quantum theory is required for its explanation.

A physical interpretation of the phase θ can be given by returning to the analogy with Bragg diffraction. It follows from our discussion that the factor $|F(\mathbf{x}_{ab})|^2$ appearing in the expression for total probability density $|\Psi|^2$ (see Eq.

(3)) is invariant to arbitrary translations of the lattice $\mathbf{p}_1$, $\mathbf{p}_2$, ...$\mathbf{p}_N$ in momentum space. Consider a pair of condensates with an initial distribution of atomic momenta satisfying Eq. (9) with $\theta = 0$, so that this distribution is one of maximum probability. Now $|F(\mathbf{x}_{ab})|^2$ is unchanged if, for each atom, the momentum is changed by the same arbitrary amount $\delta \mathbf{p}$; according to Eq. (9) the corresponding value of θ for the new distribution of momenta is given by $\theta = \mathbf{x}_{ab} \cdot \delta \mathbf{p} / 2\hbar$. This invariance property of $|F(\mathbf{x}_{ab})|^2$ is closely analogous to the invariance of the intensity of a beam of diffracted x-rays with respect to arbitrary spatial translations of the diffracting crystal lattice.

Summary

The interference pattern produced by a pair of atom lasers has been explained starting from a knowledge of the exact configuration-space wavefunction for the many-atom system. Because the identical atoms are bosons, the wavefunction must be symmetric with respect to interchanges of the atomic coordinates, including those of atoms belonging to spatially separated condensates. This exchange symmetry introduces a correlation between the atoms in the laser pair, and it is this apparent interaction (present even for an ideal gas) which leads to interference. Remarkably, the apparent interaction between atoms is significant even over the nearly macroscopic distance which separates the two lasers. This effect arises from the "Pauli Principle for bosons" (symmetric wavefunction) and is reminiscent of the corresponding apparent interaction between electrons (antisymmetric wavefunction) in an atom or molecule which places the electrons in different shells. The interference pattern produced by the atom lasers can be understood quantitatively in a relatively simple way by exploiting a formal equivalence between the expression for the quantum probability density and the expression for the intensity of the x-rays diffracted from a crystal lattice. The locations of the interference maxima are then given by the analog of the Bragg diffraction law.

References

1. Andrews, M. R.; Townsend, C. G.; Miesner, H.-J.; Durfee, D. S.; Kurn, D. M.; Ketterle, W. *Science* **1997**, *275*, 589 and references therein.
2. Hegstrom, R. A. *Chem. Phys. Lett.* **1998**, *288*, 248.
3. Kittel, C. *Introduction to Solid State Physics*; John Wiley and Sons: New York, 1986; pp 29-37; Jackson, J. D. *Classical Electrodynamics*; John Wiley and Sons: New York, 1975; pp 417-418.

Protein, DNA, and Viruses

Chapter 12

B_{12}-Dependent Methionine Synthase: A Structure That Adapts to Catalyze Multiple Methyl Transfer Reactions

Martha L. Ludwig and Rowena G. Matthews

Department of Biological Chemistry and Biophysics Research Division, University of Michigan, Ann Arbor, MI 48109

B_{12}-dependent methionine synthases from prokaryotes and from mammalian sources are large modular proteins of M_r ~140,000. In the course of catalysis and activation, three different substrates, homocysteine, methyltetrahydrofolate, and S-adenosylmethionine, transfer methyl groups to or from the enzyme-bound cobalamin cofactor. Structural and functional dissection of the *E. coli* enzyme has shown that binding determinants for each substrate are associated with different modules, implying that each of the three methyl transfer reactions requires a different but specific spatial arrangement of the domains that make up the intact enzyme. Structure determinations have demonstrated dramatic domain rearrangements in the conversion of the resting state of the cobalamin-binding fragment to its complex with the activation domain.

Introduction

Methionine synthase (MetH) is one of the two essential B_{12}-dependent enzymes found in mammalian systems. It carries out methyl transfer reactions in which the Co-C bond of the methylcobalamin cofactor (Figure 1) undergoes heterolytic cleavage with formal transfer of a methyl cation. In contrast, methylmalonyl-CoA mutase, the other mammalian B_{12}-dependent enzyme, utilizes the adenosylcobalamin cofactor and initiates substrate rearrangement by homolytic cleavage of the Co-C5' bond.

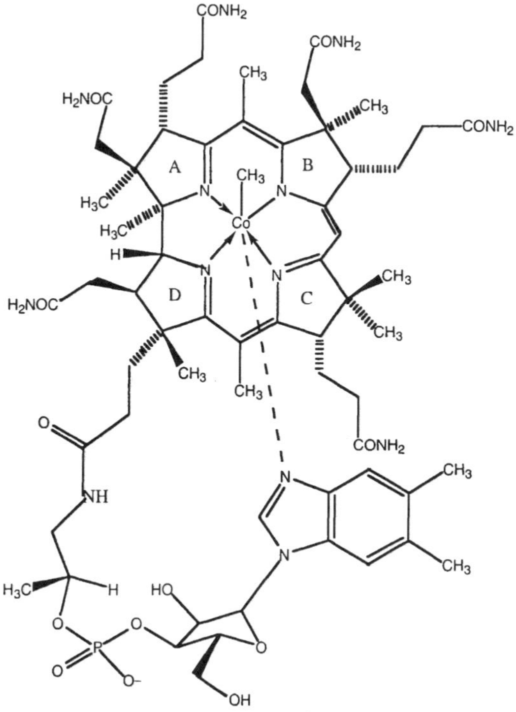

Figure 1. The cobalamin cofactor in its alkyl Co(III) form. The alkyl substituent is a methyl group in MetH and other methyltransferases, and is 5' deoxyadenosine in B_{12} dependent mutases.

The primary reaction catalyzed by methionine synthase converts homocysteine (Hcy) and methyltetrahydrofolate (CH_3H_4folate) to methionine and tetrahydrofolate (Figure 2). Occasional oxidation of the reactive cob(I)alamin intermediate produces an inactive cob(II)alamin enzyme, which is reactivated by a reductive methylation that uses S-adenosylmethionine (AdoMet) as the methyl donor and flavodoxin or a flavodoxin-like domain as an electron donor. Thus methionine synthase supports three distinct methyl transfer reactions each involving the cobalamin cofactor.

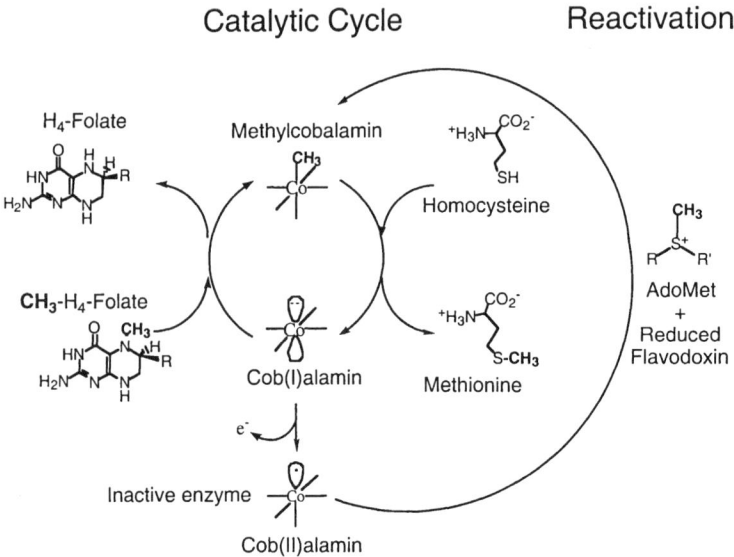

Figure 2. Catalysis and reactivation of methionine synthase. Methionine formation occurs via two half-reactions in which cobalamin serves as the intermediate methyl carrier. Reactivation is depicted in the right-hand portion of the diagram. An electron donor and AdoMet convert the inactive cob(II)alamin form of the enzyme to methylcob(III)alamin. In E. coli, flavodoxin serves as the reductant for this priming reaction (1).

Methionine synthase is comprised of four functional units arranged in a modular fashion. This organization has facilitated dissection of the molecule into substructures for correlation of structure with activity, as described in the following section. The N-terminal segment [1-353] binds an essential zinc ion and activates homocysteine; the following portion [354-649] binds the methyltetrahydrofolate substrate; and the third region [650-896] carries the B_{12} cofactor that participates in each of the methyl transfer reactions. The C-terminal domain [897-1227] houses the determinants for AdoMet binding and is essential for reactivation of the cob(II)alamin form of the enzyme. Structures have been determined for the individual cobalamin-binding (2, 3) and activation domains (4) and most recently, for the 65 kDa piece [649-1227] comprising both fragments (5). The sequence of the folate-binding module indicates that it possesses the same $(\beta\alpha)_8$ barrel fold as the corrinoid-dependent methyltransferase (AcsE) from *Moorella thermoaceticum*, solved recently by Doukov et al. (6). The structure of the N-terminal homocysteine-binding region

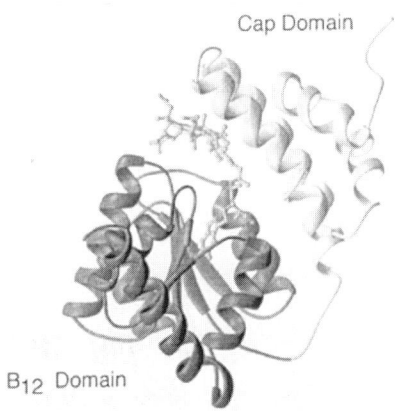
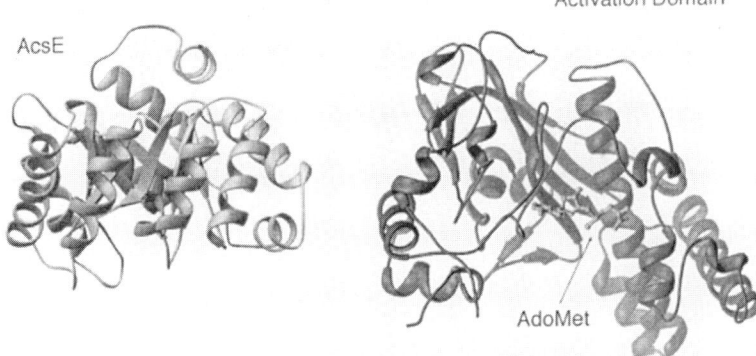

Figure 3. Three of the modules comprising methionine synthase. At the top center is the B_{12}-binding fragment [651-896], a structure with two domains, one a four-helix bundle that serves to cap the cofactor, and the other an α/β fold that interacts with the lower face of the corrin macrocycle and binds the nucleotide tail of cobalamin. Measurements of the rates of photolysis of the Co-CH_3 bond indicate that the cap domain covers the upper face of the corrin in the substrate-free form of the intact enzyme (7). On the lower right is the activation domain [897-1227] with bound AdoMet. This helmet-shaped single domain is an unusual fold with no resemblance to other well-characterized AdoMet-binding domains (8). On the lower left is the structure of the methyltransferase AcsE from Moorella thermoaceticum, which we take as a surrogate for the folate-binding domain of MetH.

is unknown, but secondary structure predictions suggest that, like the folate-binding domain, the Hcy module may include a (βα) barrel.

In order for bound cobalamin to serve as a methyl donor or acceptor in reactions with its three different substrates, domain movements must occur during the catalytic cycle and also as part of the switch between catalysis and reactivation. Tryptic digestion patterns (9, 10) and the reactivities of subpieces (11) support a model in which the tethered modules adopt a series of different domain arrangements to carry out the distinctive methyl transfer reactions. We have envisioned these rearrangements as movements of approximately rigid domains about flexible linkers. A primary strategy in our structure-function studies has been to trap and analyze the conformations that support each of the methyl transfer reactions. Examination of the structures should provide clues to the thermodynamics and mechanisms that control the distribution of conformations. One goal is to understand how cobalamin chemistry and cobalamin-protein interactions affect the selection of partners for methyl transfer. For example, the enzyme discriminates against utilization of the methyl group from AdoMet for conversion of homocysteine to methionine (10, 12). This selectivity avoids futile cycles that would result in ATP hydrolysis rather than net synthesis of methionine.

The Modules of Methionine Synthase and their Functions

Isolation and expression of stable fragments, and the ability to study their structures and activities, have proven to be powerful tools in the analysis of methionine synthase. The modular arrangement of the molecule was first recognized and exploited in experiments that released the C-terminal fragment by tryptic cleavage of the chain at residues Arg896 or Arg899 (9, 13). Separation of the fragments revealed that catalysis of methionine formation could be supported by the upstream regions [2-896] in the absence of the C-terminus, whereas the C-terminal module bound AdoMet (9) and was essential for reactivation. Functional analyses of the N-terminal regions [2-649] that are missing the B_{12}-binding domains have been possible because exogenous B_{12} can serve as an alternate substrate (11., 14) Thus the fragments or mutant fragments that bind and activate homocysteine and CH_3H_4folate can be characterized by their reactivities with exogenous methylcobalamin and exogenous cob(I)alamin respectively. These remarkable and convenient reactions with exogenous B_{12} have also been observed in all corrinoid-dependent methyl transferases that have been studied, including those from methanogens (*Archaea*) and acetogenic prokaryotes (reviewed in (15) and in (16))

The N-Terminal Fragment [2-353]: Zinc-Dependent Activation of Homocysteine

Characterization of the expressed fragments [2-353] and [2-649] and studies of site-directed mutants have established that residues 2 to ~330 constitute a module that binds and activates Hcy for reaction with either exogenous or enzyme-bound methylcobalamin (*17, 18*). The Hcy region binds a zinc ion that is required for binding and activation of Hcy (*19*). Mutation of any of three cysteine residues, Cys247, Cys310, or Cys311, results in loss of zinc binding and corresponding loss of the catalytic activity of intact MetH, and in failure of the N-terminal fragment [2-649] to convert homocysteine to methionine using exogenous methylcobalamin as a substrate (*19*).

The zinc-binding site of MetH is related to other zinc-containing proteins in which zinc activates thiol groups for methyl transfer reactions, and has been examined by EXAFS and XANES. EXAFS measurements directly demonstrate that the substrate homocysteine and its analog selenohomocysteine ligate Zn (*20, 21*). Proton release and pK measurements show that Hcy is bound to MetH as the thiolate anion at neutral pH (*22*). Other methyl transfers to thiols that are activated by Zn occur in the Ada protein (*23*), farnesyl transferase (*24*), cobalamin-independent methionine synthase (*25*), methanol-coenzyme M (mercaptoethanesulfonate) methyltransferase (*26*), methylcobamide:coenzyme M methyltransferase (*27, 28*), and betaine-homocysteine methyltransferase (BHMT) (*29*). Sequence alignments imply strong similarities between the structures of BHMT and the N-terminal Hcy-binding module of methionine synthase.

The ligation of substrate to zinc presumably lowers the thiol pK in all of these proteins, but as noted by Lipscomb and Strater (*30*), would also be expected to decrease the nucleophilicity of the thiolate. Thus the enhanced reactivity that is conferred by zinc ligation remains to be fully explained. For the case of the Ada protein where zinc is coordinated by four cysteines Lipscomb and Strater (*30*) have proposed that charge transfer from the additional cysteinate ligands helps to maintain the nucleophilic character of the reactive thiolate.

The Folate-Binding Module [354-649] and Related Methyltransferases

Alignment of the MetH sequence between residues 366 and ~ 610 with a methyltetrahydrofolate corrinoid iron-sulfur protein methyltransferase (AcsE) reveals 22% identity and 43% homology between these proteins (*6*); the same region of MetH is more distantly related to pteroate synthase (*6, 31*). Structures for pteroate synthase (*32, 33*) and the recently determined structure of AcsE (*6*) predict that the folate binding module will be a $(\beta\alpha)_8$ barrel. Model-building (*6*) suggests that the barrel, with folate near the C-termini of β-strands 1, 7, and 8, will bind B_{12} in much the same way as substrate-binding barrels interact with B_{12} in the B_{12}-dependent mutases, methylmalonyl-CoA mutase (*34*), glutamate mutase (*35*), and diol dehydratase (*36*).

The way in which MetH or AcsE activates methyltetrahydrofolate for reaction with cob(I)alamin is not fully understood. Protonation of N(5) does not occur in the binary complex of CH_3H_4folate with enzyme (*31*) but must occur at some later stage as the reaction progresses to form tetrahydrofolate.

The Cobalamin-Binding Fragment [650-896] and the Nucleotide-Off Conformation of Cobalamin

The cobalamin-binding domain is the heart of the structure of methionine synthase. It must interact and react with each of the other modules. The first piece of MetH to be studied by crystallography, the cobalamin-binding fragment revealed that the nucleotide side chain of the cofactor was buried deep in an α/β domain with the dimethylbenzimidazole (DMB) ligand replaced by a histidine from the protein (Figure 3). Displacement of the DMB ligand occurs in all the known corrinoid methyltransferases although in AcsE histidine is not coordinated to the cobalt (*15, 16*). Our current view is that the DMB tail of cobalamin is displaced because these enzymes must be able to access the 4-coordinate cob(I)alamin state (Figures 2, 4). Replacement of DMB by a protein ligand might be expected to facilitate dissociation and reassociation of the lower ligand to cobalt.

Replacement of DMB by histidine in MetH has invited speculation about the role of the histidine ligand in the methyl transfer reactions catalyzed by this enzyme. In MetH the histidine is part of a triad of residues, His759, Asp757, and Ser810, that forms a hydrogen-bonded network extending to the surface of the protein. Initial probing by mutation (*37*) indicated that His759 was "essential" for catalysis since the rate of methyl transfer from methylcobalamin to Hcy was reduced by approximately 10^5 in the His759Gly mutant. In contrast, mutations of Asp757 and Ser810 had much smaller effects, primarily on k_{cat}, which is dominated by product release.

Conservation of the triad in methionine synthases suggested its possible role as a proton conduit (*2*). Reduction of cob(II)alamin MetH to cob(I)alamin (*38*) or attack on the methylcobalamin species by Hcy (*22*) is accompanied by uptake of a proton, and proton uptake is perturbed by mutation of Asp757 (*22*). Protonation of the His-Asp pair serves to facilitate the dissociation of the histidine ligand that must occur in formation of the intermediate cob(I)alamin

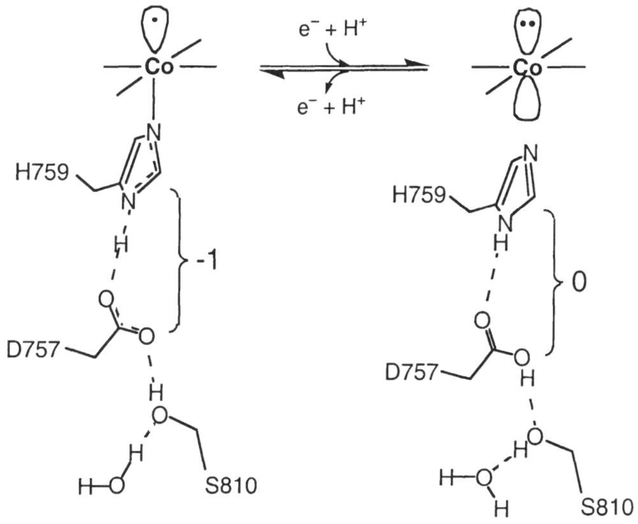

Figure 4. Protonation of the ligand triad and dissociation of His759 in the conversion of the cob(II)alamin form of MetH to the cob(I)alamin species. The net charge of −1 that is assigned to the cob(II)alamin form attributes partial imidazolate character to the histidine ligand.

species (Figure 4). It also allows the triad to signal the oxidation state or charge of the cobalt.

Contributions of the histidine ligand and other elements of the cobalamin binding site to catalysis have recently been re-examined in a different way. The discovery that methionine synthesis can be catalyzed *in trans* by a mixture of the N-terminal [2-649] and the C-terminal [649-1227] fragments of MetH permits direct comparison of the second order reactions of the substrate-loaded N-terminal piece with endogenous and with exogenous cobalamin (*17*). The second order rate constants for reactions of the N-terminal fragment with cobalamin bound to the C-terminal half of MetH are enhanced by factors of 60 to 120, relative to the reactions with exogenous cob(I)alamin or methylcobalamin. However this enhancement does not fully explain the much larger 10^5 fold effect of replacing His759 with glycine (*37*). The further impairment of the His759Gly mutant may arise from shifts in the distribution of conformations.

The Activation Domain [897-1227] and Reactivation of the Cob(II)alamin Form

The Reactivation Reaction

The simplest scheme for reactivation of the cob(II)alamin form of *E. coli* MetH involves two steps: reduction to cob(I)alamin by flavodoxin and subsequent methylation by AdoMet (Figure 1). The first step presents a thermodynamic problem, since flavodoxin potentials, even those for the semiquinone/hydroquinone equilibrium, are more positive than the MetH cob(II)/cob(I) potentials. Over a decade ago Banerjee *et al.* (*39*) showed convincingly that reduction can proceed even at rather high potential when driven by coupling with the favorable ΔG of transfer from AdoMet. Indeed these thermodynamics provide the rationale for deploying AdoMet in reactivation.

Recent investigations of the *in vitro* reactivation of cob(II) MetH by flavodoxin hydroquinone (*40*) have provided a more complete description of the reaction. Initially, only a small fraction of the enzyme is in a conformation suitable for reaction with flavodoxin, and this fraction is rapidly reduced to cob(I)alamin. For the remainder of the enzyme, reduction is preceded by a very slow step involving dissociation of the histidine 759 ligand to form a 4-coordinate cob(II)alamin species. The unligated (base-off) intermediate can then be reduced by flavodoxin, and cob(I)alamin is methylated and finally religated to His. Rate constants for formation and decay of intermediates were obtained from reactions with the mutant Asp757Glu, which is mostly 4-coordinate in the cob(II)alamin state.

The important finding from these kinetic analyses is that formation of cob(I)alamin MetH from the cob(II)alamin enzyme encounters a kinetic barrier at the step involving dissociation of the lower cobalt ligand. The dissociation is exceedingly slow, proceeding at about $0.4\ s^{-1}$ in the presence of reduced flavodoxin, relative to $27\ s^{-1}$ for turnover in catalysis. We suggest that this slow step not only involves dissociation of His, as signaled by spectral changes, but is also coupled to a larger rearrangement that generates a structure competent for methyl transfer from AdoMet.

The Reactivation Conformation

The shape of the activation domain and the location of bound AdoMet (*4*), shown in Figure 3, prompted us to model a conformation competent for the reaction of AdoMet with cob(I) MetH, in which the activation domain would displace the helical cap that covers the methyl face of the corrin in the isolated cobalamin-binding fragment (*3*). Determination of the structure of the [649-

1227] fragment has now confirmed this model, documenting the structural changes that accompany conversion to the activation conformation (Figure 5). The finding that enabled the direct structure determination was Bandarian's discovery (*17*) of a way to express the C-terminal fragment [649-1227]. Examining the covalently linked B_{12}-binding and activation modules was essential since the isolated activation domain does not support the reductive activation of cob(II) MetH *in trans* (*9*). To obtain a structure displaying the interactions of the activation domain with cobalamin, we exploited an earlier finding that mutation of His759 to glycine, which results in a "base-off" species, favors the activation-competent conformation (*37, 41*). In contrast to the wild type [649-1227] fragment, the His759Gly mutant fragment is inactive in conversion of Hcy to methionine. However it can be reduced by flavodoxin and methylated by AdoMet (Bandarian and Matthews, unpublished observations).

The structure of the cob(II)alamin form of the His759Gly fragment is depicted on the left in Figure 5. In the observed arrangement of the three domains that make up the fragment, the AdoMet site is adjacent to cobalamin, although the reactants are not quite correctly positioned for methyl transfer. The 4-helix bundle that covers methylcobalamin in the isolated B_{12}-binding fragment has been displaced. In the conversion to the "cap-off" activation complex, the cap domain rotates 62° and its center of gravity is displaced by 26 Å. Residues 650-736 of the cap are essentially undeformed in this transition; the large movement of the cap is accomplished by local conformation changes at residues 743-745 within the linker sequence between the cap and the B_{12} domain. Comparison with motions that have been analysed and codified in the MolMov (*42, 43*) and DynDom (*44, 45*) databases shows that this is a novel domain movement. The motion has both closure and twist components (*44, 45*) but its distinctive feature is that the screw axis derived from the coordinate transformation (*44, 46*) does not pass near the linker residues 743-745 or near side chains that might be regarded as hinges. We would expect similar "motions" in other situations where tethered domains undergo large translations with respect to one another. It is unclear whether these kinds of rearrangements occur by sliding mechanisms or by dissociation and reassociation of the interacting domains.

A model of the cap-on conformation of the C-terminal fragment (Figure 5, right) was constructed by a simple hinge movement of the activation domain about the 896-901 linker region. A rotation of 70° is sufficient to allow the cap domain to return to the position that it occupies in the isolated B_{12}-binding fragment. The postulated hinge movement around the linker between the activation and B_{12} domains does resemble motions that have been encountered in other enzymes where substrates must be presented to more than one site (*47, 48*). Figure 5 depicts the simplest way to open and close the activation domain: B_{12} interface. However we cannot rule out some unraveling of the chain near residue 900 that would further separate the activation and B_{12} domains.

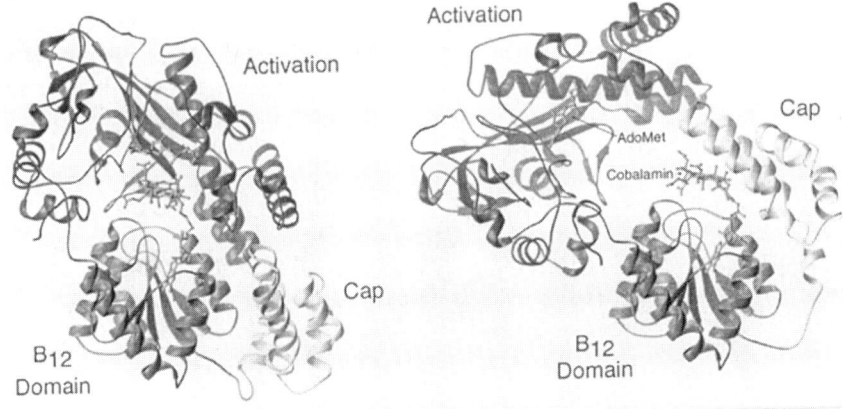

Figure 5. The structure of the activation complex (left) with the activation domain enclosing the cobalamin and part of the B_{12}-binding domain, and the cap domain at the lower right. The cobalamin cofactor, with its side chain protruding into the B_{12}-binding domain, is shown in ball-and-stick mode. AdoMet has been included at the site where it binds in the isolated activation domain (4). From cross-linking experiments (49), it is known that flavodoxin binds to this face of the gactivation complex. On the right is a model for the structure of the [649-1227] fragment in its cap-on conformation. The motions of the activation and cap domains that occur in the interconversion of conformations involve rotations around axes that are not parallel to one another.

Corrin Movement in the Switch to the Activation Conformation.

Changes in the conformation and interactions of cobalamin were an unexpected feature of the structure of the C-terminal fragment. These changes have important implications for switching between the activation complex and other conformations of MetH. As can be seen in Figure 6, the corrin ring lifts away from the B_{12} domain, tilting with respect to its position in the B_{12}-fragment. The displacement of the cofactor is driven by steric effects. Atoms in the region 1170-1175 of the activation domain act as a wedge, prying the corrin ring upward (Figure 6). In its new position the cofactor is stabilized by a number of hydrogen bonds between the amide side chains of the corrin and atoms of the activation domain. The same displacements of cobalamin would be expected in the wild type enzyme.

The changes illustrated in Figure 6 increase the distance between Cα 759 and cobalt by 2.3 Å. The observed displacement of the corrin would therefore force the dissociation of the lower histidine ligand from cobalt. The wedge thus provides a mechanism for coupling the rearrangement of protein domains to the dissociation of the histidine ligand, which is the initial step required for formation of cob(I)alamin.

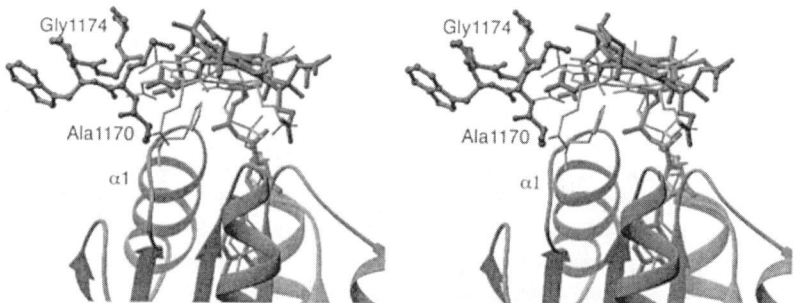

Figure 6. A stereoview showing how bound cobalamin is displaced in the activation complex. The protein structures were aligned by matching atoms from the B_{12}-binding domains (displayed as ribbons). The cofactor from the activation conformation is represented by the thicker bonds, and the peptide sequence ^{1170}Ala-Met-Trp-Pro-Gly-Ala from the activation domain is drawn in ball-and-stick mode at the left. Overlaps between the corrin in its cap-on conformation (thin bonds) and atoms of Ala 1170 and Gly 1174 are avoided by the upward movement of the corrin macrocycle.

The Distribution of Conformations in Methionine Synthase

The intact MetH enzyme must adopt several conformational states, including those in which the folate- or Hcy-binding modules are positioned for methyl transfer to or from the cobalamin, as well as the activation-competent and cap-on states that are compared in Figure 5. We expect the distribution of conformers to be linked to the oxidation state of the cobalt and to be affected by the presence of substrates or flavodoxin. The partitioning between the activation complex and other conformations of cob(II)alamin MetH is currently the best characterized of the equilibria involving domains of methionine synthase. Spectroscopic measurements of the extent of dissociation of His759 have determined that the activation conformation is disfavored in the cob(II) form of the holoenzyme by ~ 1.8 kcal/mol *(37, 40)*. Binding of flavodoxin shifts the equilibrium to favor the base-off activation form. The energetic cost of histidine dissociation from cob(II)alamin can also be estimated by comparing the binding of flavodoxin to the wild type and His759Gly enzymes, and is again found to be approximately 1.8 kcal/mol (*41*).

In contrast, the activation conformation appears to be unpopulated in methylated MetH. The methylcob(III)alamin enzyme retains the spectral signature of base-on cobalamin in the presence of flavodoxin and flavodoxin is not bound (K_d is > 70 µM) (*41*). These observations imply that ΔG for dissociation of histidine from cobalt is much more positive in methylcobalamin MetH than in the cob(II)alamin form of the enzyme.

The distribution of conformations adopted by the protein during the catalytic cycle remains to be established. We are not sure whether cap-on conformations, with the helix bundle covering cobalamin, are significant species in the presence of high concentrations of substrates. His759 is expected to dissociate from the cofactor and be protonated in the cob(I)alamin intermediate that forms in catalysis (*22*) (Figure 4). However it is not obvious that the corrin will be pried away from its binding domain as it is in the activation complex. Dissociation of His759 might proceed by a different mechanism when the Hcy- or methyltetrahydrofolate-binding domains cap the corrin. A major mystery is the occurrence of two forms of cob(I) MetH, one that reacts rapidly with CH_3H_4folate but not with AdoMet, and another in which these reactivities are reversed. These species are difficult to interconvert (*10*).

Conformational Rearrangements and the Kinetics of Methionine Synthase

An iso ping-pong mechanism that incorporates alternating domain arrangements has been proposed for MetH (*50*). In this scheme it is assumed that substrate binding or product release at the Hcy and CH_3H_4folate domains occurs only when these modules are not capping the B_{12}-binding domain. With a k_{cat} of 27 s^{-1} (*37, 51*) domain displacements in the catalytic cycle must occur at minimum rates of the order of 60 s^{-1}. These movements cannot encounter very large barriers from the interactions at the domain interfaces. In contrast, conversion to the activation conformation with its intricate interface is a very slow reaction when the enzyme exists in the cob(II)alamin state. Thus exclusion of AdoMet from the primary reaction cycle may be partly a kinetic phenomenon.

It is fascinating that a covalent tether to the B_{12} domain is needed for activation to proceed (*9*). The loss of translational freedom resulting from fusion of the domains thus appears to be an important feature of the cap-on: activation-on conversion. The contraints on other domain movements conferred by linkers may play an important role in the overall energetics and kinetics of methionine synthase.

Conversion of the C-terminal fragment to an activation-competent conformation is the first of several rearrangements of methionine synthase that we would like to study. Computations would complement the structure determinations, in the best Lipscomb tradition, by examining not only the static pictures of various conformers but also the likely pathways for interconversion of the structures (*52*) and the mechanisms for activation of bound substrates (*53*).

Acknowledgements

This research was supported by grants from the National Institutes of Health GM 16429 (MLL) and GM 24908 (RGM).

References

1. Fujii, K.; Huennekens, F. M. *J. Biol. Chem.* **1974**, *249*, 6745-6753.
2. Drennan, C. L., Huang, S., Drummond, J. T., Matthews, R. G., and Ludwig, M. L. *Science* **1994**, *266*, 1669-1674.
3. Drennan, C. L.; Dixon, M. M.; Hoover, D. M.; Jarrett, J. T.; Goulding, C. W.; Matthews, R. G.; Ludwig, M. L. In *Vitamin B_{12} and B_{12} Proteins*; Krautler, B., Arigoni, D., Golding, B. T., Eds.; Wiley-VCH: Weinheim, 1998; pp 133-156.
4. Dixon, M.; Huang, S.; Matthews, R. G.; Ludwig, M. L. *Structure* **1996**, *4*, 1263-1275.
5. Bandarian, V.; Pattridge, K. A.; Lennon, B. W.; Huddler, D. P.; Matthews, R. G.; Ludwig, M. L. *Nature Structural Biology* **2001**, in press.
6. Doukov, T.; Seravalli, J.; Stezowski, J., J.; Ragsdale, S. W. *Structure* **2000**, *8*, 817-830.
7. Jarrett, J. T.; Drennan, C. L.; Amaratunga, M.; Scholten, J. D.; Ludwig, M. L.; Matthews, R. G. *J. Bioorgan. Med. Chem.* **1996**, *4*, 1237-1246.
8. Dixon, M. M.; Fauman, E.; Ludwig, M. L. In *S-Adenosylmethionine-Dependent Methyltransferases: Structures and Functions*; Cheng, X., Blumenthal, R. M., Eds.; World Scientific: Singapore, 1999; pp 39-54.
9. Drummond, J. T.; Huang, S.; Blumenthal, R. M.; Matthews, R. G. *Biochemistry* **1993**, *32*, 9290-9295.
10. Jarrett, J. T.; Huang, S.; Matthews, R. G. *Biochemistry* **1998**, *37*, 5372-5382.

11. Goulding, C. W.; Postigo, D.; Matthews, R. G. *Biochemistry* **1997**, *36*, 8082-8091.
12. Taylor, R. T.; Weissbach, H. *Arch. Biochem. Biophys.* **1969**, *129*, 745-766.
13. Banerjee, R. V.; Johnston, N. L.; Sobeski, J. K.; Datta, P.; Matthews, R. G. *J. Biol. Chem.* **1989**, *264*, 13888-13895.
14. Taylor, R. T. *Arch. Biochem. Biophys.* **1971**, *144*, 352-362.
15. Ludwig, M. L.; Matthews, R. G. *Ann. Rev. Biochem.* **1997**, *66*, 269-313.
16. Matthews, R. G. In *Chemistry and Biochemistry of B_{12}*; Banerjee, R., Ed.; New York: Wiley, 1999.
17. Bandarian, V.; Matthews, R. G. *Biochemistry* **2001**, *40*, 5056-5064.
18. Goulding, C. W.; Matthews, R. G. *FASEB J.* **1996**, *10*, A973.
19. Goulding, C. W.; Matthews, R. G. *Biochemistry* **1997**, *36*, 15749-15757.
20. Peariso, K.; Goulding, C. W.; Huang, S.; Matthews, R. G.; Penner-Hahn, J. E. *J. Am. Chem. Soc.* **1998**, *120*, 8410-8416.
21. Peariso, K.; Zhou, Z. S.; Smith, A. E.; Matthews, R. G.; Penner-Hahn, J. E. *Biochemistry* **2001**, *40*, 987-993.
22. Jarrett, J. T.; Choi, C. Y.; Matthews, R. G. *Biochemistry* **1997**, *36*, 15739-15748.
23. Myers, L. C.; Terranova, M. P.; Ferentz, A. E.; Wagner, G.; Verdine, G. L. *Science* **1993**, *261*, 1164-1167.
24. Hightower, K. E.; Fierke, C. A. *Current Opinion in Chemical Biology* **1999**, *3*, 176-181.
25. Gonzalez, J. C.; Peariso, K.; Penner-Hahn, J. E.; Matthews, R. G. *Biochemistry* **1996**, *35*, 12228-12234.
26. Sauer, K.; Thauer, R. K. *Eur. J. Biochem.* **1997**, *249*, 280-285.
27. LeClerc, G. M.; Grahame, D. A. *J. Biol. Chem.* **1996**, *271*, 18725-18731.
28. Sauer, K.; Thauer, R. K. *Eur. J. Biochem.* **2000**, *267*, 2598-2504.
29. Millian, N. S.; Garrow, T. A. *Arch. Biochem. Biophys.* **1998**, *356*, 93-98.
30. Lipscomb, W. N.; Strater, N. *Chem. Rev.* **1996**, *96*, 2375-2433.
31. Smith, A. E.; Matthews, R. G. *Biochemistry* **2000**, *39*, 13880-13890.
32. Achari, A.; Somers, D. O.; Champness, J. N.; Bryant, P. K.; Rosemond, J.; Stammers, D. K. *Nature Structural Biology* **1997**, *4*, 490-497.
33. Hampele, I. C.; D'Arcy, A.; Dale, G. E.; Kostrewa, D.; Nielsen, J.; Oefner, C.; Page, M. G.; Schonfeld, H.-J.; Stüber, D.; Then, R. L. *J. Mol. Biol.* **1997**, *268*, 21-30.

34. Mancia, F.; Keep, N. H.; Nakagawa, A.; Leadlay, P. F.; McSweeney, S.; Rasmussen, B.; Bosecke, P.; Diat, O.; Evans, P. R. *Structure* **1996**, *4*, 339-350.
35. Reitzer, R.; Gruber, K.; Jogl, G.; Wagner, U. G.; Bothe, H.; Buckel, W.; Kratky, C. *Structure* **1999**, *7*, 891-902.
36. Shibata, N.; Masuda, J.; Tobimatsu, T.; Toraya, T.; Suto, K.; Morimoto, Y.; Yasuoka, N. *Structure* **1999**, *7*, 997-1008.
37. Jarrett, J. T.; Amaratunga, M.; Drennan, C. L.; Scholten, J. D.; Sands, R. H.; Ludwig, M. L.; Matthews, R. G. *Biochemistry* **1996**, *35*, 2464-2475.
38. Drummond, J. T.; Matthews, R. G. *Biochemistry* **1994**, *33*, 3732-3741.
39. Banerjee, R. V.; Harder, S. R.; Ragsdale, S. W.; Matthews, R. G. *Biochemistry* **1990**, *29*, 1129-1135.
40. Jarrett, J. T.; Hoover, D. M.; Ludwig, M. L.; Matthews, R. G. *Biochemistry* **1998**, *37*, 12649-12658.
41. Hoover, D. M.; Jarrett, J. T.; Sands, R. H.; Dunham, W. R.; Ludwig, M. L.; Matthews, R. G. *Biochemistry* **1997**, 127-138.
42. Gerstein, M.; Lesk, A. M.; Chothia, C. *Biochemistry* **1994**, *33*, 6739-6749.
43. Gerstein, M.; Krebs, W. *Nucleic Acids Res.* **1998**, *26*, 4280-4290.
44. Hayward, S.; Berendsen, H. J. C. *Proteins: Structure, Function, and Genetics* **1998**, *30*, 144-154.
45. Hayward, S. *Proteins: Structure, Function, and Genetics* **1999**, *36*, 425-435.
46. Cox, J. M. *J. Mol. Biol.* **1967**, *28*, 151-155.
47. Herzberg, O.; Chen, C. C.; Kapadia, G.; McGuire, M.; Carroll, L. J.; Noh, S. J.; Dunaway-Mariano, D. *Proc. Nat. Acad. Sci. U.S.A.* **1996**, *93*, 2652-2657.
48. Iwata, M.; Bjorkman, J.; Iwata, S. *J. Bioenerg. Biomembr.* **1999**, *31*, 169-175.
49. Hall, D. A.; Jordan-Starck, T. C.; Loo, R. O.; Ludwig, M. L.; Matthews, R. G. *Biochemistry* **2000**, *39*, 10711-10719.
50. Matthews, R. G. In *Enzymatic Mechanisms*; Frey, P. A., Northrup, D. B., Eds.; IOS Press, 1999; pp 155-161.
51. Banerjee, R. V.; Frasca, V.; Ballou, D. P.; Matthews, R. G. *Biochemistry* **1990**, *29*, 11101-11109.
52. Ma, J.; Sigler, P. B.; Xu, Z.; Karplus, M. *J. Mol. Biol.* **2000**, *302*, 303-313.
53. Ma, J.; Zheng, X.; Schnappauf, G.; Braus, G.; Karplus, M.; Lipscomb, W. N. *Proc. Natl. Acad. Sci. USA* **1998**, *95*, 14640-14645.

Chapter 13

Metalloproteins to Membrane Proteins: Biological Energy Transduction Mechanisms

Douglas C. Rees

Division of Chemistry and Chemical Engineering 147-75CH, Howard Hughes Medical Institute, California Institute of Technology, Pasadena, CA 91125

Structural analyses of two macromolecular systems, the metalloprotein nitrogenase and the integral membrane protein MscL (mechanosensitive channel of large conductance), are discussed within the context of energy transduction mechanisms. Nitrogenase catalyzes the ATP dependent reduction of dinitrogen to ammonia during the process of biological nitrogen fixation, while MscL is a stretch activated (mechanosensitive) channel that opens and closes in response to changes in lateral tension applied to membranes. Although nitrogenase and MscL have very different structures and functions, they both mediate the coupling of two energetic processes. From these studies, it is suggested that effective coupling of two processes by transduction proteins occurs through conformational states common to each process.

It is a special pleasure to have the opportunity to participate in this symposium in honor of William N. Lipscomb's 80th birthday. The breadth of the Colonel's interests and the significance of his contributions remain a continual source of amazement and inspiration to those of us fortunate to have been in his group. Perhaps even more remarkably, the Colonel's talents are combined with a remarkable sense of humor. One example of many provided at the symposium were the research propositions submitted by the Colonel in fulfillment of the requirements for the PhD in Chemistry at Caltech (*1*). In particular, the last prop notes:

"11.(a) Research and study at the Institute have been unnecessarily hampered by the present policy of not heating the buildings on weekends.

(b) Manure should not be used as a fertilizer on ground adjacent to the Campus Coffee Shop."

Another aspect of the Colonel's thesis that attracted considerable attention from my group was that about half of his thesis work, as well as two more props, were war-related and hence classified so that they do not appear with the deposited copy of his thesis – which immediately suggested a way to cut the writing time for the thesis or props in half.

My interest in protein structure originated while I was a graduate student in the Colonel's group. My first exposures to both macromolecular crystallography and metalloproteins were provided by my thesis research on the structural analysis of the binding of inhibitors, both small molecule and protein, to the metalloprotein carboxypeptidase A. The structure of carboxypeptidase A, established in the Colonel's group, was the first to be determined for a metalloenzyme, and this system provided a fascinating opportunity to combine structure, mechanism and metals. My future research interests in nitrogenase also grew out of my experiences in the Colonel's lab through interactions with Jim Howard of the University of Minnesota. Jim was spending his sabbatical at Harvard and through numerous discussions, he convinced me that nitrogenase was of sufficient interest that I subsequently postdoc'ed with him, and we have continued our collaboration in this area ever since.

Metalloproteins and membrane proteins represent extremely broad classifications of proteins with diverse biological roles; this article cannot begin to adequately introduce either area, much less both. Instead, one example each of a metalloprotein (nitrogenase) and a membrane protein (MscL) will be described, emphasizing the functional implications of the structural studies, particularly in the context of energy transduction mechanisms.

Energy transduction processes

Life depends on energy transduction processes that couple cellular metabolism to environmental energy sources such as light or reduced compounds. These primary energy sources must be efficiently converted into biologically usable forms such as concentration gradients or ATP, to fuel the molecular processes required for the growth and survival of organisms, including biosynthesis, transport, motility, etc. The common elements in an energy transduction process involve the coupling of an energetically favorable process to drive an energetically unfavorable process. In the case of bacteriorhodopsin, for example, the energetically favorable absorption of light is used to drive the formation of a pH gradient across the cellular membrane. Proteins mediating these interconversions represent attractive targets for structural studies since they exist in multiple conformations and often have been extensively studied by a variety of biochemical and biophysical approaches. As a result, there has been an explosion of activity in the structural characterization of transduction proteins in recent years (see (2)). Analysis of these systems indicates that a key feature of energy transduction mechanisms is that the two processes must be coupled; this can be achieved by having the biological system exclusively undergo a common, coupled transition while suppressing the independent reactions. In this fashion, "slippage" or "short-circuiting" of the transduction process will be minimized. The role of the protein in the transduction process is to kinetically facilitate the coupled reaction, while simultaneously making the uncoupled (independent) reactions kinetically unfavorable. The structural basis underlying these coupling mechanisms will be analyzed in more detail in the following sections for two specific systems: the metalloprotein nitrogenase and the membrane protein channel MscL.

Metalloproteins: Nitrogenase

Although the earth is surrounded by an atmosphere composed predominantly of dinitrogen, most organisms are unable to utilize this source for metabolic purposes. Fortunately, there are a specialized group of organisms known as nitrogen fixers or diazotrophs that are able to catalytically reduce dinitrogen to the metabolically usable form of ammonia. Organisms responsible for biological nitrogen fixation contain the nitrogenase enzyme system (reviewed in (3-8)) that consists of two component metalloproteins, the iron (Fe-) protein and the molybdenum iron (MoFe-) protein. Together, these two proteins mediate the ATP dependent reduction of atmospheric dinitrogen to ammonia. Prior to the development of the Haber-Bosch process, biological nitrogen fixation accounted for the majority of fixed nitrogen entering the biosphere (with the remainder provided by lightning and volcanic emissions); nearly a century after the development of the Haber-Bosch process, industrial ammonia formation represents one of the largest chemical industries and produces comparable amounts of ammonia to the biological process. Indeed, the introduction of the Haber-Bosch process has had a revolutionary impact on the ability of the earth to sustain substantial population increases over the past century (9).

The basic outlines of the electron flux through the nitrogenase system have been established, and involve the initial reduction of Fe-protein by low potential carriers such as ferredoxin or flavodoxin. The Fe-protein then transfers electrons to the MoFe-protein in a process that is coupled to the hydrolysis of ATP. After the appropriate numbers of electrons and protons are present in the MoFe-protein, substrate reduction can take place on the FeMo-cofactor. Despite extensive studies, the overall stoichiometry of the nitrogenase catalyzed reaction has still not been definitively established. The uncertainties are expressed in the following equation for the overall enzyme reaction (8):

$$N_2 + (6+2n)H^+ + (6+2n)e^- + p(6+2n)ATP \rightarrow 2NH_3 + nH_2 + p(6+2n)ADP + p(6+2n)P_i$$

In the "standard" model, the evolution of one molecule of dihydrogen is coupled to the reduction of one molecule of dinitrogen, and two molecules of ATP are hydrolyzed per electron transferred, so that $n=1$ and $p=2$. However, in typical experiments, larger values of both n and p are observed. There are also reports of ratios approaching 1 when an all-ferrous form of Fe-protein is used as the electron source (10).

The structures of the component proteins have been determined, both individually and complexed together. The MoFe protein is an $\alpha_2\beta_2$ heterotetramer, where the α and β subunits exhibit similar polypeptide folds that consist of three domains. This protein contains two copies each of two different types of unusual metallocenters (Figure 1), the FeMo-cofactor, the substrate reduction site, and the P-cluster, believed to participate in electron transfer from the Fe protein to the FeMo-cofactor. Each cluster contains 8 metals and associated sulfurs in distinctive arrangements, that have neither been observed in any other enzymes nor modeled synthetically. These metalloclusters are coordinated by ligands contributed by different domains of the MoFe-protein subunits. Intriguingly, the ligands are all present in a common core structure found in each domain that is composed of a four stranded, parallel β-sheet flanked by α-helices (8); a similar structural organization is also observed in the coordination of the H-cluster at the active site of iron only hydrogenases and in the so-called prismane protein. Although multiple oxidation states have been observed for the clusters, the numbers of electrons associated with a given oxidation state have not been unambiguously assigned. It appears likely, however, that the iron atoms in the as-isolated state of the P-cluster are all ferrous, and are predominantly in the ferrous form in the FeMo-cofactor.

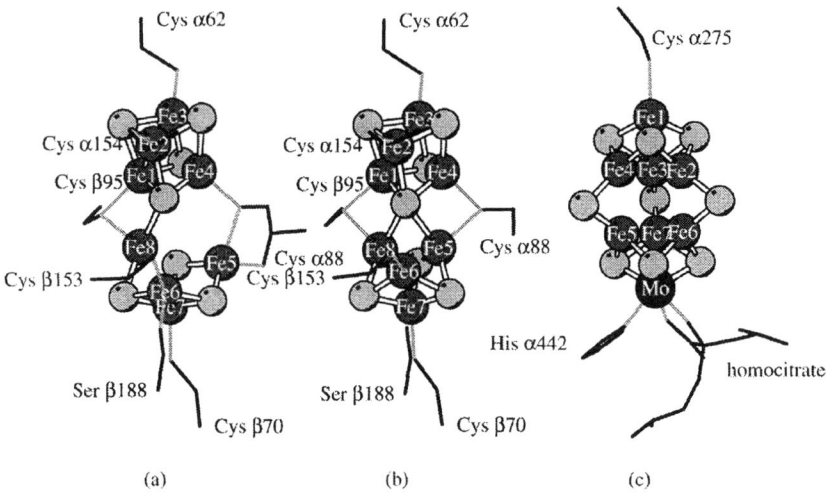

Figure 1. Structural models for the nitrogenase metalloclusters and coordinating ligands. (a) the oxidized P^{OX} state of the P-cluster. (b) the dithionite reduced P^N form of the P-cluster. (c) FeMo-cofactor. Protein Data Bank (11, 12) coordinate sets 2MIN and 3MIN were used for (a) and (b)-(c), respectively. Molecular figures in this chapter were prepared with MOLSCRIPT (13).

The Fe-protein is a dimer of two identical subunits that symmetrically coordinate a single [4Fe:4S] cluster. The isolated Fe-protein can bind MgADP or MgATP at a stoichiometry of two nucleotides/dimer, and participates intimately in the coupling between ATP hydrolysis and electron transfer to the MoFe-protein. Structurally, Fe-protein adopts a polypeptide fold characteristic of P-loop containing nucleotide binding proteins, termed "switch proteins", such as ras, G-proteins and related proteins. These similarities extend to the switch I and II regions, that are conformationally sensitive to the presence or absence of the nucleotide γ-phosphate in the G-protein family. Importantly, one of the cluster ligands, Cys 132, is contained within the switch II region of the Fe-protein and so identifies a possible mechanism for coupling ATP hydrolysis and [4Fe:4S] cluster redox behavior through conformational changes in this region. The Fe-protein cluster exhibits the unusual ability to exist in three oxidation states, including the unprecedented all-ferrous form ($[4Fe:4S]^{+0}$). The structural factors influencing this redox versatility have not been definitively established, but it has been proposed that solvent accessibility of this cluster, unusual for these typically buried clusters, may be a factor (14).

Mechanistic considerations

At the protein level, the basic mechanism of nitrogenase involves (1) complex formation between the reduced Fe-protein with two bound ATP and the MoFe-protein; (2) electron transfer between the two proteins coupled to the hydrolysis of ATP; (3) dissociation of the Fe-protein accompanied by re-reduction and exchange of ATP for ADP and (4) repetition of this cycle until sufficient numbers of electrons and protons have been accumulated so that available substrates can be reduced. Key mechanistic questions concern the chemical mechanism of the reduction of dinitrogen and other substrates, and the requirement for ATP hydrolysis in this process.

Substrate reduction is generally regarded as occurring at the FeMo-cofactor through a likely sequence of two electron transfers coupled to proton transfers (*15*). However, little experimental data is available that directly addresses these mechanistic issues, including whether N_2 binds to the FeMo-cofactor primarily at the Mo or one or more Fe sites (indeed, there is no direct experimental evidence to show that N_2 really does bind to the FeMo-cofactor); whether the reduction of any substrate (such as protons) could take place at the P-cluster; how the electron and proton transfers are coupled; and the sequence of intermediates that are generated during substrate reduction. Establishing these mechanistic features remains a priority of nitrogenase research.

Relatively speaking, the role of ATP in the nitrogenase reaction is better characterized than the chemical mechanism of substrate reduction, although many questions remain. Since the reduction of dinitrogen to ammonia is thermodynamically favored at pH 7 when coupled to the oxidation of reduced ferredoxin (*16*), ATP hydrolysis is not required to provide a thermodynamic driving force for the reaction. Instead, it seems likely that ATP hydrolysis plays a role in the kinetic mechanism of nitrogenase. Key to understanding the role of ATP in the mechanism of nitrogenase is the protein-protein complex between the MoFe-protein and Fe-protein, since it is in this species that ATP hydrolysis is coupled to interprotein electron transfer. We have determined the crystal structures of two stable forms of Fe-protein – MoFe-protein complexes (*17, 18*), prepared by either running the nitrogenase reaction in the presence of AlF_4^-, resulting in formation of the ADP•AlF_4^- stabilized complex (*19*), or through complex formation in the absence of nucleotide with the Leu 127Δ deletion mutant of the Fe-protein (*20*).

To a first approximation, the interface between the Fe-protein and MoFe-protein are similar in both complexes. The relative positions of the metalloclusters observed in these complex structures indicate that electron transfer from the Fe-protein to the FeMo-cofactor proceeds through the P-clusters (Figure 2); the edge to edge distance of ~14Å is compatible with inter-protein electron transfer rates much more rapid (~10^6 sec^{-1} (*21, 22*)) than the observed turnover time (~1 sec^{-1} per N_2 reduced). Indeed, the rate determining step with saturating amounts of substrates has been assigned to dissociation of the protein – protein complex. In both the ADP•AlF$_4^-$ and L127Δ stabilized complexes, the MoFe-protein structures are relatively constant between free and bound forms, while the Fe-proteins have undergone substantial conformational changes. A detailed comparison of the relationships between the various Fe-protein conformations has been presented (*18*).

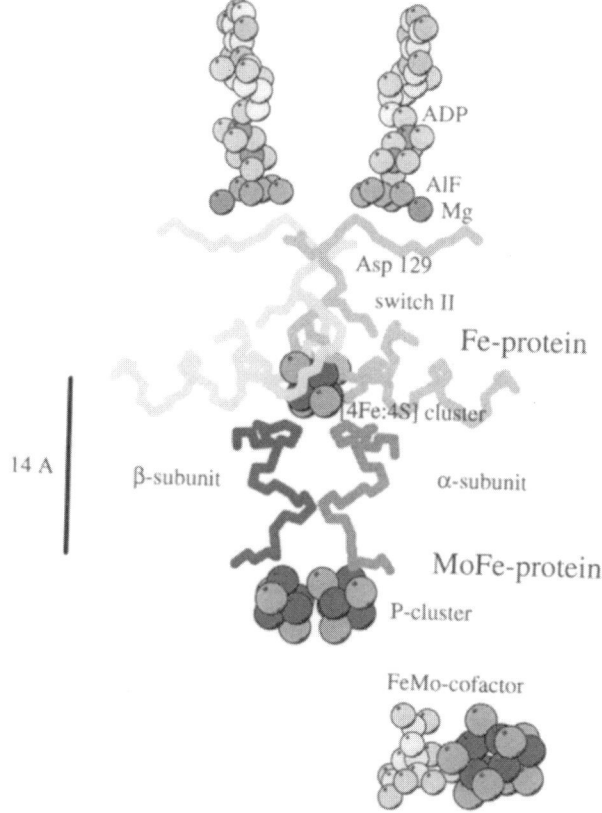

Figure 2. A slice through the ADP•AlF$_4^-$ stabilized nitrogenase complex (17) that includes the ADP•AlF$_4^-$, [4Fe:4S] cluster, P-cluster, and FeMo-cofactor, illustrating the relationship between the nucleotide binding and electron transfer sites. Relative to the isolated Fe-protein, the change in conformation of the switch II region serves to reposition the Asp 129 sidechains for nucleotide hydrolysis and also moves the [4Fe:4S] cluster closer to the P-cluster. The linkage of these conformational changes through the switch II region provides the structural basis for the coupling of ATP hydrolysis and electron transfer by nitrogenase. This figure was prepared from Protein Data Bank set 1N2C.

Energy Transduction in Nitrogenase

The similarities between Fe-protein and nucleotide switch proteins suggests that one role for nucleotide hydrolysis may be to drive a series of conformational changes that mediate the interactions between Fe-protein and MoFe-protein (see (2, 3)). One of the regions of greatest conformational variability in switch proteins between the NTP and NDP states involves the so-called switch II region. In the case of Fe-protein, the switch II region includes Asp 125, Gly 128, Asp 129 and Cys 132. Consequently, this region contains residues that interact with both the nucleotide and the cluster. Analysis of the nitrogenase complex stabilized by ADP•AlF$_4^-$ demonstrated that the conformation of switch II that is required for ATP hydrolysis is coupled to a repositioning of the [4Fe:4S] cluster such that the cluster is ~4 Å closer to the MoFe-protein, which should facilitate efficient inter-protein electron transfer by a factor of ~100 (21). As the nucleotides convert between the ATP, transition state, ADP+Pi, ADP, etc. -bound states, the conformation of the switch II region will change, necessarily modulating the electron transfer properties of the nitrogenase proteins and coupling the hydrolysis of ATP to quasi-unidirectional electron transfer from the Fe-protein to the MoFe-protein. These considerations may be critical to ensure the efficient transfer of electrons to dinitrogen by a relatively slow enzyme in the ubiquitous presence of protons that are alternative electron acceptors.

Coupling of ATP hydrolysis to electron transfer in other systems

In addition to nitrogenase, other biochemical systems have been identified that couple ATP hydrolysis to electron transfer. Protochlorophyllide reductase, that catalyzes the stereospecific reduction of the D ring during the light independent pathway for chlorophyll biosynthesis, exhibits striking similarities to the nitrogenase system, particularly in the component analogous to the Fe-protein (23). Furthermore, homologues of the Fe-protein have been identified from the genomic sequences of various microorganisms that are not nitrogen fixers, such as *Methanococcus janaschii* (24), suggesting that the ability to couple ATP hydrolysis to electron transfer may be of more widespread utility than "just" nitrogen fixation. The activator component (CompA) of 2-hydroxyglutaryl-CoA dehydratase (25) of *Acidaminococcus fermentans* that participates in glutamate fermentation provides an instructive example of a similar enzymatic activity (ATP driven electron transfer) achieved with a distinct polypeptide fold. As with the Fe-protein, CompA is a dimer of two identical subunits that together coordinate a single [4Fe:4S] cluster (26). In this case, however, the subunits are members of the actin/hexokinase/heat shock 70 family of nucleotide binding proteins, rather than the nucleotide switch protein family, and the nucleotide is bound at the interface between two domains within a subunit, rather than at the dimer interface. Despite the differences in folds, in both cases, the cluster is coordinated by both subunits in the dimer; one surface of the cluster is exposed to provide the interaction site with the second component; and the cluster and ATP hydrolysis sites are separated by ~20 Å. This arrangement indicates that the nucleotide and redox centers are not directly coupled but rather are indirectly linked through a series of conformational changes driven by ATP hydrolysis and component protein binding to ensure that unidirectional electron transfer takes place at the appropriate time between the redox partners.

Membrane Proteins: MscL

All cells are surrounded by membranes composed of phospholipid bilayers that separate the inside of the cell from the outside world. Embedded within these bilayers are integral membrane proteins that mediate the flow of molecules, information and energy into and out of the cell. One class of these membrane proteins are ion channels that facilitate the transmembrane passage of ions down their electrochemical potential gradient in a process that is central to many signaling and sensing pathways. Functionally important properties of ion channels that we would like to understand in terms of their structures include their conductance (related to the diameter of the permeation pathway), ionic selectivity (most clearly established for the KcsA potassium channel (27)), and the regulation (gating) of channel activity in response to environmental changes by conformational switching between open and closed states.

Mechanosensitive channels open and close in response to mechanical stress applied to the cell membrane or other cytoskeletal component. The best-characterized member is the prokaryotic large-conductance mechanosensitive channel (MscL) that was originally identified, isolated, and characterized in *Escherichia coli*, primarily through the efforts of Kung and co-workers (28, 29). Our interest in MscL grew out of desire to study a gated channel system to help establish the structural basis for the switching between open and closed states in response to environmental changes. For this purpose, we wanted to study a gated system that was small and prokaryotic, to hopefully simplify expression. These constraints were beautifully satisfied by MscL, perhaps uniquely so at the time this project was initiated.

MscL is a nonselective ion channel of ~2.5 nS conductance activated *in vitro* by the application of membrane tension; with a potential difference of 100 mV, this conductance is equivalent to the flow of ~10^9 ions/second across the membrane. The functional properties of MscL suggest a physiological role in the regulation of osmotic pressure in the cell. Sequence analysis and biochemical studies indicated that the MscL channel consists of a single type of subunit of ~16 kD with two transmembrane helices, and that this channel is localized within the inner cell membrane with both NH_2- and COOH- termini positioned in the cytoplasm. MscL homologs have been identified in other bacteria (30).

We took advantage of the availability of bacterial homologues of MscL to test nine different homologues for expression, purification and crystallization, with the final result that crystals of the MscL homologue from *Mycobacterium tuberculosis* were obtained that diffracted to 3.5Å resolution. The structure was subsequently solved by a combination of multiple isomorphous replacement, anomalous scattering and noncrystallographic symmetric averaging (31).

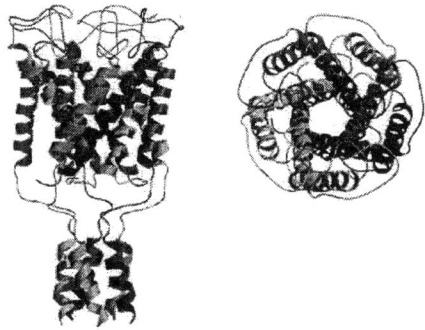

Figure 3 The Tb-MscL pentamer viewed from directions perpendicular (left) and parallel (right) to the normal of the membrane plane, respectively. The vertical bar on the left is 30Å in length and marks the approximate membrane-spanning region of these channels.

MscL is organized as a pentamer (Figure 3) composed of two domains, transmembrane and cytoplasmic, that share the same fivefold axis relating subunits within the channel. The transmembrane helices are all tilted ~30° relative to the membrane normal. The first transmembrane helix of each MscL subunit starts in the cytoplasm and crosses the membrane as the "inner" helix to form the permeation pathway of this channel, while the second helix returns to the cytoplasm across the membrane to form the "outer" helix. Hence, MscL is threaded across the membrane in the opposite manner to KcsA. The inner helices pack against each other to form a right-handed helical bundle with a crossing angle of ~-40°. Each inner helix contacts four adjacent helices (two inner and two outer), while the outer helices contact their own and an adjacent inner helix. The extracellular loop connecting the transmembrane helices exhibits extensive sequence variability within the MscL family. The cytoplasmic domain consists of a five helix bundle that extends for ~35Å away from the likely plane of the membrane-aqueous interface.

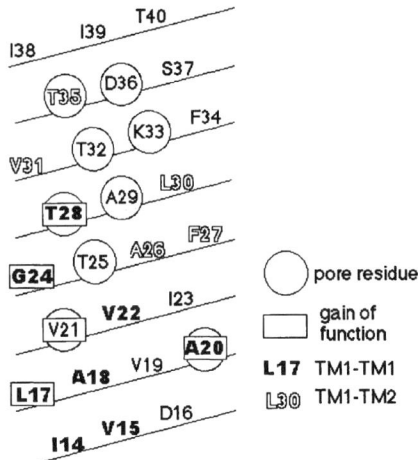

Figure 4. A helical wheel for the MscL inner helix that illustrates the positions of helix interface residues, pore residues and gain of function mutations. The sidechain of Val 21 serves to constrict the channel at the narrowest position. The correspondence between residues at the interface between TM1 helices and the gain of function mutations is evident, suggesting this interface is critical for the gating process. TM1 and TM2 are the inner and outer helices, respectively.

An important motivation for initiating the structural analysis of MscL was to understand channel gating, since MscL opens and closes in response to mechanical stresses applied directly to the membrane. The high conductance and lack of ion selectivity are consistent with a large, water-filled pore existing in the open state of the MscL channel. An important development in establishing the gating mechanism of MscL has been the identification of 'gain-of-function' mutants that display a slow or no-growth phenotype as a result of the leakage of solutes out of the cell under low osmotic strength growth conditions (32). Many of the mutations associated with severe phenotypes are located at the interface between adjacent inner helices (Figure 4), in the region of the membrane spanning domain where the pore is most restricted (at the side chain of Val 21 in the closed state). These observations suggest that contacts between inner helices play a crucial role in the gating mechanism. This interface must be rearranged, perhaps by allowing the inner and outer helices to interleave like barrel staves, to create a pore of sufficiently large diameter in the open state. More recently, a detailed model for the closed to open transition of MscL has been proposed by Sukharev and Guy based on a combination of modeling, electrophysiological and disulfide crosslinking experiments (33). This model invokes an extended rearrangement of the helices with the inner and outer helices becoming more tilted and immersed in the bilayer, while a five helix bundle formed by the N-terminal residues of MscL (not observed in the crystal structure) form the true gate that must be opened for conduction. The availability of such models will be invaluable for the design of experiments to more conclusively establish the structural organization of this state.

An interesting similarity in the channels studied to date is that the helices lining the permeation pathway pack together as right handed helical bundles with crossing angles (-40°) that are relatively steep for membrane proteins (*34*). One consequence of this relatively steep packing angle is that the contact interface between helices is localized to a fairly narrow region, which may facilitate helix-helix rearrangements associated with channel gating (*35*). As with other membrane proteins, residues buried at the interface between interacting transmembrane spanning helices are predominantly apolar, with relatively few hydrogen bonds or other polar types of contacts observed. The apolar and non-directional nature of these helix-helix contacts may also facilitate helix rearrangements associated with channel gating.

A key aspect of the function of MscL is the coupling mechanism between protein conformation and membrane stretching, which is at the heart of the energy transduction. Thermodynamically, if a protein exists in multiple states with varying cross-sectional areas, stretching the membrane will increasingly favor states with the larger cross-sectional areas. In the case of MscL, an applied tension of ~12 dynes/cm is required to open the channel, which approaches the tension needed to rupture the membrane. This suggests as a working model that the lateral pressure in the membrane bilayer clamps the channel in the closed state; when the membrane is stretched, this pressure is reduced, allowing the channel to expand to the open state as lipids are no longer as tightly packed against the channel. For a mechanism such as that proposed by Guy and Sukharev, stretching the membrane could lead to bilayer thinning, with the an associated increase in helical tilt to keep the hydrophobic residues immersed in the non-polar region of the membrane bilayer.

Concluding Remarks

Consideration of the general properties of energy transduction mechanisms indicates that there are common elements to their function. In particular, the efficient coupling of two processes by energy transduction proteins involves common conformations that are required for both processes. Since these conformations can apparently only be achieved when the complete set of substrates, ligands, etc. are bound to the transduction protein, the coupling efficiency is increased by suppression of uncoupled processes. In real systems, the coupling mechanisms are as diverse as the processes themselves. In the cases discussed in this chapter, nitrogenase is proposed to represent an example of kinetic coupling of ATP hydrolysis and electron transfer to ensure unidirectional electron transfer from the Fe-protein to the MoFe-protein, while MscL likely represents a case of thermodynamic coupling where membrane stretching favors the state of largest cross-sectional area (i.e., the open state).

Finally, as an update on the status the Colonel's last research proposition (*1*), I am pleased to note that while Caltech may not be able to provide air-conditioning this summer, the buildings are heated on weekends in the winter. As far as the use of manure is concerned, I have detected it on rare occasions, but clearly there are still more effective mechanisms for supplying fixed nitrogen to the grounds.

Acknowledgments

I would like to thank the Colonel for his advice and guidance over the years, as well as allowing us to collect the first data sets of the nitrogenase Fe-protein in his lab. None of this work would have been possible without my interactions with Jim Howard, or without the efforts of my group on the projects described from my lab. Research work on nitrogenase and MscL has been supported by NIH grants GM45162 and GM62532, respectively.

Bibliography

(1) Lipscomb, W. N., Jr. Ph.D. thesis, California Institute of Technology, Pasadena, CA, 1946.
(2) Rees, D. C.; Howard, J. B. *J. Mol. Biol.* **1999** 293, 343.
(3) Howard, J. B.; Rees, D. C. *Annu. Rev. Biochem.* **1994** 63, 235.
(4) Burgess, B. K.; Lowe, D. J. *Chem. Rev.* **1996** 96, 2983.
(5) Howard, J. B.; Rees, D. C. *Chem. Rev.* **1996** 96, 2965.
(6) Seefeldt, L. C.; Dean, D. R. *Acc. Chem. Res.* **1997** 30, 260.
(7) Smith, B. E. *Adv. Inorg. Chem.* **1999** 47, 159.
(8) Rees, D. C.; Howard, J. B. *Curr. Op. Chem. Biol.* **2000** 4, 559.
(9) Smil, V. *Enriching the Earth*; The MIT Press: Cambridge, MA, 2001.
(10) Nyborg, A. C.; Johnson, J. L.; Gunn, A.; Watt, G. D. *J. Biol. Chem.* **2000** in press,
(11) Sussman, J.; Lin, D.; Jiang, J.; Manning, N.; Prilusky, J.; Ritter, O.; Abola, E. *Acta Crystallogr.* **1998** D54, 1078.
(12) Berman, H. M.; Westbrook, J.; Feng, Z.; Gilliland, G.; Bhat, T. N.; Weissig, H.; Shindyalov, I. N.; Bourne, P. E. *Nuc. Acids Res.* **2000** 28, 235.
(13) Kraulis, P. J. *J. Appl. Cryst.* **1991** 24, 946.
(14) Strop, P.; Takahara, P. M.; Chiu, H.-J.; Hayley, C.; Angove, C.; Burgess, B. K.; Rees, D. C. *Biochemistry* **2001** 40, 651.

(15) Thorneley, R. N. F.; Lowe, D. J. *J. Biol. Inorg. Chem.* **1996** 1, 576.
(16) Alberty, R. A. *J. Biol. Chem.* **1994** 269, 7099.
(17) Schindelin, H.; Kisker, C.; Schlessman, J. L.; Howard, J. B.; Rees, D. C. *Nature* **1997** 387, 370.
(18) Chiu, H.-J.; Peters, J. W.; Lanzilotta, W. N.; Ryle, M. J.; Seefeldt, L. C.; Howard, J. B.; Rees, D. C. *Biochemistry* **2001** 40, 641.
(19) Renner, K. A.; Howard, J. B. *Biochemistry* **1996** 35, 5353.
(20) Ryle, M.; Seefeldt, L. C. *Biochemistry* **1996** 35, 4766.
(21) Gray, H. B.; Winkler, J. R. *Annu. Rev. Biochem.* **1996** 65, 537.
(22) Page, C. C.; Moser, C. C.; Chen, X.; Dutton, P. L. *Nature* **1999** 402, 47.
(23) Fujita, Y.; Bauer, C. E. *J. Biol. Chem.* **2000** 275, 23583.
(24) Bult, C. J.; White, O.; Olsen, G. J.; Zhou, L.; Fleischmann, R. D.; Sutton, G. G.; Blake, J. A.; FitzGerald, L. M.; Clayton, R. A.; Gocayne, J. D.; Kerlavage, A. R.; Dougherty, B. A.; Tomb, J.-F.; Adams, M. D.; Reich, C. I.; Overbeek, R.; Kirkness, E. F.; Weinstock, K. G.; Merrick, J. M.; Glodek, A.; Scott, J. L.; Geoghagen, N. S. M.; Weidman, J. F.; Fuhrmann, J. L.; Nguyen, D.; Utterback, T. R.; Kelley, J. M.; Peterson, J. D.; Sadow, P. W.; Hanna, M. C.; Cotton, M. D.; Roberts, K. M.; Hurst, M. A.; Kaine, B. P.; Borodovsky, M.; Klenk, H.-P.; Fraser, C. M.; Smith, H. O.; Woese, C. R.; Venter, J. C. *Science* **1996** 273, 1058.
(25) Buckel, W. *FEBS Lett.* **1996** 389, 20.
(26) Locher, K. P.; Hans, M.; Yeh, A. P.; Schmid, B.; Buckel, W.; Rees, D. C. *J. Mol. Biol.* **2001** 307, 297.
(27) Doyle, D. A.; Cabral, J. M.; Pfuetzner, R. A.; Kuo, A.; Gulbis, J. M.; Cohen, S. L.; Chait, B. T.; MacKinnon, R. *Science* **1998** 280, 69.
(28) Sukharev, S. I.; Blount, P.; Martinac, B.; Blattner, F. R.; Kung, C. *Nature* **1994** 368, 265.
(29) Sukharev, S. I.; Blount, P.; Martinac, B.; Kung, C. *Annu. Rev. of Physiol.* **1997** 59, 633.
(30) Moe, P. C.; Blount, P.; Kung, C. *Mol. Microbiol.* **1998** 28, 583.
(31) Chang, G.; Spencer, R. H.; Lee, A. T.; Barclay, M. T.; Rees, D. C. *Science* **1998** 282, 2220.
(32) Ou, X. R.; Blount, P.; Hoffman, R. J.; Kung, C. *Proc. Nat. Acad. Sci. USA* **1998** 95, 11471.
(33) Sukharev, S.; Betanzos, M.; Chiang, C.-S.; Guy, H. R. *Nature* **2001** 409, 720.
(34) Bowie, J. U. *J. Mol. Biol.* **1997** 272, 780.
(35) Rees, D. C.; Chang, G.; Spencer, R. H. *J. Biol. Chem.* **2000** 275, 713.

Chapter 14

Paradigms for Protein–Ligand Interactions

Florante A. Quiocho

Howard Hughes Medical Institute and Department of Biochemistry and Molecular Biology, Baylor College of Medicine, Houston, TX 77030

Protein-ligand recognition is the basis of biological and biochemical specificity and activity. Here we highlight key features of molecular recognition emerging from our structural and functional studies of several proteins with diverse specificity and activity. These proteins include calmodulin, a signal transducer for a variety of biological activities; the enzymes adenosine deaminase and aldose reductase; VP39, a vaccinia virus protein which act on both ends of mRNA; several initial high affinity receptors for bacterial ABC-type transport systems. Calmodulin's extraordinary versatility is reflected by its ability to activate numerous enzymes and structural proteins. Among the inhibitors that bind to adenosine deaminase are ground-state, reaction-coordinate and transition-state analogs. Because aldose reductase is a target for design of therapeutics to combat diabetic complications, about a thousand drugs have been synthesized which exhibit noncompetitive inhibition. To perform its function, VP39 must recognize the N7-methylguanosine cap of an mRNA transcript while also binding the transcript in a sequence-nonspecific manner. Because the transport receptors bind a variety of ligands (carbohydrates, peptides, tetrahedral oxyanions, etc.), they are a gold mine for the study of ligand recognition.

Calmodulin: Nonspecific But Very Tight Binding of Numerous Target Enzymes/Proteins Through Unprecedentedly Large Conformational Change

Calmodulin (CaM) is well known for its unique ability to bind to a very large number of targets (at least 30 enzymes and 20 structural proteins) in a calcium-dependent manner, and has been studied extensively in an attempt to understand how it is able to accomplish this. Ca^{2+}-activated CaM binds to these targets or their CaM-binding domains with affinities in the nanomolar range (*1*) The first crystal structure of the free form of the Ca^{2+}-CaM (*2*) offered much insight as well as a puzzle the two domains, each containing a pair of Ca^{2+} and a putative hydrophobic recognition pocket, were widely separated and twisted in *trans* with respect to each other by a long domain-linking helix. Biochemical data in conjunction with study of this structure led to a speculative model wherein this long helix had a simple hinge at a serine residue in the middle which allowed the two domains to swivel towards each other and bind targets (*1*).

We were able to determine the x-ray structure of Ca^+-CaM bound to the CaM-binding domain of smooth muscle myosin light chain kinase (MLCK) (*3*), which appeared concurrently with an NMR structure of the Ca^+-CaM complex with the skeletal muscle MLCK peptide analog (*4*). Our structure showed uncoiling of the linker helix that involved four residues changing from a helix to a strand, allowing both domains to come together in a compact *cis* configuration and embrace the target peptide closely. This dramatic conformational change, a combination of ~100° bend and ~120° twist between the two domains relative to the unbound Ca^+-CaM structure, is the largest ever observed for a protein. Each end of the target sported a large hydrophobic residue, which in the complex structure was bound in a hydrophobic pocket of a CaM domain, similar to the analogous NMR structure. This finding supported a simplistic view of CaM binding which involved two "anchoring residues" separated by twelve intervening residues of the target helix. However, we had also a well-characterized CaM-binding domain of the calmodulin-dependent multifunctional protein kinase II (CamKII) which did not contain this motif, and were able to crystallize this complex also. The structure of this complex showed that CaM was recognizing a shorter hydrophobic sequence by using a different domain-domain conformation while demonstrating a completely disordered linker helix (*5*). By now it was obvious to us that the linker was doing a lot more than simple hinging and bending, and fortunately we had available active minus-two and minus-three linker helix deletion mutants of Ca^+-Cam as well as a calmodulin containing the linker from TnC (three residues longer). All three mutant CaMs complexed and crystallized with the CaMKII domain, yielding very similar structures in spite of the altered linkers (*6*). The deletion mutants had unraveled their remaining linkers to allow correct domain positioning on the target, and the longer linker of the TnC mutant helix had folded at a radical angle at a simple hinge to accomplish the same end. Thus it was made clear that the target it is binding wholly dictates calmodulin's bound conformation, and this is managed by an expandable linker which gives the individual binding domains a very wide variety of available target-binding conformations.

Carbohydrates: Polar versus Nonpolar Interaction

The interactions between proteins and carbohydrates provide a prototypic example of how proteins specifically recognize small highly soluble organic ligands. Recognition of carbohydrates is crucial in a diverse array of process, including the transport, biosynthesis and storage of carbohydrates as an energy source, signal transduction through carbohydrate messengers, and cell-cell recognition and adhesion.

Sugars, especially those with the more common pyranoside ring structure, contain a mixture of polar and nonpolar groups of atoms. The polar groups, which constitute about 60% of the total accessible surface, are exocyclic hydroxyl groups. The nonpolar groups consist of CH bonds protruding from both faces of the sugar, which form small nonpolar clusters. The binding sites of proteins are designed to interface with these two accessible surfaces by hydrogen bonding to the peripheral hydroxyl groups and stacking aromatic sidechains against the faces with the nonpolar clusters.

Our crystallographic studies of monosaccharide and oligosaccharide binding to the D-galactose-binding protein GBP) (7), L-arabinose-binding protein (ABP) (8) and maltooligosaccharides/maltose-binding protein (MBP) (9, 10), of the ABC-transporters revealed several unusual features of the polar interactions, of which some illustrated in Figure 1. (i) Each sugar hydroxyl group is involved extensively in stable multiple cooperative hydrogen bonds (Figure 1). Cooperative hydrogen bonds generally follow the simple pattern (11, 12): $(NH)_n \rightarrow OH_S \rightarrow O$, where OH_S represents a sugar hydroxyl and $(NH)_{n=1 \text{ or } 2}$ and O correspond to donor and acceptor groups of the protein mostly from Asn, Asp, Arg, Lys (Figure 1). (ii) The vicinal sugar hydroxyls provide an ideal geometry for the formation of 'bidentate' hydrogen bonds, where the pair of hydroxyls interacts with two functional groups of planar polar residues (e.g., involving Asn 236 and Asn 91, Figure 1). (iii) The carboxylate sidechains are heavily involved in the recognition of sugar epimers and anomeric hydroxyls (Figure 1). (iv) Ordered water molecules modulates binding of multiple sugar ligands. This is exemplified by the arabinose-binding protein that binds D-galactose and D-fucose with affinities that are 2- and 40-fold, respectively, lower than L-arabinose (13). A recent review provides a more detailed description of these four features of polar interactions between proteins and carbohydrates (12).

Besides the polar interaction, another notable feature of carbohydrate binding, whose nearly universal existence has only recently come to light and whose contribution to molecular recognition only recently appreciated, is the interaction of the small nonpolar clusters from sugar faces (identified as A and B faces) with nonpolar sidechains especially from aromatic residues (recently reviewed in ref. 12). Two excellent examples of this hydrophobic interaction are portrayed by the binding of D-glucose to GBP (7) and maltooligosaccharides to the maltose-binding protein (9, 10). The glucose is sandwiched between the sidechains of Phe and Trp residues (Figure 1).

Because molecular recognition processes frequently involve both polar and nonpolar interactions, it is crucial to differentiate and experimentally assess the contribution of both types of interactions. MBP offers a unique system to make this assessment because its polar and nonpolar binding functions are segregated into two globular domains (9, 10). A bending motion of the hinge connecting the two domains modulates access to and from the sugar-binding site groove (14). The groove accommodates at least four α(1–4) linked glucose units of the maltooligosaccharides in a left-handed ribbon-like configuration. In this configuration, the unbound oligosaccharides have about three times greater accessible surface area from the polar OH groups than with the small clusters of nonpolar patches mostly from the A-faces. This segregation of polar from nonpolar groups is reflected by the distinct role that each domain plays in achieving near perfect complementarity in ligand binding in the closed form for which several structures have been determined (10). Domain I provides most of the polar residues that make cooperative hydrogen bonds with the sugar OH groups, whereas Domain II supplies all the aromatic sidechains that interface with the nonpolar patches. Given this unique system, we sought recently to identify which domain preferentially binds ligands and, thereby, uncover which interaction (polar or nonpolar) plays the major role in the initial process of molecular recognition and binding. This was accomplished by determining the structure of the maltose bound in the open form (X. Duan and F. A. Quiocho, manuscript submitted). In this structure, the maltose is bound almost exclusively in domain II that is rich in aromatic residues. This finding supports a major role for nonpolar interactions in initial ligand binding even when the ligands have significantly greater potential for highly specific polar interactions.

Another noteworthy finding of the binding of maltose to domain II in the open structure is the considerable number of unpaired buried sugar OHs making contacts with the aromatic sidechains. This finding indicates that a significant fraction of the binding interface in the open structure is non-complementary. More importantly, it further implies a smaller actual contact surface involved in purely hydrophobic interaction between domain II and maltose.

Metal: Discovery of a Zinc in Adenosine Deaminase

Adenosine deaminase (ADA) has attracted considerable attention since its existence and some aspects of its specificity were first reported about 65 years

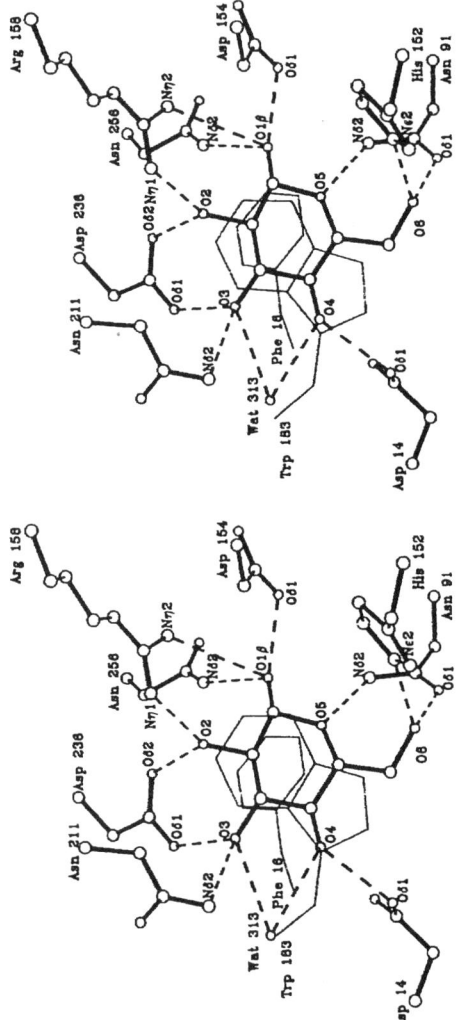

Figure 1. Stereo view of the hydrogen-bonding and stacking interactions between the D-galactose-binding protein (GBP) and D-glucose. The glucose is sandwiched between Phe 16 and Trp 183.

ago (*15*). ADA, which is present in virtually all mammalian cells and a key enzyme in purine metabolism, catalyzes the irreversible hydrolysis of adenosine to inosine and ammonia. ADA also plays key roles in a variety of biological processes (briefly reviewed in (*16*)) and is particularly crucial in the development of the lymphoid system (*17*). The crystal structure of ADA, the first for a purine/pyrimidine deaminase enzyme, determined in our laboratory (*16*) and others that followed of complexes with ground-state (*18*) and transition-state analogues (*19*) led to following major discoveries that are very much pertinent to understanding the catalytic reaction mechanism:

(i) A Zn^{2+} cofactor is present in the active site. This came as a complete surprise because of the far-ranging investigations on the activity of ADA especially in the prior two decades. It is also déjà vu once again for this author since he previously used the zinc metalloenzyme carboxypeptidase A to demonstrate enzyme activity in crystalline state in the mid 1960s in the laboratory of Prof. F. M. Richards (*20*) and participated in its structure determination in the late 1960s in the laboratory Prof. W. N. Lipscomb to whom this special issue is dedicated (*21, 22*).

(ii) Although the crystals of ADA from which the structures were determined, initially at pH 4.2 and subsequently at pH 7, were obtained with an excess purine riboside, the ligand bound in the site is 6(*S*)-hydroxy-1,6-dihydropurine riboside (HDPR) (*16, 19*). HDPR is the product of the enzyme-catalyzed stereospecific addition of a Zn^{2+}-activated water molecule or hydroxide to the C6 of the purine riboside. As HDPR represents the first half of the catalytic reaction, it is an excellent example of a "reaction coordinate analog", a term introduced by Christianson and Lipscomb (*23*) to describe "reversibly reactive substrate analogues of catalytic intermediates". HDPR has an estimated K_i of about 10^{-13} M (*24*). Our structure established the stereochemistry of the hydroxide addition and the chelation of the 6*S*-hydroxyl adduct to the Zn^{2+} as the main contributor to the extremely tight analog binding (*16, 19*).

(iii) An activated water or hydroxide is coordinated to the Zn^{2+} in the presence of 1-deazaadenosine (*18*). This finding is the first to show a zinc-activated water co-substrate in a complex with an almost perfect substrate or ground-state analog. The location of the hydroxide is fully consistent with the stereochemistry of hydroxide attack leading to the formation of HDPR described above in (ii).

(iv) The mode of binding of (8*R*)-hydroxyl-2'-deoxycoformycin (DCF) (or pentostatin, the drug name), a natural product from Streptomyces (*25*) and a potent inhibitor ($K_i \sim 10^{-13}$ M (*26*)), is very similar to that of HDPR (*19*). Pentostatin has been used to treat hairy cells leukemia (*27*) and to clinically increase the effectiveness of antitumor and antiviral analogs of adenosine. The strong inhibitory effect of DCF provided the early evidence for the theory of transition state in enzyme reaction. DCF possess a tetrahedral carbon (C8) at the position corresponding closely to C6 of adenosine that is the site of hydroxide addition in the catalytic reaction. The 8*R* isomer binds about 10^7 times stronger than the 8*S* isomer (*28*). The x-ray structure of ADA with the bound pentostatin shows that the coordination of the 8*R*-hydroxyl group to the Zn^{2+} as mainly contributing to the strong potency and very high degree of stereospecificity of inhibition (*18*).

Noncompetitive Inhibitor of Aldose Reductase: Binding in the Active Site

Aldose reductase (ALR2) is an NADPH-dependent enzyme that catalyzes the reduction of a wide variety of carbonyl-containing compounds to their corresponding alcohol, including the first and rate-limiting step of the polyol pathway of glucose metabolism. Clinical interest in ARL2 has been raised by its ability to reduce glucose to sorbitol. Enhanced flux of glucose through the polyol pathway is believed to be linked to numerous diabetic complications, including neuropathy, nephropathy, and retinopathy (*29*).

Over a thousand heterocyclic inhibitors of ARL2 have been and continue to be developed and some have shown promise in the treatment of diabetic complications (*30*). A very curious common behavior of these inhibitors, indicated by inhibition and competition studies, is that they bind at a site independent of the substrate and coenzyme (*29*). Our structure of the holoenzyme (human ARL2-NADPH complex) revealed a large active site pocket with the nicotinamide ring forming the base and several hydrophobic residues lining the sides (*31*). We next determined the first structure of the holoenzyme complexed with a potent synthetic inhibitor (zopolrestat) (*32*). Besides displaying an IC_{50} of 3 nM, zopolrestat exhibits excellent pharmacokinetic properties in humans and, at the time of the structure analysis, was in Phase II clinical studies (*30*). Kinetic studies of zopolrestat binding, like those of other inhibitors, indicate noncompetitive inhibition (*32*). The structure, to our surprise, unambiguously showed the zopolrestat binding snugly and with excellent complementarity in the hydrophobic active site pocket. This finding was subsequently duplicated in our structure determination of the same ternary complex but with FR-1, a fibroblast growth factor-induced member of the aldoketo reductase family (*33*). These two inhibitor-bound structures and subsequent others with similar results once-and-for-all invalidate the previous prevailing idea of inhibitor binding at a site other than the active site.

Cap mRNA: Recognition by "Double-Stacking" or "Cation-Double π" Interaction

The ability of proteins to discriminate alkylated from nonalkylated nucleic acids is of paramount importance in numerous biological processes, including DNA excision repair and the processing and translation of eukaryotic and viral mRNA. We have exploited the function of VP39, a vaccinia virus protein, to study the mechanism of this discrimination at the atomic level. VP39, together with the S-adenosylmethionine coenzyme, acts at the mRNA 5' end as a cap 0 (N7-methylguanosine (m^7G) (5')pppN···)-specific (nucleoside-2'-*O*-)-methyltransferase (*34*). To perform the enzymatic methylation of the 2'-hydroxyl group of the first transcribed base, VP39 must specifically recognize the m^7Gua moiety of the capped nucleotide while also binding mRNA transcript in a sequence-nonspecific manner.

We have determined at least 15 high resolution structures of VP39 complexed with m^7Guanine-capped nucleotides, dinucleotides and oligonucleotides and other methylated nucleobases (*35-38*). The series of x-ray structures of VP39 with bound m^7G-capped analogs provided paradigms for binding of a cap and an mRNA transcript in a sequence-nonspecific manner (described below). Three factors contribute to the specific recognition of the cap's m^7Gua moiety, which is inserted in a pocket or slot. (**i**) The 7-methyl groups makes a van der Waals contact or a CH···O hydrogen bond with a backbone carbonyl oxygen (Figure 2). (**ii**) All the polar groups of m^7Gua moiety are engaged in hydrogen-bonding interactions. (**iii**) The third and most striking factor is the stacking between the sandwiched m^7Gua and the side chains of Tyr 22 and Phe 180 (Figure 2). These three factors can also be elucidated from the crystal structure of the eIF4E-m^7Gpp complex (*38, 39*). The double-stacking arrangement exhibits three notable features: nearly perfect parallel alignment, substantial areas of overlap (notably the stack with Phe 180), and an optimal interplanar distance in both stacks of 3.4 Å. Data amassed from our crystallographic, biochemical and site-directed mutagenesis experiments indicate that stacking interactions between the electron-rich aromatic sidechains and the electro-deficient methylated nucleobase play a dominant role in cap recognition and binding ((*37*); also discussed in (*38*)). A "cation-π" (*40*) interpretation of the cap-protein stacking interaction has been suggested (*37, 38*) in which the positively charged nucleobase cation interacts electrostatically with the π-clouds or quadropole moment of the stacked aromatics. "Cation-double π" interaction is a more appropriate description of this distinctive double stacking arrangement.

Single-stranded RNA: Sequence Non-specific Recognition

Protein-mediated sequence-nonspecific recognition of single-stranded nucleic acids is crucial in a variety of biological processes including replication, DNA repair, RNA processing and translation. The structure of the vaccinia RNA methyltransferase VP39 bound to a 5'm^7G-capped RNA hexamer is the first to show a mechanism for non-specific recognition of an RNA (*41*). The RNA forms two short, single-stranded helices of three bases each. The first of these helices binds in the active site of VP39 solely through hydrogen bonds between the protein and the ribose-phosphate backbone. The bases of the RNA strand

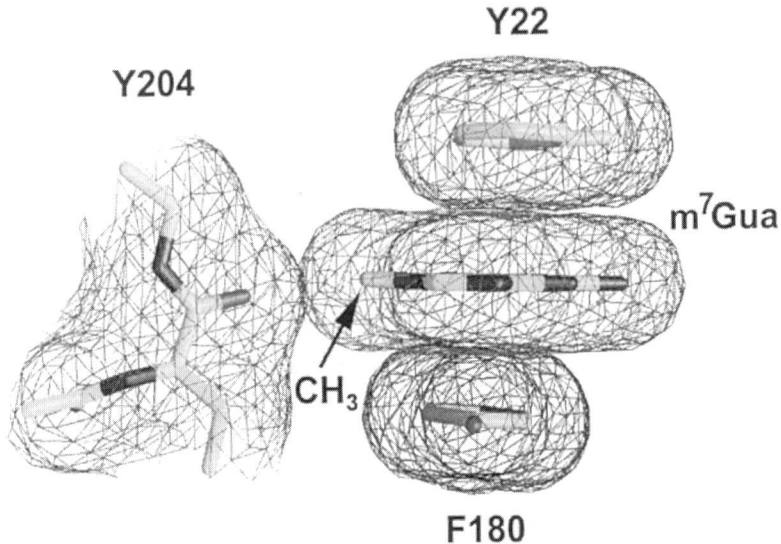

Figure 2. Double stacking interactions of Tyr 22 and Phe 204 with the guaninium and van der Waals contact of the backbone carbonyl oxygen of Tyr 204 with the methyl (CH_3) of the m^7Gua moiety of the m^7G cap analog. Figure adapted with modification from ref. 37.

stack together as trimers, but do not form any interactions with the protein. This observation suggests an intuitive mechanism for sequence non-specific nucleic acid binding, where the single-stranded RNA forms short transient helices driven by intramolecular base stacking interactions. The protein then recognizes and stabilizes the helical backbone conformation formed by this transient stacking without interacting with the bases themselves.

Charged Ligands: Prototypical Role of Local Dipolar Interactions for Exquisite Ligand Specificity and Electrostatic Balance

Proteins or biological systems very often exhibit extremely high selectivity between very similar ligands (e.g., divalent cations). One classic example of this phenomenon that our laboratory begun investigating in the 1980s is exhibited by two binding proteins of the bacterial ABC transporters for phosphate and sulfate, two structurally very similar tetrahedral oxyanions that are the principal sources for phosphorus and sulfur for a myriad of biological macromolecules (principally proteins and polynucleotides) and small molecules (notably ATP, S-adenosylmethionine, NAD, etc.). The specificity for each oxyanion is stringent; the transport system for phosphate is distinct from that for sulfate. The specificity of each oxyanion is conferred by the periplasmic phosphate-binding protein (PBP) and sulfate-binding protein (SBP). At physiological pH, PBP and SBP exhibit no overlap in specificity. This stringent specificity exhibited by each receptor has an extremely important biological implication; one oxyanion nutrient cannot become an inhibitor of transport for the other. The specificity of the PBP-dependent phosphate ABC-type transport is also shared by other phosphate transport systems in eukaryotic cells and across brush borders and into mitochondria. Discrimination between two oxyanions, as revealed by our x-ray structure analysis of PBP and SBP, appears surprisingly and fundamentally simple. It is based solely on the fact that sulfate is completely ionized at pH values above 3, whereas phosphate remains protonated at pH values below 13. How does the ligand's protonation states govern the specificity of each receptor?

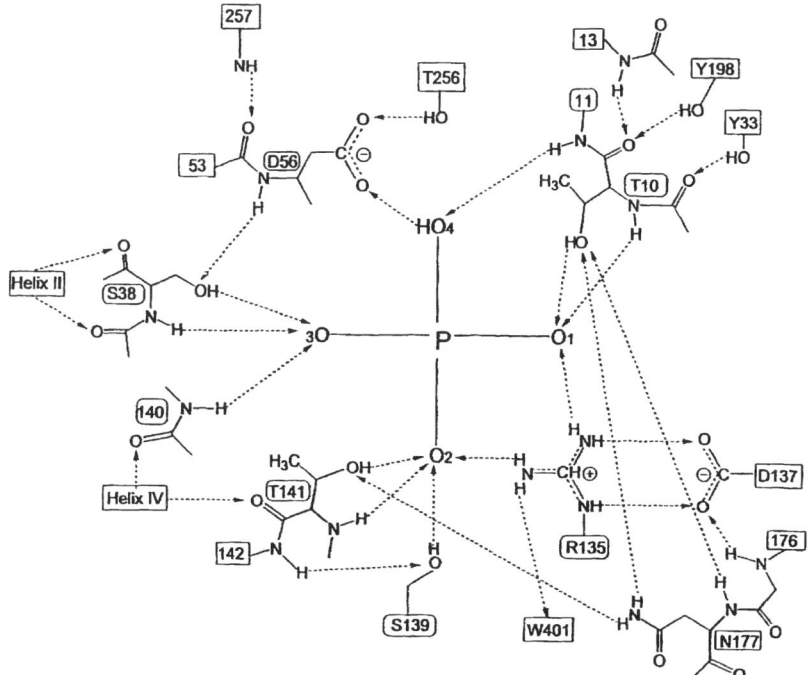

Figure 3. Hydrogen-bonding interactions between the phosphate-binding protein (PBP) and phosphate. The residues directly interacting with the phosphate are further involved in networks of hydrogen bonds.

The PBP-phosphate complex structure was determined first at 1.7 Å resolution *(42)* and later on at an ultra-high resolution of 0.98 Å *(43)*. The phosphate is completely desolvated and sequestered in the protein cleft between two domains. The phosphate makes 12 hydrogen bonds with the proteins (11 with donor groups and 1 with an acceptor group). It is noteworthy that the donor groups are mostly backbone peptide NH groups (a total of 5) and hydroxyl sidechains (a total of 4) (Figure 3). The guanidinium sidechain of Arg 135 donates 2 NH groups to the phosphate, but its full potential as a charge-coupling group to the anion is either negated or considerably diminished by the salt link with Asp 137 (Figure 3), a feature condusive to rapid nutrient release in active transport. The Asp 56, the lone acceptor group, plays two key roles in conferring exquisite specificity to PBP; it recognizes, by way of the hydrogen bond, a proton on the phosphate and presumably disallows, by charge repulsion, the binding of fully ionized sulfate dianion. The proton in this hydrogen bond may be shared between the carboxylate and phosphate oxygens.

The SBP binding site between the two lobes of the bilobate protein is also tailor-made for sulfate (44). In keeping with the high specificity of SBP for fully ionized tetrahedral oxyanions, the bound sulfate, which is also completely dehydrated and buried, accepts seven hydrogen bonds entirely from uncharged protein polar groups (mostly backbone NHs) (Figure 4). The absence of a carboxylate or a hydrogen-bond acceptor in the binding site of SBP accounts for the inability of SBP to bind phosphate.

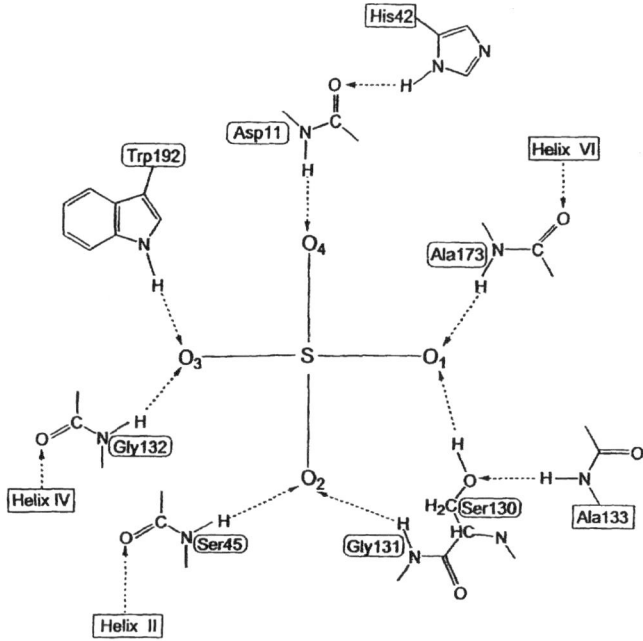

Figure 4. Schematic diagram of the hydrogen bonds from the sulfate-binding protein (SBP) to the bound sulfate.

The ability of PBP and SBP to differentiate each oxyanion ligand through the presence or absence of proton(s) is an extremely high level of sophistication in molecular recognition. The importance of complete hydrogen bonding in recognition of buried ligands is powerfully demonstrated in PBP and SBP. Despite the potential for a large number of matched hydrogen-bonding pairs, a single mismatched hydrogen bond (e.g., a sulfate providing no proton to the Asp residue of PBP and no acceptor group in SBP for a phosphate proton) represents a binding energy barrier of 6-7 kcal/mol.

A Charged-neutral Hydrogen Bond: Short Low-barrier but with A Normal Strength

The ultra high resolution (0.99 Å) refined structure of the PBP-phosphate complex is the first to establish crystallographically the formation of an extremely short hydrogen bond (2.432 ± 0.007 Å) between the Asp 56 carboxylate of PBP and phosphate (Figure 4) (*43*). Although this short hydrogen bond is within the proposed range of low barrier hydrogen bonds with estimated energies of 12-24 kcal/mol (*45*), its contribution to phosphate binding affinity has been assessed to be no better than that of a normal hydrogen bond (*43*). Thus, a unique role for short hydrogen bonds in biological systems, such as enzyme catalysis (*46, 47*), remains controversial. A hydrogen bond with an extraordinarily high strength is inconsistent with the function of a protein involved in active transport or enzyme catalysis which require relatively fast ligand translocation or release of transient intermediates in catalytic reaction.

Charged Ligands: Prototypical Role of Local Dipolar Interactions for Stabilization of Buried Charges

A novel finding of further immense importance and wide implication resulting from our studies of the sulfate- and phosphate-binding proteins in the mid 1980s is how the isolated charges of the buried anions are stabilized. The binding of the sulfate involves no counter-charged residues or cations (Figure 4) (*44*). Although Arg 135 donates two hydrogen bonds to the phosphate, its salt-linking property is usurped by Asp 137 (Figure 3) (*42*). Moreover, site-directed mutagenesis studies indicate that phosphate binding is quite insensitive to modulation of the salt link (*48*). These findings are a powerful demonstration of how a protein is able to stabilize the charges by means other than salt links. Experimental and computational studies indicate that local dipoles, including the hydrogen bonding groups (mostly backbone NH and sidechain OH groups) and the three NH groups from the first turn of α-helices, immediately surrounding the sulfate and phosphate are responsible for charge stabilization (*42, 44, 49-52*). These studies further indicated that helix macrodipoles play little or no role in charge stabilization of the anions (*50, 51*). However, the first and last turns of helices are a rich source of local dipoles for charge stabilization. The same principle of charge stabilization and ion recognition as well by local dipoles also applies for the following buried uncompensated ionic groups in other proteins: the Arg 151 of the arabinose-binding protein (*49*), the zwitterionic leucine ligand bound to the leucine/isoleucine/valine-binding protein (*49*), the potassium in the specificity filter of the potassium channel pore (*53*), Arg 56 of synaptobrevin-II in a SNARE complex (*54*), and Lys 79 of AP180 of the endocytosis machinery (*55*). Dipolar interactions are ideal not only for molecular recognition but also for the functions of proteins/enzymes in processes which require relatively fast ligand or ion movement (active transport or ion channel systems) or transient stabilization of intermediates in enzyme catalysis.

Acknowledgment

The work in the author's laboratory was supported in part by grants from NIH and Welch Foundation. The author is an Investigator in the Howard Hughes Medical Institute. He is also indebted to colleagues who have contributed immensely to the work described above. He thanks G. Hu, Y. Mao, W.E. Meador, A. Nickitenko and N.K. Vyas for assistance in preparing the manuscript.

References

1. Means, A. R; VanBerkum, M. F.; Bagchi, I.; Lu, K. P.; Rasmussen, C. D. *Pharmacol. Ther.* **1991**, *50*, 255-270.
2. Babu, Y. S.; Bugg, C. E.; Cook, W. J. *J. Mol. Biol.* **1988**, *204*, 191-204.
3. Meador, W. E.; Means, A. R.; Quiocho, F. A. *Science* **1992**, *257*, 1251-1255.
4. Ikura, M.; Clore, G. M.; Gronenborn, A. M.; Zhu, G.; Klee, C. B.; Bax, A. *Science* **1992**, *256*, 632-638.
5. Meador, W. E.; Means, A. R.; Quiocho, F. A. *Science* **1993**, *262*, 1718-1721.
6. Meador, W. E.; George, S. E.; Means, A. R.; Quiocho, F. A. *Nature Struct. Biol.* **1995**, *2*, 943-945.
7. Vyas, N. K.; Vyas, M. N.; Quiocho, F. A. *Science* **1988**, *242*, 1290-1295.
8. Quiocho, F. A.; Vyas, N. K. *Nature* **1984**, *310*, 381-386.

9. Spurlino, J. C.; Lu, G. Y.; Quiocho, F. A. *J. Biol. Chem.* **1991**, *266*, 5202-5219.
10. Quiocho, F. A.; Spurlino, J. C.; Rodseth, L. E. *Structure* **1997**, *5*, 997-1015.
11. Quiocho, F. A. *Pure App. Chem.* **1989**, *61*, 1293-1306.
12. Quiocho, F. A.; Vyas, N. K. In *Bioorganic Chemistry: Carbohydrates;* Hecht, S. M. Ed.; Oxford University Press, Oxford, England, 1999; pp 441-457.
13. Quiocho, F. A.; Wilson, D. K.; Vyas, N. K. *Nature* **1989**, *340*, 404-407.
14. Sharff, A. J.; Rodseth, L. E.; Spurlino, J. C.; Quiocho, F. A. *Biochemistry* **1992**, *31*, 10657-10663.
15. Schmidt, G. *Z. Physiol. Chem.* **1932**, *208*, 185-191.
16. Wilson, D. K.; Rudolph, F. B.; Quiocho, F. A. *Science* **1991**, *252*, 1278-1284.
17. Kredich, N. M.; Hershfield, M. S. In *The metabolic basis of inherited disease* Scriver, C. R.; et al., Eds.; McGraw-Hill, New York, 1989; pp 1045-1075.
18. Wilson, D. K.; Quiocho, F. A. *Biochemistry* **1993**, *32*, 1689-1694.
19. Wang, Z.; Quiocho, F. A. *Biochemistry* **1998**, *37*, 8314-8324.
20. Quiocho, F. A.; Richards, F. M. *Proc. Natl. Acad. Sci. U.S.A.* **1964**, *52*, 833-839.
21. Steitz, T. A; Ludwig, M. L; Quiocho, F. A.; Lipscomb, W. N. *J. Biol. Chem.* **1967**, *242*, 4662-4668.
22. Quiocho, F. A.; Lipscomb, W. N. *Adv. Protein. Chem.* **1971**, *25*, 1-78.
23. Christianson, D. W.; Lipscomb, W. N. *Acc. Chem. Res.* **1989**, *22*, 62-69.
24. Kati, W. M.; Wolfenden, R. *Biochemistry* **1989**, *28*, 7919-7927.
25. Woo, P. W. K; Dion, H. W; Lange, S. M; Dahl, L. F.; Durham, L. J. *J. Heretocycl. Chem.* **1974**, *11*, 641-643.
26. Agarwal, R. P.; Spector, T.; Parks, R. E. Jr. *Biochem. Pharmacol.* **1977**, *26*, 359-367.
27. Spiers, A. S. D.; Moore, D.; Cassileth, P. A.; Harrington, D. P.; Cummings, F. J.; Neiman, R. S.; Bennett, M. J.; O'Connell, M. J. *N. Engl. J. Med.* **1987**, *316*, 825-830.
28. Schramm, V. L.; Baker, D. C. *Biochemistry* **1985**, *24*, 641-646.
29. Kador, P. F.; Sharpless, N. E. *Mol. Pharmacol.* **1983**, *24*, 521-531.
30. Sarges, R.; Oates, P. J. *Prog. Drug. Res.* **1993**, *40*, 99-161.
31. Wilson, D. K.; Bohren, K. M.; Gabbay, K. H.; Quiocho, F. A. *Science* **1992**, *257*, 81-84.
32. Wilson, D. K.; Tarle, I.; Petrash, J. M.; Quiocho, F. A. *Proc. Natl. Acad. Sci. U. S. A.* **1993**, *90*, 9847-9851.
33. Wilson, D. K.; Nakano, T.; Petrash, J. M.; Quiocho, F. A. *Biochemistry* **1995**, *34*, 14323-14330.
34. Barbosa, E.; Moss, M. *J. Biol. Chem.* **1978**, *253*, 7692-7702.
35. Hodel, A. E.; Gershon, P. D.; Shi, X.; Quiocho, F. A. *Cell* **1996**, *85*, 247-256.

36. Hodel, A. E.; Gershon, P. D.; Shi, X.; Wang, S. M.; Quiocho, F. A. *Nat. Struct. Biol.* **1997**, *4*, 350-354.
37. Hu, G.; Gershon, P. D.; Hodel, A. E.; Quiocho, F. A. *Proc. Natl. Acad. Sci. U.S.A.* **1999**, *96*, 7149-7154.
38. Quiocho, F. A.; Hu, G.; Gershon, P. D. *Cur. Opin. Struct. Biol.* **2000**, *10*, 78-88.
39. Marcotrigniano, J.; Gingras, A.-C.; Sonnenberg, N.; Burley, S. K. *Cell* **1997**, *89*, 951-961.
40. Dougherty, D. A. *Science* **1996**, *271*, 163-168.
41. Hodel, A. E.; Gershon, P. D.; Quiocho, F. A. *Mol. Cell.* **1998**, *1*, 443-447.
42. Luecke, H.; Quiocho, F. A. *Nature* **1990**, *347*, 402-406.
43. Wang, Z.; Luecke, H.; Yao, N.; Quiocho, F. A. *Nat. Struct. Biol.* **1997**, *4*, 519-522.
44. Hibbert, F.; Emlsey, J. *J. Adv. Phys. Org. Chem.* **1990**, *26*, 255-379.
45. Gerlt, J. A.; Gassman, P. G. *Biochemistry* **1993**, *32*, 1943-1952.
46. Cleland, W. W.; Kreevoy, M. M. *Science* **1994**, *264*, 1887-1890.
47. Pflugrath, J. W.; Quiocho, F. A. *Nature* **1985**, *314*, 257-260.
48. Yao, N.; Ledvina, P. S.; Choudhary, A.; Quiocho, F. A. *Biochemistry* **1996**, *35*, 2079-2085.
49. Quiocho, F. A.; Sack, J. S.; Vyas, N. K. *Nature* **1987**, *329*, 561-564.
50. Åqvist, J.; Luecke, J.; Quiocho, F. A.; Warshel, A. *Proc. Natl. Acad. Sci.* **1991**, *88*, 2026-2030.
51. He, J. J.; Quiocho, F. A. *Prot. Sci.* **1993**, *2*, 1643-1647.
52. Ledvina, P. S.; Tsai, A. L.; Wang, Z.; Koehl, E.; Quiocho, F. A. *Prot. Sci.* **1998**, *7*, 2550-2559.
53. Doyle, D. A.; Cabral, J. M.; Pfuetzner, R. A.; Kuo, A.; Gulbis, J. M.; Cohen, S. L.; Cahit, B. T.; MacKinnon, R. *Science* **1998**, *280*, 69-77.
54. Sutton, R. B.; Fasshauer, D.; Jahn, R.; Brunger, A. T. *Nature* **1998**, *395*, 347-353.
55. Mao, Y.; Chen, J.; Maynard, J. A.; Zhang, B.; Quiocho, F. A. *Cell* **2001**, *104*, 433-440.

Chapter 15

Structures of the Central Dogma of Molecular Biology

Thomas A. Steitz

Departments of Molecular Biophysics and Biochemistry and Chemistry, Howard Hughes Medical Institute, Yale University, New Haven, CT 06520–8114

Working in the laboratory of Bill Lipscomb, the Colonel, as a graduate student, I learned several principles for constructing a research program. The most important among these is to select a significant research question and then investigate all aspects of it, rather than hopping from one little problem to another. This approach is particularly important in X-ray crystal structure analysis which can be applied to any molecule that crystallizes. The Colonel set out to understand chemical bonding in the boron hydrides, and the majority of structures determined in Gibbs laboratory when I was there in the 1960s were of various boron containing compounds. Further, the structure of a particular boron hydride was not the end of the story but rather one step in a continuous cycle that included theoretical calculations, synthesis of new compounds and further structure determinations. The fruits of this approach to understanding bonding in boron hydrides have been amply recognized. I believe there is an important message in these studies for the field of structural biology and for those who now advocate investing time and money into structural genomics, solving the structures of any protein gene product that crystallizes without reference to its biological function.

About 25 years ago I decided to embark on a study of the proteins and nucleic acids that function in the replication and recombination of the genome and in the expression of genes into RNA and protein – the molecules involved in the process that Francis Crick called the Central Dogma of Molecular Biology (1). In the 1950s, long before all of the critical molecular players had been identified, Crick summarized the information flow in biology: DNA is copied into DNA, DNA can be transcribed into RNA and the sequences encoded in RNA can be translated into protein (Fig. 1). We have now established the structures of proteins and nucleic acids involved in all of these steps, many of them caught in the act of executing their biological function.

DNA replication

Although DNA replication involves many proteins, even in a simple bacterial phage like T4, the central player in replication is DNA polymerase itself. Possibly the earliest enzymatic activity to appear in evolution was that of the polynucleotide polymerase, the ability to replicate the genome accurately being a prerequisite for evolution itself. These enzymatic scribes must faithfully copy the sequences of the genome into daughter nucleic acids or the information contained within would be lost. Thus, some mechanisms of assuring fidelity are required. Further, all classes of polynucleotide polymerases must be able to translocate along the template being copied, usually processively, as synthesis proceeds.

We determined the first crystal structure of a DNA polymerase, that of the Klenow fragment of *E. coli* DNA polymerase I (pol I), in 1985 (2). This structure showed the surprising fact that the polymerizing active site and the editing active site were located on different domains separated by some 35 Å. However, the structure of the apo enzyme gave no accurate insights concerning the way in which DNA was either copied or edited. The polymerase domain, however, did show an overall architectural feature that appears to be shared in all polymerase from various families whose structures have been subsequently determined. It had a shape that can be compared with that of a right hand and having domains called "thumb", "palm", and "fingers".

We were subsequently able to obtain the crystal structure of an editing complex showing duplex DNA snuggled up against the thumb with the frayed 3' end of single-stranded DNA extending into the exonuclease active site. In a series of studies that included mutagenesis, various metal ions and several phosphorothioate substrates, we were able to clearly establish that the mechanism of this exonuclease involved two divalent metal ions separated by about 3.8 Å (3,4,5,6). One metal ion appears to activate the attacking water molecule by lowering its pKa while the second metal ion appears to function to stabilize the leaving oxyanion. That this hydrolytic mechanism involves only the use of divalent metal ions properly positioned by carboxylate groups, led to the suggestion that such a two metal ion mechanism of hydrolysis might be used in ribozyme involved in the cleavage of RNA such as the group I intron ribozyme (7).

The structures of binary and ternary complexes containing primer-template DNA with and without dNTP bound to the polymerase active site have

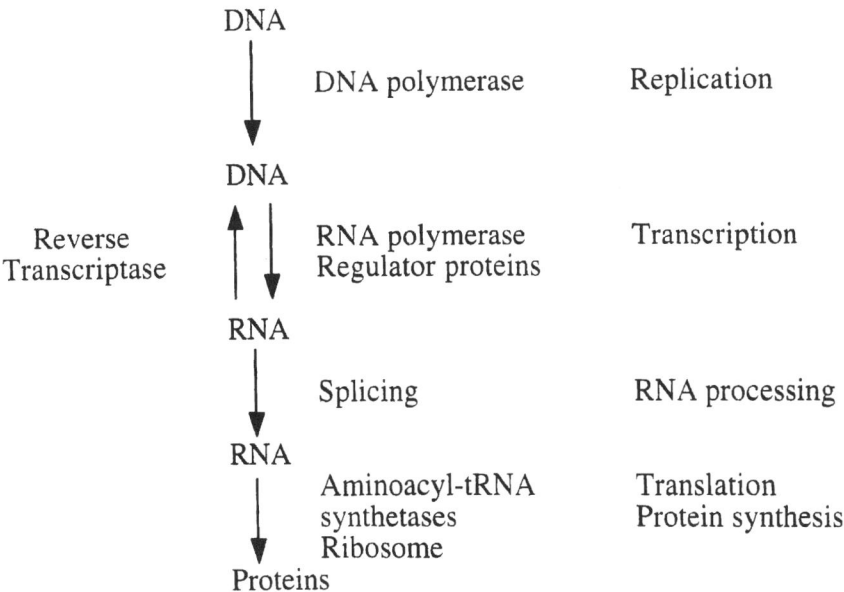

Figure 1. Crick's Central Dogma of Molecular Biology

been successfully obtained in the DNA polymerase I family more recently (8,9,10). A complex with primer-template DNA bound to *Thermus aquaticus* DNA polymerase showed the duplex DNA positioned up against the thumb approximately where duplex DNA was found in editing mode but not precisely superimposed (Fig. 2). A true ternary complex involving primer-template DNA and deoxynucleoside triphosphate was achieved subsequently with T7 DNA polymerase by Tom Ellenberger and coworkers (9). This complex showed that the significant structural reorganization had occurred and that this structural reorganization was very likely an important part of maintaining the fidelity of nucleotide addition. The fingers domain had rotated and was now tightly packed against the base of the incoming triphosphate and the template base with which it was paired. A mismatched base pair would not fit and it was proposed that this substrate induced conformational change, which is necessary for subsequent catalytic steps would not occur if the incoming deoxynucleoside triphosphate forms a mismatched base pair.

More recently, we have determined the structures of a DNA polymerase and its complexes from phage RB69, which is a member of the pol II family and a homologue of the human DNA polymerase α (11,12,13). The structures of the apo enzyme as well as those of both editing (Fig. 3) and polymerizing complexes show dramatic conformational changes in the enzyme that are induced by the binding of substrates. As in the case of the pol I family, the formation of a ternary complex between primer-template DNA, deoxynucleoside triphosphate in the enzyme showed a large reorientation of the fingers domain that interacts with a newly formed base pair (13). This domain rotates by 60° compared to its position in the unliganded enzyme. Perhaps equally unexpected is the difference in the orientations of the bound DNA in the editing and in the polymerizing complexes. The duplex portion of the DNA's product is rotated by 40° in the editing complex compared to its position in the polymerizing complexes. The thumb, which is in contact with the DNA, also changes position, seemingly to guide the reorientation of the DNA from polymerizing to editing mode.

Structural studies as well as sequence comparisons among polymerases strongly suggest the hypothesis that the phosphoryl transfer reaction of all polymerases is catalyzed by a two metal ion mechanism that was originally proposed by analogy to the well studied two metal mechanism in the 3' exonuclease reaction (14). It is perhaps of interest to note that such a mechanism, which involves only the properties of two correctly positioned divalent metal ions, could easily be used by an enzyme made entirely of RNA and thus could function in an all RNA world. The fidelity of DNA synthesis appears to arise from two sources. First, "enforced" Watson-Crick interactions at the polymerase active site increases the accuracy of the incorporation step (9,13). Second, there is a competitive editing at the 3' exonuclease active site that removes misincorporated nucleotide (3,5). When nucleotides are misincorporated, further polymerization is retarded, the duplex DNA is destabilized, and the more energetically favored single-stranded 3' terminus shows enhanced binding to the exonuclease active site where the offending nucleotide is removed.

Transcription

The transcription of the genes encoded in DNA into mRNA is accomplished by DNA dependent RNA polymerase that can recognize specific promoter sequences in DNA where they initiate RNA transcription by melting the DNA duplex and, unlike DNA polymerase, initiate RNA synthesis primed by a single nucleotide. At the stage of initiating a new transcript, RNA polymeases cycle abortively making numerous short transcripts before they are able to enter an elongation phase during which completely processive synthesis results in a complete mRNA transcript. The feature of transcription that probably attracts the most attention is the regulation of these polymerases to activators and depressors.

Catabolite Gene Activator Protein

We established the first crystal structure of a transcription regulator, the *E. coli* catabolite gene activator protein (CAP) (15), which activates transcription by RNA polymerase at specific promoters. This structure was followed almost immediately by that of the λ phage cro protein (16), which is a repressor. Our first attempts to model build a CAP complex with DNA proved incorrect. The Colonel always quoted Linus Pauling as saying that if one never made a mistake in science, one would never make a great discovery. Of course, just making mistakes would not lead to great discoveries either.

Several years later we were able to determine the structure of CAP complexed with cAMP and about 30 base pairs of a specific DNA sequence to which it binds at the lac promoter (17). This complex showed a DNA that was kinked by the protein in two places, each by more than 40° (Fig. 4). This unprecedented observation of severely bent DNA wrapping around a protein accounted for our inability to correctly model the complex earlier and explained the large size of the known CAP binding site on DNA.

T7 RNA Polymerase

Our structural studies of an RNA polymerase from phage T7 began in the early 1980s and have now led to structural insights into the mechanisms by

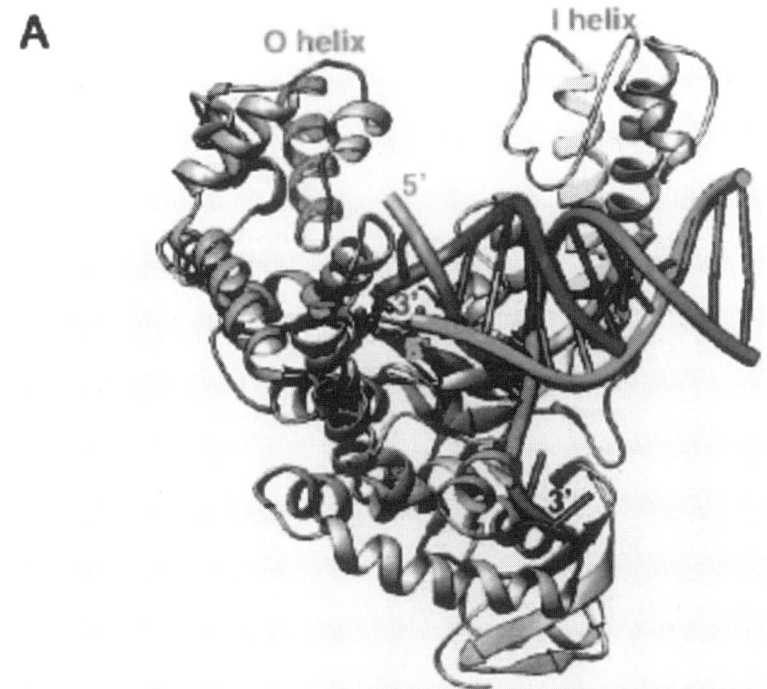

*Figure 2. The shuttle mechanism of editing in DNA polymerase.
(A) Superposition of DNAs bound in editing and in polymerizing modes. To orient the two DNAs, the polymerase domains of the Klenow fragment editing complex and the Taq polymerase synthetic complex were superimposed, and the DNA from the editing complex was added to the Taq polymerase-DNA complex. The 3' end of the primer strand in polymerizing mode is duplex and lies near three catalytically important carboxylates in the polymerase active site. The 3' end of the primer strand in editing mode is single-stranded and lies in the 3'-5'-exonuclease domain active site. (B) The shuttling model for polymerase editing proposes that the equilibrium between the 3' end of the primer strand being bound as a single strand in the exonuclease active site (right) and bound as duplex at the polymerase active site (left) is shifted toward the editing mode by mismatched base pairs, which destabilize duplex DNA and retard addition of the next nucleotide. The shuttling of the 3' end between the two active sites is fast compared with the rate of next nucleotide addition.
(Adapted with permission from reference 28)*

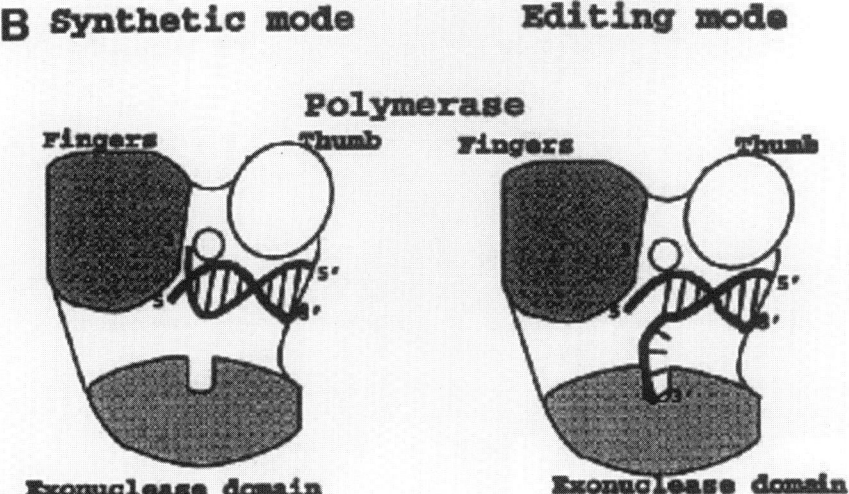

Figure 2. *Continued.*

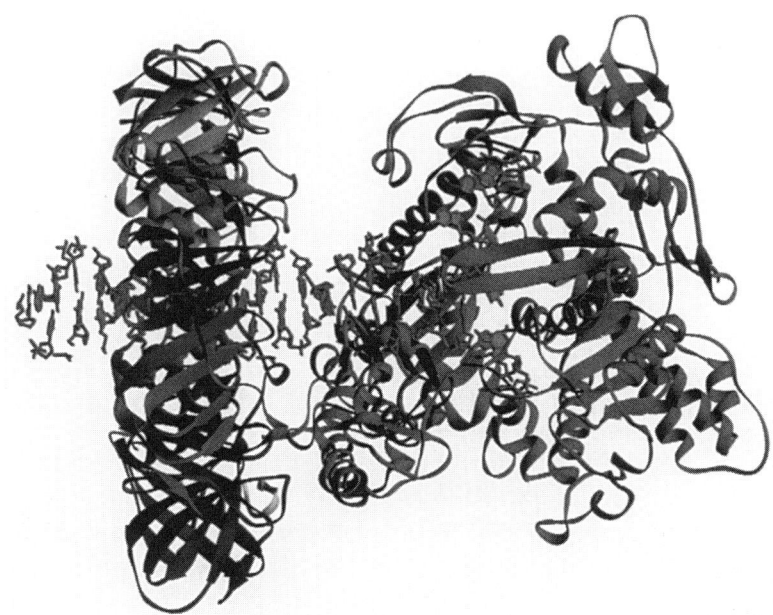

Figure 3. *View of the docked editing and sliding clamp complexes. The model was made by conjoining the RB69 DNA polymerase editing structure with the structure of the sliding clamp structure complexed with a peptide consisting of the 11 C-terminal residues of the polymerase. Using their areas of overlap as guides, it is possible to extend the B-DNA of the DNA polymerase editing structure through the central channel of the sliding clamp and then orient the clamp via the bound polymerase peptide.*
(Adapted with permission from reference 12)

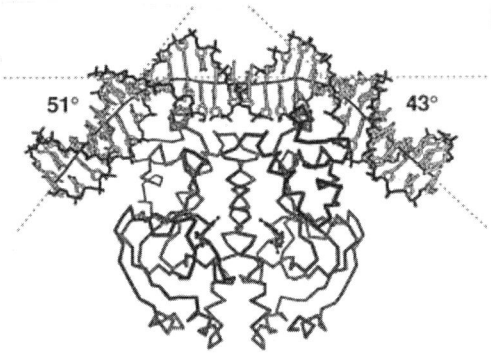

Figure 4. *Structure of the CAP-DNA complex shows angles of DNA bending in the CAP-DNA complex. The DNA helix axis as defined by the program "Curves" is shown as a black line running down the middle of the DNA helix.*
(Adapted with permission from reference 17)

which this polymerase can initiate specifically at a promoter sequence, denature duplex DNA to form an open bubble complex, transcribe the DNA into RNA in the initiation phase and undergo a structural transition to an elongation phase of transcription (18,19). Furthermore, we have been able to provide a structural basis for understanding how a transcriptional repressor works in this system (20; Yin and Steitz, unpublished). The T7 RNA polymerase is one of the simplest DNA-dependent polymerase enzymes, a single polypeptide chain of about 100,000 molecular weight, and is capable of transcribing a complete gene without the need for additional proteins. The structure of a complex with promoter (18) shows the DNA being recognized by an anti-parallel β-loop interacting in the major groove and an N-terminal domain that is novel to the RNA polymerase is also involved in promoter recognition and DNA melting. The structure of a transcribing T7 RNA polymerase initiation shows that synthesis of RNA in the initiation phase leads to the accumulation, or "scrunching" of the template in the enclosed active site pocket (19). The structure of this enzyme with 30 base pairs of duplex DNA and 17 n.t of RNA caught in the act of an elongation phase of synthesis shows some astounding transformations to the polymerase structure (W. Yin and T.A. Steitz, unpublished). Work not yet finished shows that the polymerase caught in the elongation phase interacts differently with upstream duplex DNA and with the heteroduplex RNA-DNA hybrid. It appears that the repressor (T7 phage lysozyme) works by stabilizing the initiation phase polymerase structure and preventing the transition to the elongation phase structure. We have established the crystal of T7 RNA polymerase complexed with T7 phage lysozyme (20), an inhibitor, and observe it to bind to the side of the polymerase opposite its active site interacting with a domain whose structure changes dramatically upon entering the elongation phase; the T7 lysozyme binding site no longer exists in the elongation state structure (Fig. 5). Since the polymerase domain is a homologue of T7 DNA polymerase, in the pol I family of DNA polymerases, we have been able to completely understand the differences between a DNA-dependent DNA polymerase and an DNA-dependent RNA polymerase. The additional functions exhibited by an RNA polymerase are performed by two insertions in the polymerase domain and an added domain.

HIV Reverse Transcriptase

Unanticipated by Crick was the existence of an enzyme capable of copying RNA into DNA - reverse transcriptase. HIV reverse transcriptase (RT) copies the HIV genomic RNA into a DNA duplex which is subsequently inserted into the human host's DNA. HIV RT is the target of many anti-AIDS drugs, including AZT and many non-nucleotide inhibitors such as nevirapine.

We determined the first structure of HIV RT, which was complexed with nevirapine, and discovered a stunningly asymmetric dimer (21). In spite of

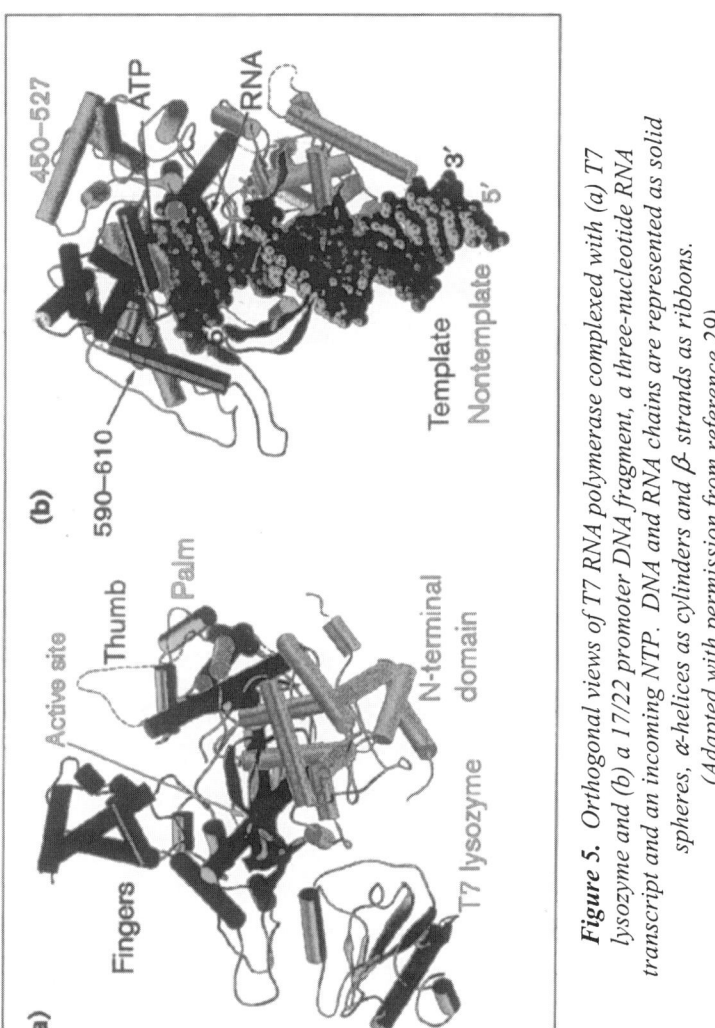

Figure 5. Orthogonal views of T7 RNA polymerase complexed with (a) T7 lysozyme and (b) a 17/22 promoter DNA fragment, a three-nucleotide RNA transcript and an incoming NTP. DNA and RNA chains are represented as solid spheres, α-helices as cylinders and β-strands as ribbons. (Adapted with permission from reference 29)

the two subunits having the same sequence, the four domains they have in common are arranged completely differently (Fig. 6). The polymerase domain of the 66 kilodalton subunit has a large cleft analogous to that of the Klenow fragment, however, the 51 kD subunit has no such cleft due to the different arrangement of subdomains. The effect of residues whose mutation results in drug resistance can be largely understood in terms of this structure.

This structure is being used by ourselves and others to embark upon structure based drug design with the goal of obtaining novel drugs that are effective against AIDS.

Protein Synthesis

The messenger RNA transcribed by the RNA polymerase is translated into protein on the ribosome. In order to decode the message and allow the addition of correct amino acids on a growing polypeptide chain, specific amino acids have to be attached to a specific tRNA molecule. The job of aminoacylating a tRNA containing an anticodon specifying a particular amino acid with that cognate amino acid is accomplished by specific aminoacyl-tRNA synthetases. There are 20 synthetases, one for each amino acid, and they must correctly identify both the tRNA containing the appropriate anticodon as well as the amino acid and utilize ATP to attach these amino acids to the 3' end of the tRNA in an activated form.

Aminoacyl-tRNA Synthetases

We were the first to determine the co-crystal structure of an aminoacyl-tRNA synthetase complexed with its cognate tRNA and ATP which was also the first complex between a protein and a specific RNA molecule (22,23). The structure of *E. coli* glutamine tRNA complexed with the *E. coli* glutaminyl-tRNA synthetase (Fig. 7) immediately showed how this synthetase is able to recognize the specific tRNA. The 3 bases of the anticodon loop are splayed out and are interacting with specific pockets on the enzyme. Furthermore, there are interactions in the minor groove of the acceptor stem making the enzyme specific for several base pairs in the acceptor stem of tRNA. Finally, the binding site for the amino acid contains hydrogen bond donors and acceptors that are properly positioned to allow only the binding of the amino acid glutamine and to exclude amino acids that are similar, such as glutamate (24).

242

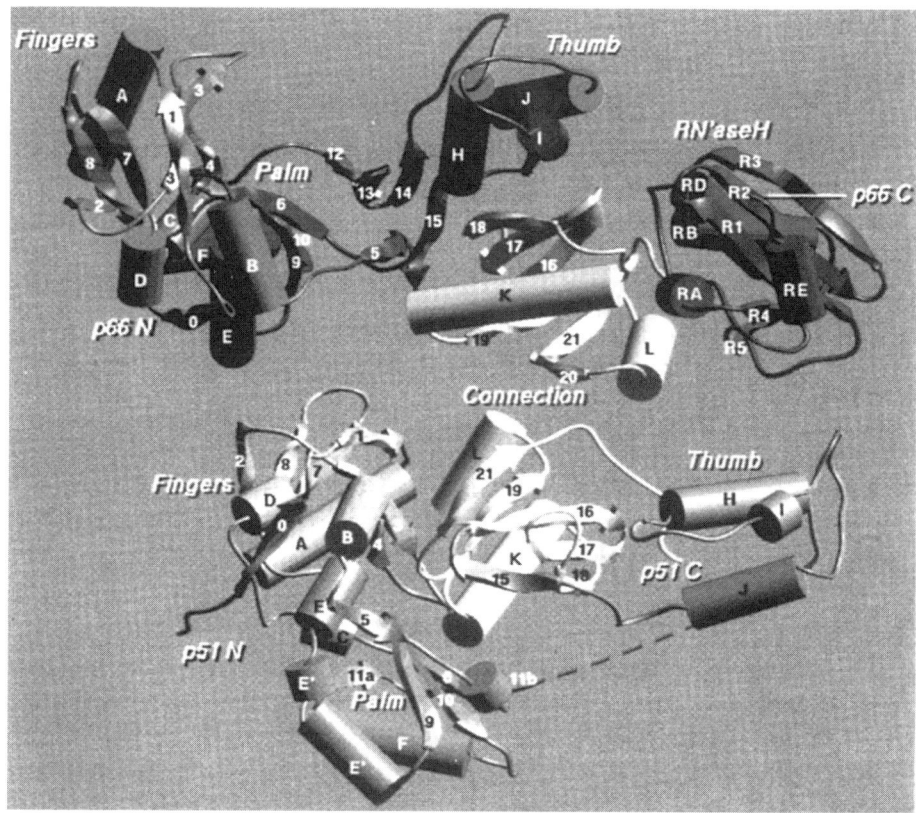

Figure 6. Schematic drawing of the polypeptide backbone of the RT heterodimer. α- helices and β- strands are represented by tubes and arrows, respectively. The p66 (upper) and p51 (lower) subunits are pulled apart in the vertical direction to make the interaction surfaces clear.
(Adpated with permission from reference 30)

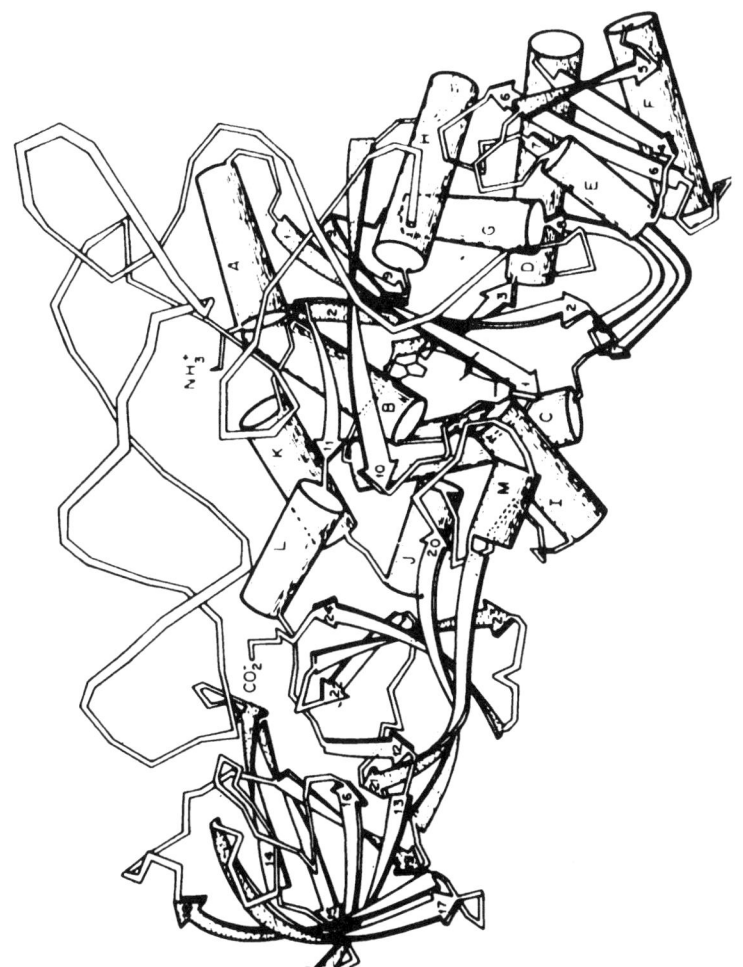

Figure 7. *GlnRS complexed with tRNA^{Gln} and ATP. For the protein, α-helices are represented as tubes sequentially lettered and β-strands as arrows sequentially numbered, both from the amino terminus. (Adpated with permission from reference 23)*

Ribosome

In the last step of the gene expression pathway, genomic information encoded in messenger RNAs is translated into protein by a ribonucleoprotein called the ribosome. The prokaryotic ribosome (MW approximately 2.6×10^6 daltons) is about 2/3 RNA and 1/3 protein and consists of 2 subunits the larger of which is approximately twice the molecular weight of the smaller. The small subunit mediates the interaction between the mRNA codon and the tRNA anticodon, an interaction on which the fidelity of translation depends. The large subunit, which sediments at 50S in prokaryotes, includes the activity that catalyzes peptide bond formation - peptidyl transferase.

We began our studies of the structure of the large ribosomal subunit isolated from *Haloarcula marismortui* in the fall of 1995 and successfully determined its atomic structure in the summer of 2000 (25). While crystals of this large ribosomal subunit had been grown earlier in Ada Yonath's laboratory, these crystals contained a number of significant pathologies that had rendered them resistant to crystal structure determination. These pathologies included 10 micron thin crystals, formation of multiple crystals, extreme non-isomorphism from one crystal to another and a most insidious problem, twinning. We were able to successfully overcome each of these problems and produce an electron density map at 2.4 Å resolution. Data collection, which was done at the APS synchrotron beamline ID 19, involved the collection of 6 million reflections for the 2.4 Å resolution native data set which took only a couple of hours. This rate of data collection is about 10^5 times faster than the rate of data collection from crystals of carboxypeptidase A using diffractometers and laboratory X-ray sources in the mid 1960s when I was a graduate student in the Colonel's lab. Furthermore, we were able to refine the positions of approximately 100,000 atoms to generate a crystallographic free R-factor of 0.22 producing a structure that is significantly more accurate than the structure of carboxypeptidase as established in the 1960s in spite of its being approximately 50 times larger. The technology of macromolecular X-ray crystallography has advanced dramatically in the past 35 years.

The structure of the large ribosomal subunit that we were able to obtain includes 2,833 of the subunit's 3,045 nucleotides and 27 of its 31 proteins (Fig. 8). The domains of its RNAs all have irregular shapes and fit together in the ribosome like the pieces of a three dimensional jig saw puzzle to form a large, monolithic structure. Proteins are abundant everywhere on its surface except in the active site where peptide bond formation occurs and the surface where it contacts the small subunit. Most of the proteins stabilize the structure by interacting with several RNA domains often using idiosyncratically folded extensions that reach into the subunit's interior.

The ribosomal 23S RNA is a large polyanion and it is held in a compact configuration by RNA-RNA, RNA-protein and RNA-metal ion

245

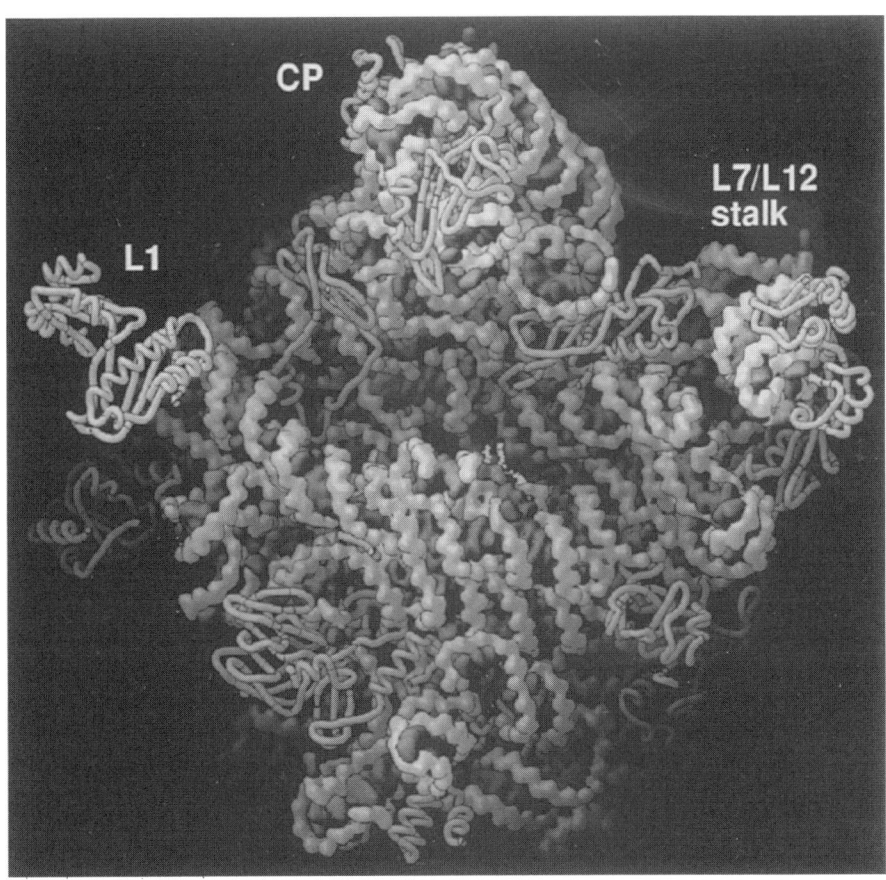

Figure 8. The H. marismortui *large ribosomal subunit in the rotated crown view. The L7/L12 stalk is to the right, the L1 stalk is to the left, and the central protuberance (CP) is at the top. In this view, the surface of the subunit that interacts with the small subunit faces the reader. The RNA is shown in a space-filling surface representation and the protein backbones as worms. (Adapted with permission from reference 27)*

interactions. The RNA-RNA interactions that orient elements of secondary structure to form compact tertiary structures are of two major kinds. The first involves Watson-Crick base pairs between single stranded bases that are located in remote positions in the secondary structure. There are about 100 such interactions. A more prevalent and perhaps energetically more important RNA-RNA interaction involves what we have termed the A-minor motif (26). Adenosines are unusually well conserved throughout the 23S and 5S RNA and are most frequently found in stretches that are designated as single strand in a secondary structure representation. These patches of adenosine are seen to be interacting in the minor groove of RNA helices. We have identified four different types of adenosine interactions in the minor groove, the most prevalent of which we call the type I A minor interaction. In this interaction the adenosine is making snug hydrogen bonding contacts with a G-C base pair via the minor groove. There are usually 2 to 6 stacked A's making such interactions. In total, there are approximately 180 A minor interactions.

The proteins may play the most important role in stabilizing the overall rigidity of the ribosome structure which they do by crosslinking RNA helices. The total surface area of the 27 proteins interacting with RNA is 40 times larger than the surface area of glutaminyl tRNA synthetase interacting with glutaminyl-tRNA. Perhaps the most unexpected aspect of these protein structures is the existence of extended polypeptides that snake in among the RNA helices and only form a specific structure in the presence of the RNA. These extensions are highly basic, highly conserved in sequence and also contain a significant fraction of proline and glycine residues as compared with the globular domains.

We have also been able to bind substrates, intermediates and product analogues to the large ribosomal subunit and have used these complex structures to establish that the ribosome is indeed a ribozyme as well as address the catalytic properties of its all RNA active site (27). All the substrate and product analogues are contacted exclusively by conserved ribosomal RNA residues from domain V of 23S rRNA. There are no protein side chain atoms closer than about 18 Å to the peptide bond being synthesized. The ribosome is able to facilitate the catalysis of peptide bond formation first of all by precisely orienting the α amino group of the amino acid to be added in a position that is adjacent to the carbonyl carbon of the ester linked growing polypeptide chain that it is to attack in a manner that is optimal for peptide bond formation. This positioning of the CCA ends of the tRNA molecules carrying the new amino acid (the A-site tRNA) and the growing polypeptide chain (the P-site tRNA) is largely achieved by Watson-Crick hydrogen bonding interactions between the 3' penultimate cytosines of these tRNA molecules that are bound in the A-site and the P-site by the so-called A-loop and P-loop of ribosomal RNA. All of the substrate positioning in the active site is done by 23S rRNA. We have also observed that one specific base, A2486, is positioned in such a way that it's N3 is hydrogen bonded to the α amino group of the amino acid that is attacking the carbonyl carbon of the ester linked peptidyl-tRNA. We have suggested that the chemical component to the mechanism of peptide bond synthesis may resemble the reverse of the acylation step in the serine proteases with the base of 2486 playing the same general acid base role as Histidine57 and chymotrypsin.

The ribosome is one of the major targets of antibiotics, and many of the clinically important antibiotics target the large ribosomal subunit. We have been able to make crystalline complexes between the 50S subunit and 7 antibiotics and from the structures of these complexes we are able to derive their mechanisms of action (J. Hansen, P.B. Moore and T.A. Steitz, unpublished). Among the more interesting of the antibiotics are the macrolides which appear to bind in the polypeptide exit tunnel just below the peptidyl-transferase active site. The macrolides appear to inhibit polypeptide synthesis by blocking the egress of the growing polypeptide from the tunnel, a sort of molecular constipation. The existence of these complex structures at high resolution appears to be sufficient to form the basis of structure based drug design. Indeed, on the basis of the structures of the 50S ribosomal subunit and these antibiotic complexes, we have founded a small biotech company called Rib-X Pharmaceuticals, Inc. to exploit the potential of these crystal structures for the design of novel, more powerful antibacterial drugs.

When I began my graduate studies on the crystal structure of carboxypeptidase with the Colonel in the 1960s it never occurred to me that there was any possibility that X-ray crystallography would ever be medically useful. It only appeared to be powerful tool to peer into the inner workings of biological macromolecules. However, some 35 years later, it looks as if this powerful tool of basic research is going to become important in the battle against bacterial diseases.

Acknowledgments - The research summarized here was supported, in part, by NIH grants (GM-22778, GM-57510, AI43896).

References

1. Crick, F.H.C. *Symposium Experimental Biology* **1958**, *12, 138.*
2. Ollis, D.L., Brick, P., Hamlin, R., Xuong, N.G., Steitz, T.A. *Nature* **1985**, *313, 762.*
3. Freemont, P.S., Friedman, J.M., Beese, L.S., Sanderson, M.R., Steitz, T.A. *Proc. Natl. Acad Sci USA* **1988** , *85, 8924.*
4. Derbyshire, V., Grindley, N.D.F., Joyce, C.M. *EMBO J* **1991**, *10, 17.*
5. Beese, L.S., Steitz, T.A. *EMBO J* **1991**, *10, 25.*

6. Brautigam, C.A., Steitz, T.A. *J Mol Biol* **1998**, *277, 363.*
7. Steitz, T.A., Steitz, J.A. *Proc. Natl Acad Sci USA* **1993**, *90, 6468.*
8. Eom, S.H., Wang, J., Steitz, T.A. *Nature* **1996**, *382, 278.*
9. Doublié, S., Tabor, S., Long, A.M., Richardson, C.C., Ellenberger, T. *Nature* **1998**, *391, 251-258.*
10. Kiefer, J.R., Mao, C., Braman, J.C., Beese, L.S. *Nature* **1998**, *391, 304.*
11. Wang, J., Sattar, A.K.M.A., Wang, C.C., Karam, J.D. *Cell* **1997**, 89, 1087.
12. Shamoo, Y., Steitz, T.A. *Cell* **1999**, *99, 155.*
13. Franklin, M.C., Wang, J., Steitz, T.A. *Cell* **2001**, 105, 657.
14. Steitz, T.A. *Curr Opin Struct Biol* **1993**, *3, 31.*
15. McKay, D.B., Steitz, T.A. *Nature* **1981**, *290, 744.*
16. Anderson, W.F., Ohlendorf, D.H., Takeda, Y., Matthews, B.W. *Nature* **1981**, 290, 750.
17. Schultz, S.C., Shields, G.C., Steitz, T.A. *Science* **1991**, *253, 1001.*
18. Cheetham, G.M.T., Jeruzalmi, D., Steitz, T.A. *Nature* **1999**, *399, 80.*
19. Cheetham, G.M.T., Steitz, T.A. *Science* **1999**, *286, 2305.*
20. Jeruzalmi, D., Steitz, T.A. *EMBO J.* **1998**, *17, 4101.*
21. Kohlstaedt, L.A., Wang, J., Friedman, J.M., Rice, P.A., Steitz, T.A. *Science* **1992**, *256, 1783.*
22. Rould, M.A., Perona, J.J., Söll, D., Steitz, T.A. *Science* **1989**, *246, 1135.*
23. Rould. M.A., Perona, J.J., Steitz, T.A. *Nature* **1991**, *353, 213.*
24. Rath.,V.R., Silvian, L.F., Beijer, B., Sproat, B.S., Steitz, T.A. *Structure* **1998**, 6, 439.
25. Ban, N., Nissen, P., Hansen, J., Moore, P.B., Steitz, T.A. *Science* **2000**, *289, 905.*
26. Nissen, P, Ippolito, J.A., Ban, N., Moore, P.B., Steitz, T.A. *Proc Natl Acad Sci USA* **2001**, *98, 4899.*
27. Nissen, P., Hansen, J., Ban, N., Moore, P.B. and Steitz, T.A. *Science* **2000**, *289, 930.*
28. Steitz, T.A. *J Biol Chem* **1999**, *274, 17395.*
29. Cheetham, G.M.T., Steitz, T.A. *Curr Opin Struc Biol* **2000**, *10, 117.*
30. Wang, J., Smerdon, S.J., Jäger, J., Kohlstaedt, L.A., Rice, P.A., Friedman, J.M., Steitz, T.A. *Proc Natl Acad Sci USA* **1994**, *92, 7242.*

Chapter 16

Analysis of Intradomain Signaling in the Multifunctional Protein CAD Using Novel Hybrids and Chimeric Molecules

David R. Evans and Hedeel I. Guy

Department of Biochemistry and Molecular Biology,
Wayne State University School of Medicine, Detroit, MI 48201

CAD is a 1.5 Mda complex that catalyzes the first three steps in *de novo* pyrimidine biosynthesis in mammalian cells. The protein consists of six copies of a 243 kDa polypeptide that is organized into 15 domains, subdomains and linkers each with a specific catalytic or regulatory function. Most of these domains have been subcloned and expressed in *E. coli* where they fold into stable, fully functional proteins. While each domain functions autonomously, interdomain signaling modulates the reactions occurring on different domains to ensure that biosynthesis proceeds in a coordinated fashion. Insights into the signaling mechanism have been provided by the analysis of several hybrid and chimeric molecules constructed by combining domains and subdomains of the mammalian, yeast and *E. coli* proteins in novel ways.

Multifunctional proteins are large polypeptides that have several catalytic or regulatory functions consolidated on a single polypeptide chain. They represent a common class of proteins, especially in eukaryotic organisms, and are thought to have evolved by ancestral gene duplication, translocation and fusion events. Our interest in complex proteins comprised of multiple interacting components originated with the fascinating structural and biochemical studies of aspartate transcarbamoylase carried out under the thoughtful guidance and inspiration of Professor William Lipscomb during the seventies.

CAD Structure and Function

CAD catalyzes the initial steps (Figure 1) of the *de novo* pyrimidine biosynthetic pathway in mammalian cells, in which glutamine, ATP, bicarbonate and aspartate are converted to dihydroorotate (*1*). Dihydroorotate is subsequently oxidized by mitochondrial dihydroorotate dehydrogenase and then converted in two steps to UMP by a second multifunctional protein that has PRPP transferase and decarboxylase activities.

GLN	glutamine $\longrightarrow$	glutamate + NH_3
CPS	NH_3 + 2 ATP + HCO_3^- $\longrightarrow$	carbamoyl phosphate + 2ADP + P_i
ATC	carbamoyl phosphate + aspartate $\longrightarrow$	carbamoyl aspartate + P_i
DHO	carbamoyl aspartate $\longrightarrow$	dihydroorotate + H_2O

Figure 1. Reactions Catalyzed by CAD

CAD is also the major locus of regulation of *de novo* pyrimidine biosynthesis, controlling the flux of metabolites through the pathway in response to changes in the demand for pyrimidine nucleotides. CAD was first isolated by Coleman et al. (2) from an overproducing mammalian cell line selected by resistance to the potent aspartate transcarbamoylase inhibitor, N-phosphonacetyl-L-aspartate (PALA). These authors found that the protein contained multiple copies of a large polypeptide chain (>220 kDa) that catalyzed glutamine-dependent carbamoyl phosphate synthetase (Gln-CPSase), aspartate transcarbamoylase (ATCase) and dihydroorotase (DHOase) activities. The availability of these cell lines opened up this area for further investigation.

Domain Structure of CAD

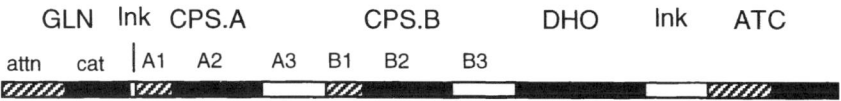

Figure 2. Domain Structure of CAD. Each domain, submdomain and the two major linkers (lnk) are represented by segments approximately proportional to their size. Carbamoyl phosphate synthesis involves the concerted action of the GLN domain and the two CPS subdomains, CPS.A and CPS.B. The second and third steps of the pathway are catalyzed by the DHO and ATC domains.

Determination of the Domain Structure

Early controlled proteolysis experiments (*3, 4*) clearly showed that the CAD polypeptide was organized into discrete functional domains. The complete domain structure of the polypeptide was mapped using three approaches. Proteolytic fragments were isolated, partially sequenced and their function was determined (*6-8*). The CAD cDNA (*5, 9-12*) was sequenced and the identity of the domains and the domain junctions were determined by comparison with the sequence of monofunctional bacterial proteins. More recently, many of the domains and subdomains were cloned and characterized (*13-21*). The polypeptide (Figure 2) consists of 2225 residues organized into four major functional domains that are further divided into 15 subdomains and linkers (*8*). The 40 kDa GLN domain catalyzes the hydrolysis of glutamine and the transfer of ammonia to the synthetase domains. The CPS domain consists of two homologous 60 kDa subdomains (*22*), CPS.A and CPS.B that together catalyze carbamoyl phosphate formation from ammonia, ATP and bicarbonate. The ATCase domain, which catalyzes carbamoyl aspartate synthesis, is connected to the remainder of the polypeptide by a long hydrophilic linker. The 44 kDa DHO domain, that catalyzes the cyclization of carbamoyl aspartate to form dihydroorotate, is spliced directly onto the end of CPS.B. While higher and lower oligomers are present, the major species is the hexamer (*23*) of the 243 kDa CAD polypeptide.

Carbamoyl Phosphate Synthesis

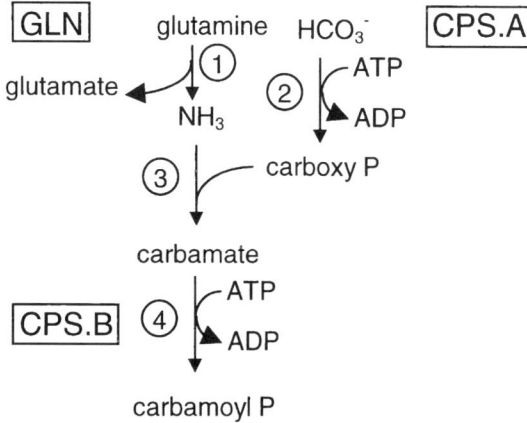

Figure 3. Partial Reactions in Carbamoyl Phosphate Synthesis

The synthesis of carbamoyl phosphate is a complex process (Figure 3) involving four partial reactions (24-25) and the concerted action of three domains. Glutamine is hydrolyzed by the GLN domain, glutamate is released and ammonia is transferred to CPS.A. At the same time, bicarbonate reacts with ATP to form the activated intermediate, carboxy phosphate, at the active site of CPS.A. Carboxy phosphate and ammonia react spontaneously (reaction 3) and the resulting carbamate is transferred to CPS.B where it is phosphorylated by a second ATP (reaction 4) to form carbamoyl phosphate. The partial reactions can be individually assayed as a glutaminase (reaction 1), HCO_3^--dependent ATPase in the absence of a nitrogen donating substrate (reaction 2) and carbamoyl phosphate-dependent ATP synthesis (the reverse of reaction 4). All of the intermediates are labile and must be sequestered within the complex. NH_3, not ammonium ion, is the substrate and would be protonated if released. Carboxy phosphate has a half life of 70 ms in aqueous solution while carbamate dissociates to NH_3 and CO_2 with a half life of approximately 28 ms (26).

Structural Studies

The large size and polymorphic oligomeric structure of CAD has made it difficult to obtain diffraction quality crystals. However, the strong sequence similarity made it possible to model the structure of the major domains using the *E. coli* x-ray structures (26, 30, 69) as tertiary templates. The GLN-CPS subdomain structure predicted by the biochemical studies was clearly visible in the *E. coli* CPSase x-ray structure (26). Since the intermediates are labile, many investigators thought that the active sites of the GLN, CPS.A and CPS.B would be in close proximity to ensure efficient utilization of these metabolites. The x-ray structure conclusively showed that this was not the case. The active sites are far apart and connected via a narrow intramolecular tunnel. Thus, the intermediates are not released from the complex but rather pass through the interior of the molecule a total distances of about 96 Å. There is also good evidence (27-29) that there is channeling of carbamoyl phosphate between its site of synthesis on CPS.B to the ATC domain, although the mechanism remains unknown. Lipscomb and his associates (30) have solved the structure of *E. coli* ATCase and its complex with many substrate analogs and effectors. The ATCase domain of CAD (31), closely resembles the catalytic subunit of the *E. coli* enzyme, a trimer composed of 34 kDa polypeptides. The structure of *E. coli* DHOase has also been recently determined (69), although it is a type II DHOase that differs in size and sequence from the type I DHOase found in CAD (32).

Nevertheless, the overall tertiary fold of the mammalian domain is likely to be similar to its bacterial counterpart. While the structure of the CAD domains is known with some certainty, there is no information regarding their arrangement within the CAD complex.

Regulation of CAD Activity

The reaction catalyzed by CAD CPSase is the initial rate limiting step of the *de novo* pyrimidine biosynthetic pathway (*33*). The CPSase activity is feedback inhibited by the end product UTP and allosterically activated by PRPP, a substrate for purine biosynthesis as well as a subsequent step in the pyrimidine pathway (*1*). Moreover, protein kinase A phosphorylates two sites on the CAD polypeptide (*34*). One site is located within the CPS B3 subdomain and the other in the DHO ATC interdomain linker (Figure 2). Phosphorylation has little effect on the activity but the response to allosteric ligands is appreciably modulated. UTP inhibition is virtually abolished (34) and the binding of PRPP is also diminished (*21*). Thus, PKA mediated phosphorylation alters the response to allosteric effectors in a way that would be expected to have opposing effects on pyrimidine biosynthesis in the cell. In contrast, the phosphorylation of a single site in CPS A1 by MAP kinase (*35*) promotes the synthesis of carbamoyl phosphate. The catalytic activity is unchanged but UTP inhibition is abolished and PRPP activation increases about 5-fold. The interplay between these two signaling cascades is likely to be responsible for the cell cycle dependent activation and down regulation of pyrimidine biosynthesis. Neither ATCase nor DHOase activities are regulated in the mammalian complex.

Interdomain Signaling

While each of the individual CAD domains and subdomains can fold and function autonomously, there is extensive interdomain signaling that modulates their function. The binding of substrates and effectors to one domain can profoundly influence the activity of remote regions within the molecule. The discovery that the CAD domains could be cloned and expressed with retention of function and could be reconstituted into fully functional complexes that closely resembled the parent molecule, made it possible to construct some interesting hybrid and chimeric proteins that have revealed a great deal about the structural organization and interdomain communication in the CAD complex.

Glutaminase Domain (GLN)

Organization of the GLN domain

The glutaminase domain consists of two subdomains initially identified by sequence comparisons of several CPSases and other amidotransferases (36-38). The carboxyl half of the domain is homologous to the trpG or triad-type amidotransferases (39) and contains several conserved residues implicated in catalysis. The amino half has a unique sequence that does not resemble any other protein in the database. The GLN subdomains were clearly identifible in the *E. coli* CPSase x-ray structure (26).

Mechanism of Glutamine Hydrolysis

The mechanism of glutamine hydrolysis is reminiscent of the thiol proteases. A thioester intermediate (Figure 4) is formed with the release of ammonia and then hydrolyzed in the subsequent step regenerating the active site for another round of catalysis (40-43). The formation and breakdown of the thioester involves the participation of a catalytic triad.

$$\text{GLN-SH} + \text{gln} \underset{}{\overset{K_s}{\rightleftharpoons}} \text{GLN-SH.gln} \xrightarrow{k_3} \text{GLN-S-glu} \xrightarrow{k_4} \text{GLN-SH} + \text{glu}$$
$$\downarrow$$
$$NH_3$$

Figure 4. Mechanism of Glutamine Hydrolysis. The γ-glutamyl thioester is formed (k_3) with the release of NH_3 and hydrolyzed (k_4) in a subsequent step.

In CAD, the members of the catalytic triad were identified by sequence comparisons (11, 12) and mutagenesis (44). Cys252 forms a thioester intermediate by nucleophilic attack on the γ-carbonyl of glutamine. His336 serves as a general base that activates the nucleophile and also participates in the hydrolysis of the thioester. The primary role of Glu338 is to optimally position His336. These residues have the correct juxtaposition in *E. coli* CPSase (26) and the CAD GLN model structure.

Coupling of Glutamine Hydrolysis and Bicarbonate Activation

The hydrolysis of glutamine and the activation of bicarbonate are parallel reactions that must be coupled to insure that they proceed in a coordinated fashion. Control implies a need for an attenuation mechanism to allow the up and down regulation of catalytic activity. In the absence of other substrates needed for carbamoyl phosphate synthesis, glutamine hydrolysis proceeds very slowly (*24, 41, 42*). Bicarbonate alone or ATP alone does not have an appreciable effect on the turnover (Table I), although bicarbonate increases the K_m for glutamine 6-fold. Only when both substrates bind to the CPS domain is the GLNase activity fully realized. The K_m for glutamine is the same as in the absence of substrates, but the k_{cat} increases 14-fold. In mammals, the intracellular concentration of bicarbonate is likely to be constant and saturating, so a more physiological comparison may be the effect of ATP on the enzyme saturated with bicarbonate. Under these circumstances, ATP induces a 38-fold increase in the apparent second order rate constant, k_{cat}/K_m.

Table I. Kinetic Parameters for Glutamine Hydrolysis

Protein	Ligands	K_m	k_{cat}	k_{cat}/K_m
		μM	s^{-1}	$M^{-1}s^{-1}$
CAD	none	95	0.14	1470
CAD	ATP	76	0.28	3670
CAD	HCO_3^-	599	0.32	530
CAD	ATP, HCO_3^-	96	1.92	20,000

In the absence of ATP and bicarbonate, the thioester intermediate could be isolated by chromatography or acid precipitation of the protein. However, the intermediate does not accumulate when the other substrates needed for carbamoyl phosphate synthesis are present. Presteady state kinetics of CAD (*44, 45*) and *E. coli* CPSase (*46*) indicate that substrate binding to CPS.A primarily increases the rate of breakdown of the thioester intermediate. The non-hydrolyzable ATP analogue, 5'-adenylylimidodiphosphate (AMP-PNP), also activates the GLN domain indicating that the conformational change that up-regulates glutamine hydrolysis is driven by the ATP binding energy, not the hydrolysis of the nucleotide. Thus, when bicarbonate and ATP are limiting, the thioester accumulates tying up the active site and preventing further rounds of catalysis and the futile hydrolysis of glutamine.

Construction of a Mammalian E. coli Hybrid

The CAD GLN domain (16) was cloned and expressed in an *E. coli* strain lacking an endogenous CPSase GLN subunit. The recombinant plasmid could complement the defect in the host strain, but when isolated the mammalian GLN domain was found to have barely detectable glutaminase activity because of a high K_m for glutamine and a low k_{cat} (Table II). This apparently contradictory result was explained when it was discovered that the mammalian GLN domain forms a fully functional hybrid with the *E. coli* CPSase synthetase subunit.

Table II. Kinetics of Glutamine Hydrolysis of the Mammalian GLN- *E. coli* CPS Hybrid

Protein	Ligands	K_m	k_{cat}	k_{cat}/K_m
		μM	s^{-1}	$M^{-1}s^{-1}$
GLN^M	none	4270	0.02	4.2
$GLN^M\ CPS^E$	none	90	0.31	3,433
$GLN^M\ CPS^E$	ATP, HCO_3^-	92	3.15	34,200

When the purified mammalian GLN domain (GLN^M) is mixed with the *E. coli* CPSase synthetase subunit (CPS^E), a stable 1:1 stoichiometric complex is formed (Figure 5). Upon association, the k_{cat} for glutamine hydrolysis increases 17-fold and the K_m decreases 47-fold suggesting that interdomain interactions are important for optimal activity of the GLN domain. The most interesting observation was that the functional linkage between the GLN and CPS subdomains is intact. In the presence of saturating ATP and bicarbonate, the activity of the GLN domain increased 10-fold without any significant change in the K_m^{gln}. The kinetic parameters of the hybrid in the presence of ATP and bicarbonate are comparable to those of the native *E. coli* enzyme. The coupling of glutamine hydrolysis and carbamoyl phosphate synthesis and the conservation of the interdomain linkage in the hybrid suggest that the subunit interfaces must be nearly identical in the eukaryotic and prokaryotic proteins.

Cloning of the GLN Subdomains

The substructure of the CAD GLN was investigated by cloning and expression of the two halves of the domain (18). The 21 kDa carboxyl half was found to be hyperactive and designated the catalytic subdomain (GLN^{cat}). The

k_{cat} for the hydrolysis of glutamine is 347-fold higher (5.7s^{-1}) and the K_m 40-fold lower (Km = 110 µM) than the isolated full length GLN domain. Thus, all of the residues involved in catalysis and glutamine binding are located within the subdomain. The catalytic subdomain does not form a stable complex with the *E. coli* CPS subunit. However, titration of the *E. coli* subunit with increasing amounts of the mammalian GLN subdomain restores glutamine dependent CPSase. Maximum activity was observed at a molar ratio of GLN^{cat}/CPS of 12 indicating that the catalytic subdomain interacts weakly with the CPS domain (Figure 5). Moreover, ATP and bicarbonate do not affect the glutaminase activity of the GLN^{cat} indicating that the interdomain linkage is not present.

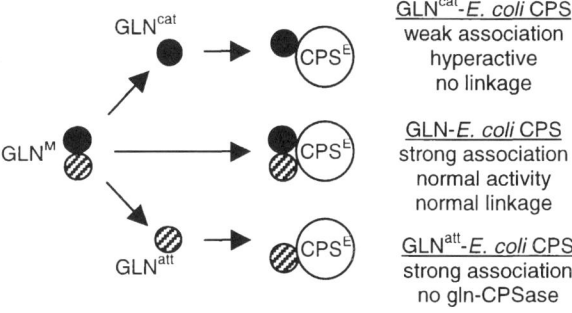

Figure 5. Mammalian GLN-E. coli CPS Hybrids

The amino half of the GLN domain was also expressed as a soluble compactly folded protein in *E. coli*. As expected, the 19 kDa subdomain had no catalytic activity but could form a stable complex with the *E. coli* CPSase synthetase subunit (Figure 5), indicating that it is crucial for the formation of the GLN CPS complex. This region of the molecule, now designated the attenuation subdomain, suppresses the high intrinsic catalytic activity of the catalytic subdomain. Excision of the attenuation subdomain from GLN results in a 14,000-fold increase in the apparent second order rate for glutamine hydrolysis.

The attenuation domain is likely to be an important element of the linkage by suppressing the breakdown of the thioester intermediate and preventing the enzyme from embarking on another round of catalysis. The binding of ATP and bicarbonate to CPS.A induces a conformational change that is transmitted to the attenuation subdomain with which it intimately interacts and disrupts the interactions between the attenuation and catalytic subdomains, thus relieving inhibition. The thioester once formed rapidly breaks down and the enzyme turns over.

Attenuation of the GLN Activity

Replacement of Cys252 in the GLN domain with alanine completely abolishes catalytic activity. Mutants of His336 and Glu338 also appreciably reduce the glutamine dependent activity of the mammalian *E. coli* hybrid, but significantly, the small residual activity is not stimulated by ATP and bicarbonate (*44*). This observation suggested that the catalytic residues are involved in the interdomain signaling. The only residue of the attenuation subdomain close to the active site of the catalytic subdomain is Ser44 (Figure 6). Its proximity to the catalytic triad suggested that it might be a catalytic residue. However, replacement of Ser44 with alanine resulted in a 10-15-fold increase in the k_{cat}, a rate enhancement similar to the change induced by the binding of ATP and bicarbonate to the CPS domain in CAD. The mutant was no longer responsive to ATP and bicarbonate, an observation that supports the idea that the attenuation domain suppresses the catalytic activity of the catalytic subdomain. Moreover, glutamine hydrolysis and bicarbonate activation are no longer coordinated in the mutant. The ratio of glutamine hydrolysis to carbamoyl phosphate formation, which is 1:1 in the wild type protein, is 3:1 in the mutant and the excess ammonia leaks out of the complex. We conclude that Ser44 in the attenuation domain is not a catalytic residue in the usual sense but rather is an essential element in the regulatory linkage that phases glutamine hydrolysis and carbamoyl phosphate synthesis.

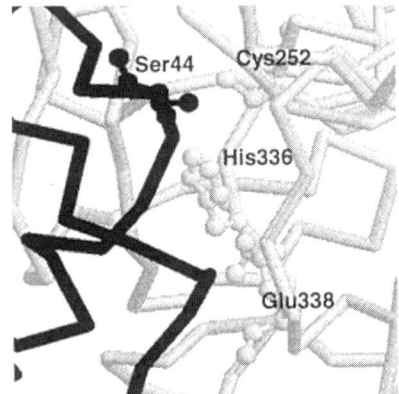

Figure 6. Active Site of CAD GLN Domain

CPS Subdomains

Cloning and Expression of CPS.A and CPS.B

Although CPS.A and CPS.B have for the most part identical tertiary folds (26), there is good evidence that the subdomains are specialized in the sense that CPS.A catalyzes the activation of bicarbonate, while CPS.B phosphorylates carbamate (47-52). The most compelling proof (53) came from mutagenesis studies showing that the introduction of a disabling mutation into CPS.A abolishes the bicarbonate-dependent ATPase, while CPS.B domain mutants cannot synthesize ATP from carbamoyl phosphate and ADP. We (19) cloned the GLN-CPS.A domain of CAD and, as a control, GLN-CPS.B (Figure 7).

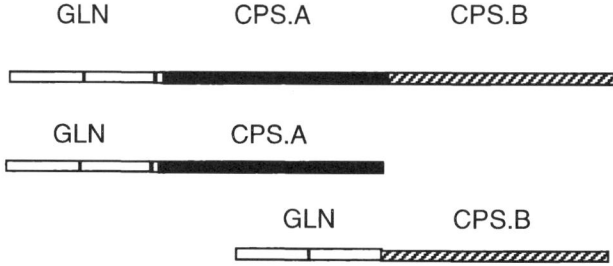

Figure 7. Cloning of the CPS.A and CPS.B Domains

The expectation was that GLN-CPS.A would catalyze the bicarbonate dependent ATPase, hydrolyze glutamine and, if we were fortunate, would transfer ammonia to CPS.A for carbamate formation. GLN-CPS.B was expected to catalyze carbamoyl phosphate dependent ATP synthesis and hydrolyze glutamine non-productively. The striking result was that each of the purified half molecules could catalyze both partial reactions and, most surprisingly, both ammonia and glutamine dependent carbamoyl phosphate synthesis (Table III). The rate of the overall reaction using either ammonia or glutamine as a nitrogen donating substrate is comparable to the values observed for intact CAD. Moreover, the rate of ATP hydrolysis in the presence of glutamine is two fold greater than the rate of carbamoyl phosphate formation, an observation consistent with the 2:1 stoichiometry for the biosynthetic reaction.

Table III. Activity of the Isolated CPS Subdomains

Activity	CAD	GLN-CPS.A	GLN-CPS.B
	(μmol/min/mg)		
NH_3-CPSase	0.263	0.201	0.368
Gln- CPSase	0.180	0.164	0.245
ATPase - gln	0.112	0.169	0.241
ATPase + gln	-	0.328	0.482

Hypothesis

How can this result be rationalized in view of the strong evidence that the CPS.A and CPS.B subdomains are specialized? The simplest explanation is that the half molecule dimerizes (Figure 8). In the CPS.A dimer, one molecule of CPS.A occupies the position of CPS.B in the native molecule. The dimer is symmetrical and the monomers are functionally equivalent. In any given catalytic cycle (Figure 8A), glutamine, ATP and bicarbonate bind to one of the

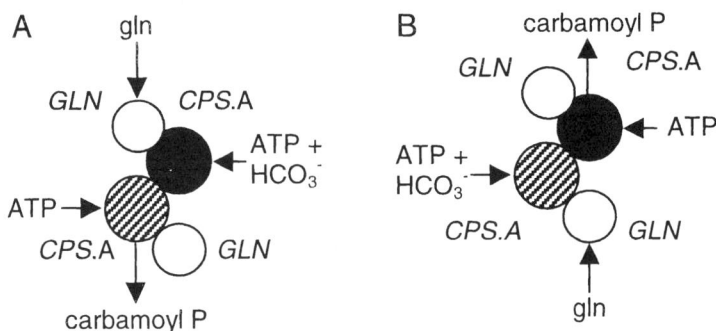

Figure 8. Carbamoyl Phosphate Synthesis by GLN-CPS.A

monomers (solid sphere) and catalyze the formation of carboxy phosphate. The other monomer (hatched sphere) catalyzes the phosphorylation of carbamate by default. In a subsequent catalytic cycle (Figure 8B), the other monomer may bind the substrates and the roles are reversed.

The tenets of this model are 1) GLN-CPS.A forms dimers 2) each monomer catalyzes both partial reactions but only the dimer can catalyze the overall reaction, 3) one of the monomers takes the place of CPS.B in the native molecule, 4) the function of the CPS subdomain in the native molecule depends on its juxtaposition to the GLN domain.

Pressure Induced Dissociation of the Hybrid Molecules

The isolated CPS subdomains were shown (19) by gel permeation chromatography and chemical crosslinking to be dimers. Moreover, CPS.A and CPS.B have an identical tertiary fold. Unpublished modeling studies showed that one CPS.A monomer should be able to assume the same juxtaposition relative to the other as CPS.B in the native enzyme. To determine whether dimer formation is a prerequisite for catalysis of the overall reaction and whether the monomers could catalyze the partial reactions, a method was needed to gently dissociate the dimer without disrupting the structure of the monomer. We found (54) that when subjected to elevated pressure, the dimer reversibly dissociated (Figure 9).

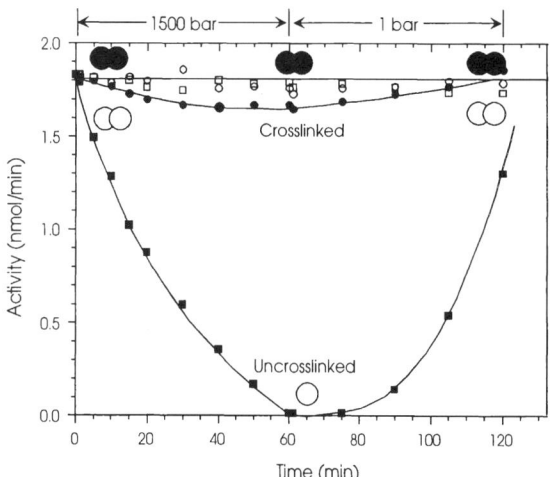

Figure 9. Pressure induced dissociation of the GLN-CPS.B dimer. The uncrosslinked dimer (open spheres) was subjected to 1500 bar (-■-) or incubated at atmospheric pressure (-□-). The crosslinked dimer (filled spheres) was also subjected to 1500 bar (-●-) or atmospheric pressure (-○-).

When subjected to 1500 bar, the GLN-CPS.B dimer gradually lost NH_3-CPSase activity (Figure 9) and chemical crosslinking showed that the dimer dissociated into monomers. When atmospheric pressure was restored, the activity was recovered in a concentration dependent manner. There was no loss of activity when the sample was incubated at atmospheric pressure. If the sample was first crosslinked and then subjected to pressure, there was no dissociation and little loss of catalytic activity. Moreover, the monomer, produced at high pressure, could catalyze both ATP dependent partial reactions. As predicted by the hypothesis, the monomers can catalyze the partial reactions but the dimer is required for the overall synthesis of carbamoyl phosphate.

The GLN Domain determines the function of the CPS domains

The hypothesis states that the proximity to the GLN domain determines the reaction catalyzed by the CPS subdomain. The GLN domain is in intimate contact with CPS.A but there are no direct interactions with CPS.B. According to this interpretation, CPS.A catalyzes the bicarbonate dependent ATPase reaction because of this proximity to the source of ammonia, while CPS.B, which is far from the GLN domain, catalyzes the phosphorylation of carbamate. To test the idea that the GLN domain directs the function of the CPS domain, the mutagenesis experiments previously conducted with *E. coli* CPSase (53) were repeated with the same results. However, if the GLN domain is removed and a disabling mutation is introduced into CPS.A, rather than selectively abolishing the ATPase activity, both the ATPase and ATP synthetase reactions were inhibited by approximately 50%. These results suggest that, in the absence of the GLN domain, CPS.A and CPS.B of the intact *E. coli* CPSase can catalyze both ATP dependent partial reactions. However, when the CPS subunit is associated with the GLN domain, CPS.A is specialized for the activation of bicarbonate while CPS.B phosphorylates carbamate. Thus, we have the interesting situation in which the role of two functionally equivalent domains is determined by their juxtaposition relative to a third domain in the complex.

Other Rearranged Molecules

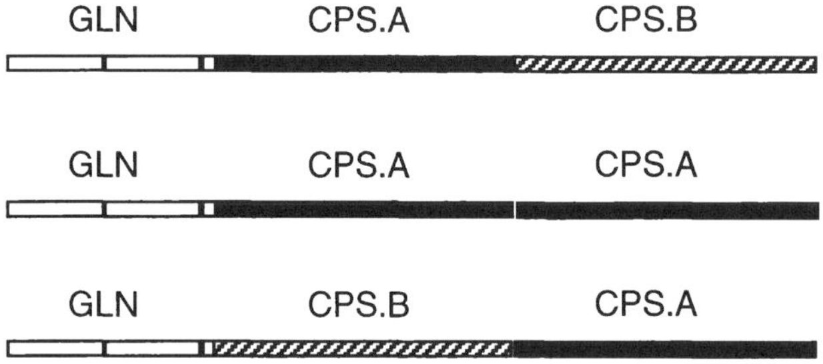

GLN   CPS.A   CPS.B

GLN   CPS.A   CPS.A

GLN   CPS.B   CPS.A

Figure 10. Rearranged CPSases

Further evidence that the subdomains are functionally equivalent came from domain swapping experiments (Figure 10) that produced two unusual constructs. One molecule consisted of the CAD GLN domain fused to two tandem copies of CPS.A and a second in which the location of the CPS subdomain were switched to give a molecule in the arrangement GLN-CPS.B-CPS.A. Both of these constructs were fully functional CPSases. Interestingly, only the latter species retains sensitivity to allosteric effectors.

The functional equivalence of the CPS subdomains appears to be a general phenomenon occurring in mammalian, *E. coli* (*19*) and yeast (*55*) CPSases. The CPS from *Aquifex aeolicus* may be a naturally occurring analog of the mammalian half molecules. Is this organism, there is no full length CPSase such as that found in *E. coli* and other bacterial species. Instead, there are two genes each a homologue (*68*) to one of the two major CPS subdomains and a separate gene encoding a GLN domain homologue. We have recently cloned and expressed these three polypeptides in *E. coli*. As for CPSase from other species, the isolated CPS.A and CPS.B subdomains can catalyze both ATP-dependent partial reactions. *A. aeolicus* GLN, CPS.A and CPS.B subunits associate to form a 1:1:1 stoichiometric complex that catalyzes glutamine-dependent CPSase activity. Unlike the situation in other CPSases, the subdomains do not associate into functional homodimers that can catalyze the overall reaction, a necessary requirement to ensure the correct posttranslational assembly of the constituent domains.

Further Dissection of the CPS subdomains

CPS.A and CPS.B are each comprised of three smaller subdomains (Figure 2), A1, A2, A3 and B1, B2, B3, respectively. Separately cloned A2 and B2 are catalytic subdomains (*20*) that can catalyze the formation of carbamoyl phosphate from NH_3, ATP and bicarbonate. While these species dimerize, they lack intermolecular tunnels and have a catalytic mechanism similar to carbamate kinases that synthesize carbamoyl phosphate by the phosphorylation of carbamate formed chemically from ammonia and bicarbonate in solution. The designation of A2 and B2 as catalytic subdomains is consistent with the x-ray structure of the *E. coli* enzyme that showed ADP and an ATP analogue bound to these bilobal subdomains (26, 70). The function of A3 is unknown, while B3, as discussed below, is the major locus of regulation. Comparison of the kinetics of A1-A2 and A2 suggest that A1 is an attenuation subdomain that suppresses the catalytic activity of A2. As in the case of the GLN domain, the coordination of reactions occurring on the GLN, CPS.A and CPS.B requires a mechanism that attenuates the activity of each catalytic domain. It is likely that A1 and B1 serve this function in the CPS.A and CPS.B subdomains.

Regulatory Subdomain

E. coli and CAD CPSase have a homologous sequence and structure, catalyze the same series of reactions by the same mechanism, but the mode of regulation is very different. The *E. coli* enzyme is inhibited by UMP and activated by ornithine and IMP (56). The mammalian enzyme is subject to allosteric regulation by UTP and PRPP and in addition is controlled by PKA and MAP kinase phosphorylation. Several lines of evidence (57-64) suggested that the B3 region at the extreme carboxyl end of CPS.B binds allosteric effectors. In the case of CAD, B3 was identified (65) as a regulatory subdomain by constructing an interesting chimeric molecule.

Construction of the Mammalian *E. coli* chimera

A domain swapping experiment (65) was conducted in which the postulated regulatory domain of the *E. coli* CPS *car*B gene was deleted and replaced with the corresponding region of the CAD cDNA. The chimeric protein (Figure 11, R1) was expressed at high levels in *E. coli* but was aggregated. Nevertheless, a kinetic analysis of the properties was revealing. The steady state kinetic parameters were similar to the *E. coli* CPSase as expected since the catalytic domains are derived from the bacterial enzyme. However, the regulation of the molecule closely resembles the mammalian enzyme. The chimera is no longer sensitive to UMP, but is modulated by the mammalian allosteric effectors, UTP and PRPP.

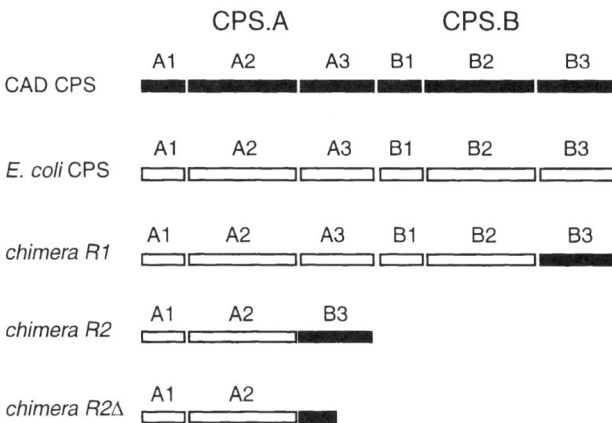

Figure 11. Mammalian E. coli Regulatory Chimeras

Unexpectedly, the R1 chimera had sensitivity to the *E. coli* allosteric activator, ornithine. However, the x-ray structure (*26*) subsequently showed that ornithine is not bound exclusively to the regulatory subdomain, but rather bridges the catalytic and regulatory subdomains. These experiments clearly demonstrated that the mammalian allosteric effectors bind to B3, a separately folded regulatory subdomain located at the carboxyl end of CPS.B.

Regulation of CPS.A

There is kinetic evidence (*66-67*) that the target of allosteric effectors is CPS.B. Since the CPS subdomains are functionally and structurally equivalent, we were curious as to whether CPS.A could be placed under allosteric control. To determine whether this occurs, a second chimeric molecule (R2) was constructed (*21*) in which the mammalian regulatory domain (B3) replaced the A3 subdomain in *E. coli* CPS.A (Figure 11). The control mechanisms (Figure 12) are nearly the same as that observed for the mammalian CAD. The chimera is inhibited by UTP and activated by PRPP although the affinity for the latter ligand is somewhat lower than in the native molecule. While chimera R1 was catalytically active, its aggregation made it a poor substrate for protein kinase A. The second chimera, R2 is monodisperse and can be readily phosphorylated. Protein kinase A abolishes UTP inhibition and reduces the affinity of the protein for PRPP, the same effect observed in CAD.

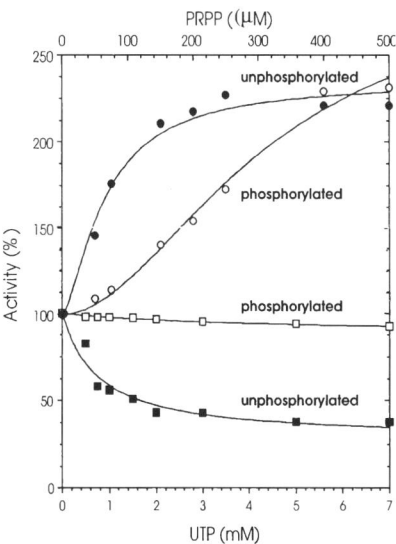

Figure 12. Control Mechanism of the Regulatory Chimera R2

A deletion mutant (Figure 11) was constructed in which the carboxyl half of the regulatory domain in the R2 chimera was deleted. The resulting construct bound allosteric inhibitors nearly as well as the parent molecule, but the allosteric transitions were abolished. Thus, the amino half of the regulatory subdomain binds allosteric effectors, while the carboxyl half is crucial for transmitting the allosteric signal to the catalytic domain. This result is consistent with the x-ray structure of *E. coli* CPS and the CAD model that indicates that most of the interactions between B3 and B2 (or A2 in the case of the chimera) involve residues in the carboxyl half of the regulatory domain.

These experiments suggest that the regulatory subdomain is an exchangeable ligand binding module that can control both CPS.A and CPS.B subdomains. CPS.B is not unique in its sensitivity to allosteric effectors. There has been no differentiation in structure or function that have rendered it subject to control, rather it is the locus of regulation in the native molecule as a consequence of its proximity to the regulatory domain.

Conclusions

Studies of the hybrid and chimeric molecules of CAD domains and other complex proteins make it possible to draw a few general principles about the interdomain interactions.

- Multifunctional proteins consist of a hierarchy of autonomously folded structural domains and subdomains each with a specific function.
- The domains function autonomously but there is interdomain signaling within the complex that modulates their function.
- The activity of the catalytic domains is often up and down-regulated by interactions with specific attenuation domains.
- The function of the domain is often determined by its juxtaposition to other domains in the complex.

This research was supported by grants from the National Institutes of Health GM47399 and GM/CA60371 and the National Science Foundation MCB-9810325.

References

1. Jones, M. E. *Ann. Rev. Biochem.* 1980, *49*, 253.
2. Coleman, P.; Suttle, D.; Stark, G. *J. Biol. Chem.* 1977, *252*, 6379.

3. Mally, M. I.; Grayson, D. R.; Evans, D. R. *Proc. Natl. Acad. Sci. U. S. A.* 1981, *78,* 6647.
4. Davidson, J. N.; Rumsby, P. C.; Tamaren, J. *J. Biol. Chem.* 1981, *256,* 5220.
5. Davidson, J. N.; Niswander, L. A. *Proc. Natl. Acad. Sci. U. S. A.* 1983, *80,* 6897.
6. Grayson, D. R.; Evans, D. R. *J. Biol. Chem.* 1983, *258,* 4123.
7. Kelly, R. E.; Mally, M. I.; Evans, D. R. *J. Biol. Chem.* 1986, *261,* 6073.
8. Kim, H.; Kelly, R. E.; Evans, D. R. *J. Biol. Chem.* 1992, *267,* 7177.
9. Simmer, J. P.; Kelly, R. E.; Scully, J. L.; Grayson, D. R.; Rinker, A. G., Jr.; Bergh, S. T.; Evans, D. R. *Proc. Natl. Acad. Sci. U. S. A.* 1989, *86,* 4382.
10. Simmer, J. P.; Kelly, R. E.; Rinker, A. G., Jr.; Zimmermann, B. H.; Scully, J. L.; Kim, H.; Evans, D. R. *Proc. Natl. Acad. Sci. U. S. A.* 1990, *87,* 174.
11. Simmer, J. P.; Kelly, R. E.; Rinker, A. G., Jr.; Scully, J. L.; Evans, D. R. *J. Biol. Chem.* 1990, *265,* 10395.
12. Bein, K.; Simmer, J.; Evans, D. R. *J. Biol. Chem.* 1991, *266,* 3791.
13. Musmanno, L. A.; Maley, J. A.; Davidson, J. N. *Gene* 1991, *99,* 211.
14. Zimmermann, B. H.; Evans, D. R. *Biochemistry* 1993, *32,* 1519.
15. Williams, N. K.; Peide, Y.; Seymour, K. K.; Ralston, G. B.; Christopherson, R. I. *Protein Eng.* 1993, *6,* 333.
16. Guy, H. I.; Evans, D. R. *J. Biol. Chem.* 1994, *269,* 7702.
17. Davidson, J. N.; Jamison, R. S. *Adv. Exp. Med. Biol.* 1994, *370,* 591.
18. Guy, H. I.; Evans, D. R. *J. Biol. Chem.* 1995, *270,* 2190.
19. Guy, H. I.; Evans, D. R. *J. Biol. Chem.* 1996, *272,* 13762.
20. Guy, H. I.; Bouvier, A.; Evans, D. R. *J. Biol. Chem.* 1997, *272,* 29255.
21. Sahay, N.; Guy, H. I.; Xin, L.; Evans, D. R. *J. Biol. Chem.* 1998, *273,* 31195.
22. Nyunoya, H.; Lusty, C. J. *Proc Natl Acad Sci U S A* 1983, *80,* 4629.
23. Lee, L.; Kelly, R. E.; Pastra-Landis, S. C.; Evans, D. R. *Proc. Natl. Acad. Sci. U. S. A.* 1985, *82,* 6802.
24. Meister, A. *Adv. Enzymol. Relat. Areas. Mol. Biol.* 1989, *62,* 315.
25. Anderson, P. M. In Nitrogen Metabolism and Excretion; Walsh, P.J., Wright, P.A., Ed.; Evolutionary and Ecological Perspectives, CRC Press, New York, NY, 1995; pp 33-55.
26. Thoden, J. B.; Holden, H. M.; Wesenberg, G.; Raushel, F. M.; Rayment, I. *Biochemistry* 1997, *36,* 6305.
27. Christopherson, R. I.; Jones, M. E. *J. Biol. Chem.* 1980, *255,* 11381.
28. Guy, H. I.; Evans, D. R. *Adv Exp Med Biol* 1994, *370,* 729.

29. Irvine, H. S.; Shaw, S. M.; Paton, A.; Carrey, E. A. *Eur. J. Biochem.* 1997, *247*, 1063.
30. Lipscomb, W. N. *Adv. Enzymol. Relat. Areas Mol. Biol.* 1994, 68, 67.
31. Scully, J. L.; Evans, D. R. *Proteins* 1991, *9*, 191.
32. Fields, C.; Brichta, D.; Shepherdson, M.; Farinha, M.; O'Donovan, G. A. *Paths to Pyrimidines* 1999, *7*, 49.
33. Chen, J. J.; Jones, M. E. *In: Srere PA, Estabrook RW, ed. Microenvironments and metabolic compartmentation. New York, Academic Press,* 1978, 211.
34. Carrey, E. A.; Campbell, D. G.; Hardie, D. G. *Embo J.* 1985, *4*, 3735.
35. Graves, L. M.; Guy, H. I.; Kozlowski, P.; Huang, M.; Lazarowski, E.; Pope, R. M.; Collins, M. A.; Dahlstrand, E. N.; Earp III, H. S.; Evans, D. R. *Nature* 2000, *403*, 328.
36. Nyunoya, H.; Lusty, C. J. *J. Biol. Chem.* 1984, *259*, 9790.
37. Piette, J.; Nyunoya, H.; Lusty, C. J.; Cunin, R.; Weyens, G.; Crabeel, M.; Charlier, D.; Glansdorff, N.; Pierard, A. *Proc. Natl. Acad. Sci. U. S. A.* 1984, *81*, 4134.
38. Nyunoya, H.; Broglie, K. E.; Lusty, C. J. *Proc. Natl. Acad. Sci. U. S. A.* 1985, *82*, 2244.
39. Zalkin, H. *Methods Enzymol* 1985, *113*, 263.
40. Anderson, P. M. *Comp. Biochem. Physiol. [B]* 1976, *54*, 261.
41. Chaparian, M. G.; Evans, D. R. *J. Biol. Chem.* 1991, *266*, 3387.
42. Lusty, C. J. *FEBS Lett.* 1992, *314*, 135.
43. Mareya, S. M.; Raushel, F. M. *Biochemistry* 1994, *33*, 2945.
44. Hewagama, A.; Guy, H. I.; Chaparian, M.; Evans, D. R. *Biochim. Biophys. Acta.* 1998, *1388*, 489.
45. Hewagama, A.; Guy, H. I.; Vickrey, J. F.; Evans, D. R. *J. Biol. Chem.* 1999, *274*, 28240.
46. Miles, B. W.; Raushel, F. M. *Biochemistry* 2000, *39*, 5051.
47. Britton, H. G.; Rubio, V.; Grisolia, S. *Eur. J. Biochem.* 1979, *102*, 521.
48. Boettcher, B. R.; Meister, A. *J. Biol. Chem.* 1980, *255*, 7129.
49. Powers-Lee, S. G.; Corina, K. *J. Biol. Chem.* 1987, *262*, 9052.
50. Kim, H. S.; Lee, L.; Evans, D. R. *Biochemistry* 1991, *30*, 10322.
51. Potter, M. D.; Powers-Lee, S. G. *J. Biol. Chem.* 1992, *267*, 2023.
52. Alonso, E.; Rubio, V. *Eur. J. Biochem.* 1995, *229*, 377.
53. Post, L. E.; Post, D. J.; Raushel, F. M. *J. Biol. Chem.* 1990, *265*, 7742.
54. Guy, H. I.; Schmitt, B.; Herve, G.; Evans, D. R. *J. Biol. Chem.* 1998, *273*, 14172.
55. Serre, V.; Guy, H. I.; Penverne, B.; Lux, M.; Rotgeri, A.; Evans, D. R.; Herve, G. *J. Biol. Chem.* 1999, *274*, 23794.
56. Anderson, P. M.; Marvin, S. V. *Biochem. Biophys. Res. Commun.* 1968, *32*, 928.

57. Rodriguez-Aparicio, L. B.; Guadalajara, A. M.; Rubio, V. *Biochemistry* 1989, *28,* 3070.
58. Rubio, V.; Cervera, J.; Lusty, C. J.; Bendala, E.; Britton, H. G. *Biochemistry* 1991, *30,* 1068.
59. Cervera, J.; Conejero-Lara, F.; Ruiz-Sanz, J.; Galisteo, M. L.; Mateo, P. L.; Lusty, C. J.; Rubio, V. *J. Biol. Chem.* 1993, *268,* 12504.
60. Bueso, J.; Lusty, C. J.; Rubio, V. *Biochem. Biophys. Res. Commun.* 1994, *203,* 1083.
61. Czerwinski, R. M.; Mareya, S. M.; Raushel, F. M. *Biochemistry* 1995, *34,* 13920.
62. Cervera, J.; Bendala, E.; Britton, H. G.; Bueso, J.; Nassif, Z.; Lusty, C. J.; Rubio, V. *Biochemistry* 1996, *35,* 7247.
63. McCudden, C. R.; Powers-Lee, S. G. *J. Biol. Chem.* 1996, *271,* 18285.
64. Bueso, J.; Cervera, J.; Fresquet, V.; Marina, A.; Lusty, C. J.; Rubio, V. *Biochemistry* 1999, *38,* 3910.
65. Liu, X.; Guy, H. I.; Evans, D. R. *J. Biol. Chem.* 1994, *269,* 27747.
66. Braxton, B. L.; Mullins, L. S.; Raushel, F. M.; Reinhart, G. D. *Biochemistry* 1992, *31,* 2309.
67. Braxton, B. L.; Mullins, L. S.; Raushel, F. M.; Reinhart, G. D. *Biochemistry* 1996, *35,* 11918.
68. Ahuja, A.; Purcarea, C.; Guy, H.I.; Evans, D.R. *J. Biol. Chem.* 2001, Sep 26, [epub ahead of print].
69. Thoden, J. B.; Phillips, G. N., Jr.; Neal, T. M.; Raushel, F. M.; Holden, H. M. *Biochemistry* 2001, *40,* 6989.
70. Thoden, J. B.; Wesenberg, G.; Raushel, F. M.; Holden, H. M. *Biochemistry* 1999, 38, 2347.

Chapter 17

A New Engine for Cleaving Nucleic Acid

Kurt L. Krause and Mitchell D. Miller

Department of Biology and Biochemistry, University of Houston, Houston, TX 77204-5513

Rapid and accurate cleavage of nucleic acid material, such as DNA and RNA, is a fundamentally important biochemical process in all living organisms carried out by enzymes called nucleases. We present here a discussion and comparison of two nucleases of widely disparate properties but who share a newly described nuclease active site geometry. *Serratia marcescens*, a pathogenic Gram negative bacterium produces an enzyme that presents a new paradigm in nucleases. Its fold is new, it is capable of very rapid, and relatively non sequence dependent cleavage of several types of DNA and RNA. On the other hand, I-*Ppo*I is a homing endonuclease which is encoded by a group I intron in the rRNA genes of *Physarum polycephalum (1)*. It also possesses a new fold, but it only slowly cleaves DNA at a very specific 15 bp site. Both enzymes possess the same active site geometry, but no other structural homology. The structural basis for their different properties is due to two main factors, the nature of their interaction with substrate and the presence of a mobile metal in I-*Ppo*I endonuclease that must migrate into position prior to catalysis.

Background on Nucleases

Nucleases are a class of enzymes that breakdown and process DNA and RNA in all living organisms. They differ greatly in their substrate specificity, ranging from restriction endonucleases that cleave only at a particular nucleotide sequence to broad specificity nucleases that cut a wide swath through DNA and RNA *(2)*. Some broad specificity nucleases function simply to recycle sources of nucleic acid material for cellular function, but other more important roles for these enzymes have been found in DNA replication, DNA repair, and apoptosis. This report focuses on using X-ray structural studies to uncover the chemical mechanism and structural-function relationships for two nucleases with widely differing structures, but with active site and mechanistic similarities, the endonuclease from *Serratia marcescens* and the homing endonuclease from I-*Ppo*I.

Nucleases accomplish a difficult chemical reaction. The oxygen-phosphate cleavage reaction they catalyze destroys a linkage that is ordinarily quite stable *(3)* (Fig. 1). For example, cleavage of dimethylphosphate in 1M sodium hydroxide in the absence of enzyme catalysis has a 15 year half-time *(4)*. Living organisms can't wait that long for nucleic acid breakdown, but nucleases can help meet the demand for rapid nucleic acid processing. For example, acceleration of nucleic acid cleavage by Staphylococcal nuclease over the uncatalyzed reaction is thought to exceed a factor of 10^{16} *(5)*.

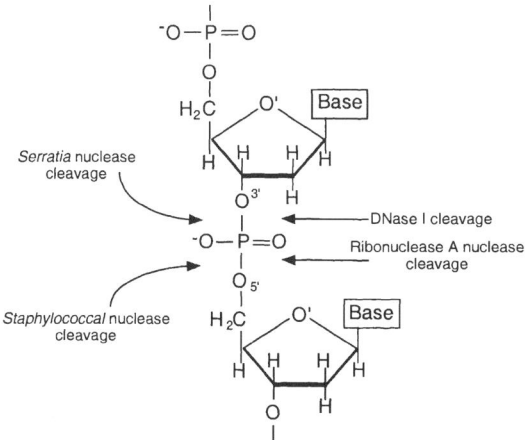

Figure 1. Cleavage sites in common nuclease reactions.

Mechanisms of enzyme catalyzed phosphodiester cleavage studied to date often follow an "in-line" associative S_N2 mechanism *(6)*. This mechanism typically begins with a nucleophilic attack at the phosphorous atom and proceeds through a pentacoordinate phosphorane intermediate. Enzymes can facilitate this reaction by stabilizing the high energy phosphorane, which carries a formal charge of minus 2, by using adjacent positively charged residues or divalent cations. This charge stabilization is thought to confer an acceleration factor of 10^7, but does not account for the remaining acceleration of 10^9 *(7)*. This additional acceleration is thought to occur primarily through general base activation of the nucleophile and general acid catalysis of the leaving group.

Within this common framework individual nucleases studied to date differ greatly in their choice of catalytic residues, metal requirements and in their geometry of interaction with substrate. Detailed knowledge of the active site geometry of nucleases is usually very helpful for the most complete understanding of their mechanism, and this is most often accomplished using x-ray crystallographic methods.

Background on Serratia family of endonucleases

Serratia marcescens is a member of the clinically important Klebsielleae tribe that contains *Klebsiella, Enterobacter, Serratia*, and *Hafnia* *(8)*. Historically Serratia has been an important bacteria for several unique reasons. It produces a red pigment during normal growth that has been used to track growth and distribution of bacteria. It it thought to be responsible for ancient reports of "blood" appearing on the host in churches. Because is was originally thought to be non-pathogenic, it was used in biological warfare experiments to test the possible airborne distibution of bacteria in cities. Now it is known as an important cause of nosicomial infections, particularly in children and immunosuppressed patients *(9, 10)*. Notably, *Serratia marcescens* produces a number of extracellular protein products during infection and cultivation including an endonuclease *(11)*, two proteases *(12)*, two lipases *(13, 14)*, and two chitinases *(15)*. Some of these enzymes, are clearly virulence factors and they contribute to the pathogenicity of *Serratia* infections. The *Serratia* endonuclease has been suggested to play a role in the virulence of the *Serratia* infections, but no strong evidence for this has yet been found. It is of particular interest because of its broad specificity, high activity and chemical stability.

The Serratia nuclease has homologs in eukaryotes and prokaryotes

The *Serratia* endonuclease belongs to a group of major sugar non-specific nucleases that have been found in several different species from both prokaryotic and eukaryotic organisms *(16)* (Table I). All of the nucleases of this group are sugar non-specific; several of them show a preference for cleavage in GC rich regions of DNA *(17, 18)*. Recent reports have implicated endonuclease

G in certain human diseases caused by impaired mitochondrial DNA replication *(19, 20)*.

Table I. Amino acid alignment of nucleases homologous to the *Serratia* endonuclease.

```
                  90            100           110           120           130
H. sapiens     GFDRGHLAAAANHRW-SQKAMDDTFYL-SKVAPQVTH-LNQNAWNNLEKYSRSLTR
B. tarus       GFDRGHLAAAANHRW-SQKAMDDTFYL-SNVAPQVPH-LNQNAWNNLEKYSRSLTR
M. musculus    GFDRGHLAAAANHRW-SQRAMDDTFYL-SNVAPQVPH-LNQNAWNNLERYSRSLTR
S. cerevisiae  GYDRGHQAPAADAKF-SQQAMDDTFYL-SNMCPQVGEGFNRDYWAHLEYFCRGLTK
S. pombe       GYDRGHQVPAADCKF-SQEAMNETFYL-SNMCPQVGDGFNRNYWAYFEDWCRRLTS
C. elegans     GFDRGHLA-AAGNHRKSQLAVDQTFYL-SNMSPQVGRGFNRDKWNDLEMHCRRVAK
Anabaena sp.   GYDRGHIAPSADRTK-TTEDNAATFLM-TNMMPQTPD-NNRNTWGNLEDYCRELVS
B. burgdorferi GYDRGHIVSSADMSF-SENAMKDTYFL-SNMSPQKSE-FNSGIWLKLEKLVREWAI
S. marcescens  KVDRGHQAPLASLAGVSDWESL-NYL--SNITPQKSD-LNQGAWARLEDQERKLID
T. brucei      GLSRGHLA-AAQFHKSSTVELAQTFNMNANTVPQDMT-MNAVDWLRLENLTRKLRR
S. pneumoniae  AVDRGHLLGYALIGGLDGFDASTSNP--KNIAVQTAW-ANQAQAEYSTGQNYYESK
consensus      gfdRGHla--A-----s---m--tf-l-snv-pQ-----N---w--le---r-l--
```

(Adapted from reference*(21)*)

General properties of the Serratia nuclease

The *Serratia* nuclease is a homodimer composed of a two polypeptide chains 245 residues long each with a molecular weight of 26,700 *(11)*. When incubated with substrate, the enzyme is known to catalyze the production of 5' monophosphate terminated oligonucleotides, i.e. the 3' O–P bond is broken during the reaction *(22-24)* (Fig. 1). Magnesium is required for activity and the pH optimum is 8.

The enzyme cleaves nucleic acids very rapidly, with a catalytic rate almost 15 times faster than DNase I *(25)*. It shows long term stability at room temperature and is active in the presence of many reducing and chaotropic agents *(26, 27)*. Because of these unique properties we undertook its crystal structure solution in our laboratory.

Crystallographic studies of Serratia endonuclease

The *Serratia* endonuclease is well suited for X-ray studies. Crystals can be grown from PEG or ammonium sulfate and diffract to very high resolution (beyond 0.88 Å at 100 K). The nuclease crystallizes in space group $P2_12_12$ with one homodimer (490 amino acids / 53.4 kDa) in the asymmetric unit of the cell (a = 106.7 Å, b = 74.5 Å, c = 68.9 Å).

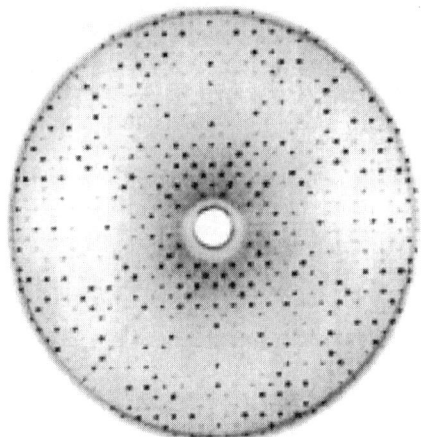

Figure 2. Precession photograph of the 0KL zone from crystals of Serratia *endonuclease demonstrating strong pseudo-centering.*

We solved the structure using multiple isomorphous replacement, but our initial attempts at a structure solution were hindered by space group ambiguity and pseudo-centering. We were able to solve Patterson functions in I222 and P222, as well as the correct space group P2$_1$2$_1$2. Following phase refinement, our density maps had a well-defined protein solvent boundary even in incorrect orientations of the correct space group. Finally in order to obtain a buildable electron density map we had to transfer the crystals from ammonium sulfate to PEG. After we completed refinement of the structure at 2.1Å we calculated that our pseudo-centered cell could be transformed into a body-centered lattice by shifting one monomer by only 1.4 Å and rotating it by only 1° about its centroid. Another indication of degree of pseudo-centering can be obtained from viewing the "herringbone" pattern of spots in precession photographs (Fig. 2). All of the information about how the monomers differ is contained in the very few weak reflections that violate the centering pattern.

Serratia nuclease at 2.1 Å

Inspection of the *Serratia* endonuclease structure reveals that its fold differs from other nucleases whose structures are known *(28)*. In the SCOP protein classification system it is listed as the only member of its family *(29)* In the CATH protein classification system, it is listed as the only member of its tertiary group *(30)*. The dimeric enzyme is almost cylindrical with overall dimensions of 30 Å x 35 Å x 90 Å (Fig. 3). Each monomer consists of a central six-stranded anti-parallel β-sheet flanked by α-helices on both sides. One long helix, termed

H1, dominates one face of each monomer and is located opposite a long region of random coil. The dimer interface is circular and nearly flat *(31)*.

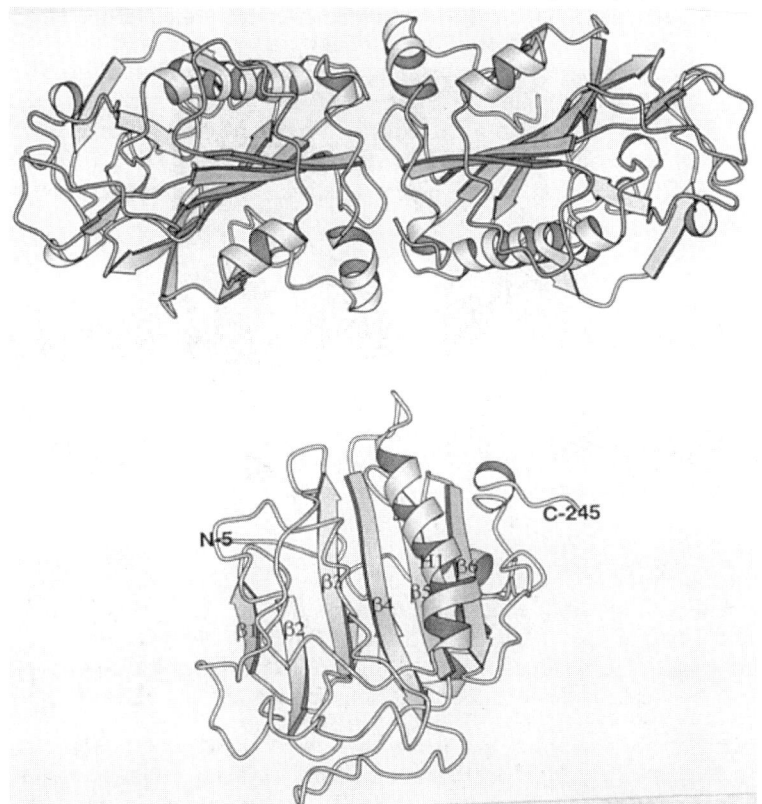

Figure 3. Top. Ribbon diagram of the nuclease dimer. The two active sites are located along the long helices on opposite sides of the molecule. Bottom. A ribbon diagram of one monomer is shown for clarity.

At the time we solved the crystal structure of *Serratia* nuclease, comparative sequence data, mutagenesis data and chemical modification data were not available. After we completed the building and refinement of the *Serratia* nuclease structure, visual inspection did not reveal the active site location. Since there were no other nuclease structures to use to help locate the active site, we analyzed the structure in terms of its electrostatic surface potential. Inspection of electrostatic maps calculated in GRASP *(32)* revealed that one side of the nuclease monomer displayed a broad area of positive electrostatic potential while the opposite side displayed negative electrostatic potential. By noting the size and charge of helical DNA we were able to determine that the active site for each monomer was located between a prominent helix, called H1, and a long region of coiled protein in the vicinity of His 89.

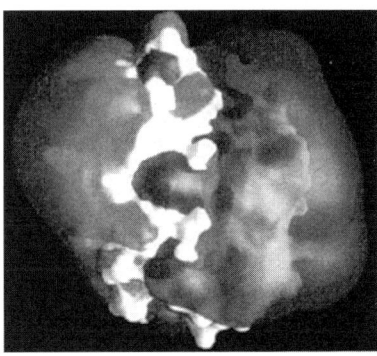

Figure 4. Electrostatic potential surface superimposed on a van der Waals surface of Serratia *nuclease. The positive surface is to the right, while the negative surface extends to the left.*

The location of the active site of *Serratia* nuclease we proposed is in agreement with all other experimental analysis which has appeared including that based on multiple sequence alignments, kinetics and mutagenesis studies *(25, 33, 34)*. The residues we cited in our electrostatic analysis have been shown to be mechanistically important, including His 89, Glu 127 and Asn 119.

Structure of the catalytic site

The active site of *Serratia* nuclease is located between the long helix H1 (116-135) and a coil region extending from residues 50 to 114. This coiled region contains His 89, a residue that is essential for catalysis. Using magnesium soaked nuclease crystals we identified the metal binding location within the active site. The Mg^{2+} atom is coordinated by Asn 119 and 5 solvent water molecules arranged in an octahedron. His 89 and Glu 127 are in close proximity to the Mg^{2+} but do not directly interact with the metal ion. Refined density for all of the waters is clear and their B-factors range from 12.7 $Å^2$ to 17.5 $Å^2$. Asn 119 is the only protein ligand of the magnesium ion (Fig. 5). Residues that are in contact with the metal-water cluster are highly conserved across species (Table I).

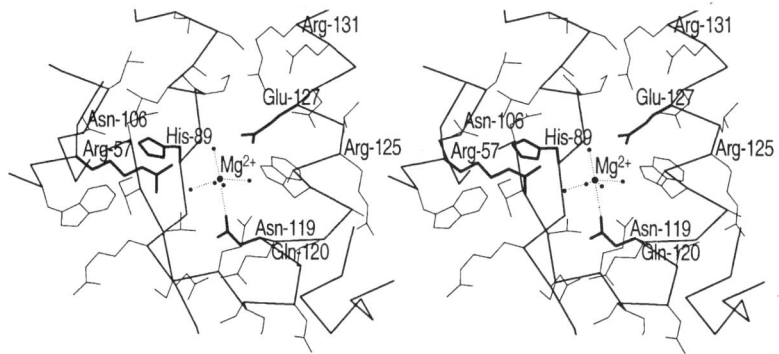

Figure 5. Stereo diagram of the Serratia *nuclease active site. The α-carbon tracing identifies local secondary structure and displays the side chains of residues thought to be catalytically important.*

Mechanism of Serratia nuclease

Based on our crystal structure and relevant biochemical data *(25, 33, 34)* we have proposed two possible catalytic schemes (Fig. 6). They agree on their choice of general base, likely catalytic groups and make-up of the catalytic metal site. They differ in that one proposes a nucleophilic role for cluster bound water.

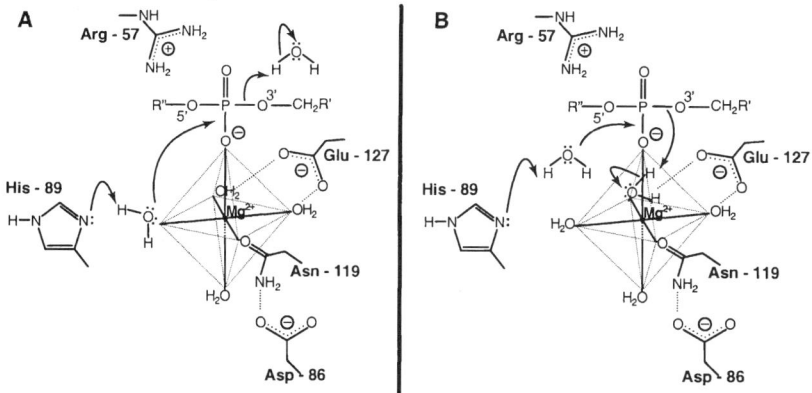

Figure 6. Proposed mechanisms for nucleotide cleavage by Serratia *nuclease. (Adapted from reference 21)*

Getting to this point in our study of *Serratia* endonuclease represented a significant advance to us because we thought we understood the structural basis for the extremely rapid binding and catalysis carried out by the *Serratia* enzyme. We were surprised to learn that another unrelated protein, called I-*Ppo*I endonuclease, shared the same geometry in its active site, and to learn that, in contrast to *Serratia*, its rate of catalysis was slow, and very sequence specific. Coincidentally, before the structure of I-*Ppo*I was known we had already begun a collaboration on I-*Ppo*I nuclease with Professor Ron Raines, University of Wisconsin.

Background on I-*Ppo*I endonuclease

I-*Ppo*I is a homing endonuclease which is encoded by a group I intron in the rRNA genes of the slime mold, *Physarum polycephalum (1)*. This intron contains a ribozyme and an open reading frame which is self-spliced and translated to yield the functional nuclease *(35)*. This homing endonuclease promotes the insertion of its own intron into intronless alleles by catalyzing double-strand breaks near potential insertion sites. Subsequent double-strand break repair inserts a copy of the intron and converts the site to an intron containing copy. I-*Ppo*I belongs to the His-Cis box class of homing endonucleases *(see review 36)*. In this class, the histidine and cystine residues form tight zinc binding sites which are believed to help stabilize the structure. Zinc co-purifies with the enzyme even after extensive dialysis. *(37)*.

I-*Ppo*I recognizes a incompletely symmetric 15 bp sequence. It is tolerant of many substitutions in its recognition sequence; however, it is intolerant of insertions or deletions. Palindromic sequences derived from either half site of the wild type recognition sequence are cleaved with the same efficiency as the wild type sequence *(38)*.

```
C T C T C  T T A A ↓G G T A G C
G A G A G ↑A A T T  C C A G C G
```

*Figure 7. Cognate cleavage site for I-*Ppo*I endonuclease*

Like the *Serratia* endonuclease, the enzyme cleaves the 3' O–P bond in DNA. Also like the *Serratia* endonuclease, the enzyme prefers magnesium over other divalent metal ions. I-*Ppo*I endonuclease is a domain swapped dimer with 163 amino acids per monomer. In the absence of salt, its k_{cat}/K_M appraoches 10^9. This number is dominated by the K_M value which is close to 1×10^{-11}M. The k_{cat} value is highly dependent on ionic strength, but without salt it has been reported to be 0.046 min^{-1} *(39)*.

The structure of the enzyme has been reported by the Stoddard group in several forms. Two forms have been solved with DNA bound. One is a product complex with magnesium bound (PDB id 1A73), and one is a substrate complex, without magnesium, (PDB id 1A74) *(40)*. The structure of I-*Ppo*I is a dimer, with each monomer interacting primarily with one-half of the cognate cleavage site (Fig. 8). The C-terminus contains a domain swapped loop. The active site residues contain two important residues, His 98 and Asn119. Asn 119 is the sole residue that interacts directly with the catalytically essential magnesium ion. Both active sites of the monomer are oriented to cleave both strands at the same time. Recently, another structure appeared of I-*Ppo*I without magnesium or DNA bound *(41)*. In our group, we obtained crystals of I-*Ppo*I with and without magnesium, but without DNA bound. We describe the structure solution and results below. A complete discussion of the structure solution is included to illustrate the method we used to get past a "stalled" refinement.

Structural studies on the I-*Ppo*I endonuclease – magnesium complex

Crystallization and X-ray data collection.

I-*Ppo*I endonuclease was kindly provided by Prof. Ron Raines, University of Wisconsin. Crystals of the I-*Ppo*I endonuclease were grown at 4 °C by vapor diffusion against 19 % PEG 4000, 0.2 M lithium sulfate, 0.1 M tris, pH 8.3. Drops were prepared by mixing 4 µl of 10.7 mg/ml I-*Ppo*I in 10 mM sodium acetate, 25 mM NaCl, pH 5.0 with 4 µl of the reservoir solution. One homodimer of the endonuclease (35.6 kDa / 326 amino acids) crystallizes in space group $P2_12_12_1$ with a=58.58 Å, b=61.10 Å, c=111.70 Å. Several crystals were transferred to a well containing 100 µl of 19 % PEG 4000, 0.2 M lithium sulfate, 0.1 M tris, 2 mM zinc acetate, 5 mM magnesium chloride, pH 8.3. After 1 day in the magnesium/zinc stabilization solution, a single crystal of dimension 0.4 x 0.4 x 0.3 mm^3 was transferred to a cryoprotection solution containing 30 % PEG 4000, 10 % PEG 400, 2 mM zinc acetate, 5 mM magnesium chloride, pH 8.3 and flash frozen in a 100 K cryostream (Oxford Cryosystems). Data to 1.8 Å resolution were collected at 100 K on a Rigaku R-axis IIc image plate detector using graphite monochromated Cu Kα radiation from a rotating anode generator operating at 50 kV / 90 mA collimated to 0.5 mm. Data were processed using DENZO version 1.7b and merged with scalepack *(42)*. The 120,674 observations of 36,097 reflections merged with an R-factor of 3.9% on I (15% for the 1.8-1.86 Å resolution shell). The completeness to 1.8 Å is 95 % (85 % I > $2\sigma_I$) with the 1.8-1.86 Å shell being 80 % (52 % I > $2\sigma_I$) complete. Diffraction from this crystal showed three prominent ice rings. Data in resolution bins 3.60 - 3.75, 2.23 - 2.28, and 1.89 - 1.93 Å were omitted from the crystallographic refinement.

Another data set was collected from a crystal which was not soaked with zinc and magnesium. This 0.35 x 0.3 x 0.25 mm^3 crystal was transferred from the mother liquor to a cryoprotection solution containing 16 % PEG 4000, 15 % PEG 400, 0.17 M lithium sulfate, 85 mM tris, pH 8.3 and immediately mounted in a cryoloop and cooled in a 95 K nitrogen gas cyrostream (Cryo Industries of America). Data to 1.9 Å resolution were collected with a Bruker 2K CCD detector with Cu Kα radiation from a Bruker M06Xce operating at 50 kV / 90 mA collimated to 0.5 mm and monochromated using Göbel Mirrors. Frames, 0.125° in width, were collected with detector distance was 60 mm with $2\theta = 15°$. Two scans were collected to insure full coverage of reciprocal space with the first scan covering 125° and the second scan 37.5°. The data was processed and merged with SMART/SAINT *(43)*. The 98,583 observations of 31,631 reflections merged with and R-factor of 6.2 % on I (23.5 % for the 1.9-1.97 Å shell). This data set is 99 % complete to 1.9 Å with 76 % of data having $I > 2\sigma_I$.

Molecular replacement and refinement

A molecular replacement solution was found with AMoRe *(44)* using, as a search model, a monomer from the 2.2 Å structure of I-*Ppo*I with the domain swapped carboxy terminus tail removed (residues 2-144 from monomer A of PDB id 1A74) *(40, 45)*. The search was conducted versus 15 - 3.5 Å resolution data set taken from a crystal collected at room temperature to 2.2 Å. (This crystal was grown under the same conditions described above and was mounted in a glass capilliary without soaking in cryoprotection or metal ion solutions.) The solution was then inspected in O *(46)* to look for symmetry clashes and the model was expanded to include the C-terminal tail from each monomer.

The initial R-factor after rigid body minimization of the AMoRe solution was 36.5 % for data 8 - 3 Å. The structure was subjected to both slow cooling *(47)* and torsion angel dynamics *(48)* simulated annealing runs in X-PLOR 3.851 *(49)*. The domain swapped junction (residues 141-150 of each monomer) was omitted from of the model to prevent model bias from obscuring the connections. The force field used was based on the Engh and Huber values *(protein_rep.param, 50, 51)* and included repulsive non-bonded forces without electrostatic terms. After each cycle, the structure was evaluated and rebuilt using the graphics program O *(46)*. Throughout refinement, the cross validation statistic, R$_{free}$, *(52)* was followed to monitor whether the addition of parameters to the model (or changes to the restraints) was justified. The same test set was used for all data sets with all programs and contained 5 % of the reflections.

After several rebuilding cycles and the addition of 38 water molecules, the R-factor decreased to 22.9 % and the R_{free} to 27.7 % for the 18,877 reflections between 20 and 2.2 Å with $F > 2\sigma_F$. A bulk solvent mask was included. At this point the 2.2 Å room temperature data set was replaced with the data from the 1.9 Å cryo-cooled data set and the 1.8 Å cryo-cooled data set collected from a magnesium and zinc soaked crystal.

Upon cryo-cooling, the unit cell dimensions of the I-*Ppo*I crystals shrunk 0.6-1 Å on each edge (e.g. from a=59.19 Å to a=58.58 Å, from b=62:13 Å to b=61.10 Å, and from c=112.69 Å to c=111.70 Å for the metal soaked crystal). Simple rigid body minimization followed by Powell minimization of the partially refined room temperature structure or the AMoRe solution did not did not drop the R_{free} below 40 % for either cryo-cooled data set. The structures were then subjected to torsion angle dynamics protocols *(48)* at several temperatures. Both starting models, with either data set, failed to produce a structure with an R_{free} below 39 % for data between 8 and 2 Å. In addition, the resulting maps did not look very good.

The beta version of the warpNtrace procedure in ARP/wARP version 5 was then employed to improve the quality of the maps *(53, 54)*. This procedure alternates between ARP, real space evaluation of atoms including the addition and removal of atoms and real space refinement, and reciprocal space refinement using the CCP4 program Refmac *(55)*. After every tenth cycle, the resulting map is auto-traced, and the auto-traced structure is then subjected to additional cycles of the ARP / Refmac refinement.

After 40 cycles and 4 auto-tracings using ARP/wARP, the structure converged with the R and R_{free} values of 19 and 24 % for data 20 - 1.8 Å. Five structures from various points in the procedure were selected and averaged with an averaged structure produced by the X-PLOR torsion angle dynamics runs using the wARP averaging option. The averaged structure factors (R=26 % and R_{free}=29 %) were used to calculate an FOM weighted 2Fo-<Fc> map using the averaged phases. This map was much improved over the individual and averaged torsion angle dynamics maps. The main chain fragments from the warpNtrace model were then merged with the starting structure in O. Side chains were auto built using lego_auto_sc in O followed by automatic real-space rotomer selection (rsr_rotomer). This structure was then easily rebuilt to optimize side chain positions and fix loops which were deleted by the warpNtrace procedure. The rebuilt structure, which still had approximately 150 protein atoms omitted and no solvent, had R of 33.1 % and R_{free} of 34.6 % for the 25,473 reflections between 8 and 2 Å with $F > 2\sigma_F$. Positional and grouped B-factor refinement (2 / residue) with a bulk solvent correction and overall anisotropic temperature factor dropped R to 28.8 % and R_{free} to 32.4 % for the 33,637 reflections between 50 and 1.8 Å with $F > 2\sigma_F$. This procedure took about 6 hours of CPU plus 2 hours of manual rebuilding and was able to escape a local minima which multiple torsion angle dynamics runs did not.

The bulk solvent parameters and overall anisotropic temperature factor were updated at the beginning of each refinement cycle. After several cycles of refinement and rebuilding, we were able to satisfactorily refine these two structures. Care was taken to avoid biasing the metal site electron density as the structure was refined.

Table 2. Final Refinement Statistics for I-*Ppo*I endonuclease structures

Parameter	*without Mg*	*With Mg*
R-factor	19.8 %	20.2 %
Free R-factor (5%)	25.0 %	25.2 %
Resolution (Å)	40 - 1.9	40 - 1.8
No. Reflections (>2σ)	28,213	30,360
No. Protein atoms	2494	2494
No. Solvent	285	284
rms Bonds	0.006 Å	0.008 Å
rms Angles	1.3°	1.4°
Average B (Å^2)		
Main Chain	16.5	20.6
Side Chain	19.1	23.2
Solvent	22.7	27.5
Bulk Solvent Parameters		
mean density (e$^-$/Å^3)	0.34	0.31
B-factor (Å^2)	84	55

Parallel refinements were carried out with both the 1.9 Å data set without magnesium and the 1.8 Å data set from the crystal soaked in magnesium and zinc. The same test set was used for R_{free} calculations during all steps with all structures. These two structures were refined to crystallographic R-factors of 20 % with good geometry (Table 2). The final models included 2494 protein atoms, 284 solvent water molecules (the 5 mM magnesium soaked crystal had two magnesium ions and 283 solvent water molecules). Several validation tools were used to assess the quality of the structure and yielded results consistent with those reported for our previously published structures. These tools included PROCHECK *(56)*, WHATCHECK *(57)*, OOPS *(58)*, and O's residue real space correlation function *(46)*.

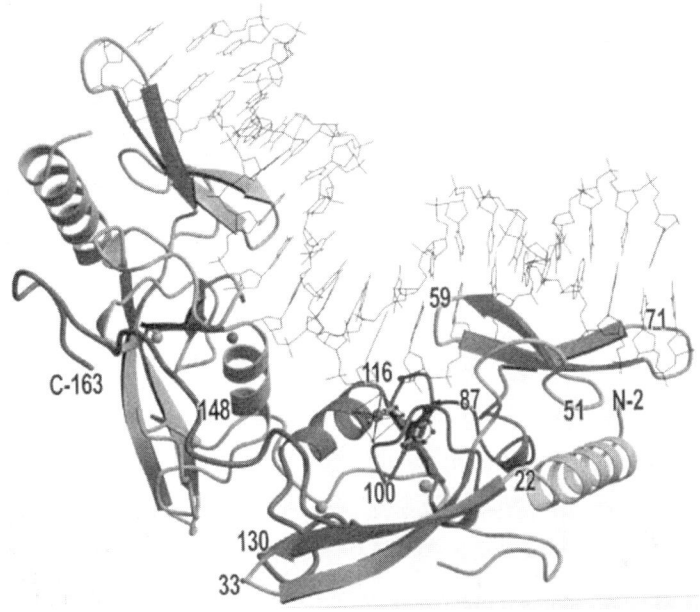

Figure 8. Ribbon diagram showing the I-PpoI dimer structure complexed with DNA as reported by Stoddard (PDB 1A73). One monomer has residue numbers and shows Asn119, His 98, and the magnesium-water cluster(40).

Analysis I-*Ppo*I endonuclease structures

We solved the structure of I-*Ppo*I without DNA using the amino-terminal domain of a monomer from the structure of I-*Ppo*I determined in the presence of cognate DNA as a search model (PDB id 1A74, *40*). We refined this molecular replacement solution against data collected from crystals in the presence and absence of 5 mM magnesium chloride in order to locate the magnesium binding site. These two structures were refined in parallel in order to minimize any differences which are the result of different interpretations of similar conformations. What follows below is a description of our I-*Ppo*I structures and a comparison to themselves, to other I-*Ppo*I structures in the Protein Data Bank and to the *Serratia* endonuclease structure. Superposition of I-*Ppo*I structures was accomplished using O, LSQMAN *(59)*, and X-PLOR. Δ phi / Δ psi measurements and buried surface analysis and hinge angle calculations were also made in X-PLOR (Figure 8).

Differences between I-PpoI with and without magnesium bound.

There are very few differences between the structures of I-*Ppo*I we solved with and without magnesium bound. In fact after superposition, the polyalanine root mean square difference (rmsd) (0.13 Å) is less than the rmsd between the two monomers in the asymmetric unit (see below). Even the water sites in the region of the metal binding site are similar. Magnesium binding only shifts the location of solvent water sites slightly. Since the two structures are so similar, we will focus on the I-*Ppo*I structure without DNA in the presence of Mg^{2+}.

Differences between the two monomers.

ϕ and ψ are on average 10° different between monomer A and B in the asymmetric unit. χ_1 is on average of 20° different between the two monomers. The average B-factors in monomer A are 2 $Å^2$ higher than for monomer B. Most of this increase is localized to the DNA recognition β-sheet (residues 54-76). This result is somewhat surprising since monomer A has much clearer density for the metal bound to His 101 and Asn 123. Only in the region just after the recognition strands, from residue 85-92, are the B-factors significantly higher than in monomer A. Monomer B has an additional 200 $Å^2$ of surface area buried in crystal contacts. This is due to the packing in the unit cell. A:A crystallographic pairs make crystal contacts parallel to the X axis forming a semi-continuous chain of monomers, while B:B crystallographic pairs interact parallel to the Y axis. The DNA binding loop from monomer B makes contact with resides near residue 20 of a symmetry related monomer B. Monomer A does not make such a contact and the DNA binding loop is more solvent exposed.

*Differences between I-*Ppo*I with and without DNA bound.*

There are both local and global changes observed in I-*Ppo*I upon DNA binding. There is a 6 ° change in the dimer hinge angle between the structures with DNA bound (125 °) and those without DNA (131 °) (Fig. 9). This number is consistent with the 5 ° change reported by the Stoddard group *(41)*. Dimerization of the free enzyme buries 4859 $Å^2$ of surface or 23.8 %, which is about 60 $Å^2$ less than the surface area buried by the dimer with DNA bound. Thus, there is little change in the amount of surface area buried in the dimer interface even though there is a sizable change in the angle relating the two monomers.

285

Figure 9. Superposition of I-PpoI *structures without DNA (thin, our structure) and with DNA (1A73; thick)(40). Note the change in hinge angle (6°) upon DNA binding.*

In addition, the DNA recognition beta-sheet (residues 54-76) shifts upon binding DNA. This is visible in Figure 9 where the active site residues and Mg^{2+} ion were superimposed. In our structure without DNA bound, these strands are less well-ordered with B-factor averages which are larger than for the DNA bound structures. The ΔB values between the free enzyme and the product complex are 9.5 for all atoms, 4.25 for the main chain and 12.75 for the side chain protein atoms involved in DNA binding *(60)*.

Surprising changes in the active site.

Perhaps the most surprising change between the I-*Ppo*I structures occurs in the magnesium site. In the structure refined from the crystals which were soaked in the presence of 5 mM magnesium chloride, we found an octahedrally coordinated magnesium ion bound to His 101 and Asn 123 along with four water molecules. In this position, the magnesium ion is located 3.8 Å from the catalytic metal site as revealed in the I-*Ppo*I product complex (Fig. 10). This metal is bound primarily to only one monomer, monomer A. We do see an octahedral complex in the same location in monomer B but the density is much weaker and is apparently only partially occupied. After refinement with a magnesium coordination sphere in monomer B, the B-factor for the magnesium is 40 Å^2, which is four times that in monomer A.

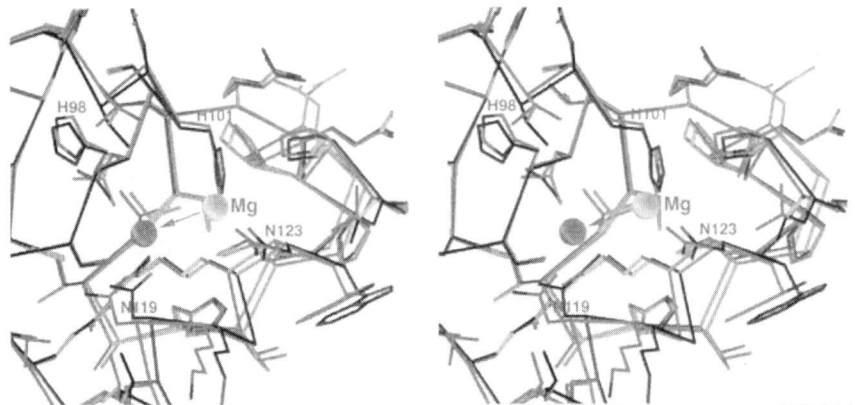

Figure 10. Superposition of the I-PpoI – Mg^{2+} structure without DNA (dark) with the I-PpoI - Product complex (1A73; light). Note the 3.8 Å shift of the magnesium site when DNA is bound.

Comparison of I-*Ppo*I nuclease to *Serratia* endonuclease.

I-*Ppo*I and *Serratia marcescens* endonuclease share a similar active site and mechanism in spite of a lack of sequence or tertiary homology *(21, 61, 62)*. Superposition of the *Serratia* endonuclease with magnesium *(*PDB id 1QAE, *21)* and I-*Ppo*I bound to product DNA and magnesium *(*PDB id 1A73, *40)* reveals significant local overlap between the two active sites including the catalytic metal and its coordination sphere. To create this superposition, the structure of I-*Ppo*I endonuclease (PDB id 1A73) was superimposed on the Serratia endonuclease structure with the program O. Residues 89, 108-113, 116-119 and the Mg^{2+} – water cluster from Serratia endonuclease were aligned with residues 98, 108-113, 116-119 and the Mg^{2+} – water cluster from I-*Ppo*I. Outside of the immediate vicinity of the active site, there is little structural similarity between the two enzymes. Only one monomer from each homodimer can be superimposed at a time. (Fig. 11). The Mg–coordination water clusters from these enzymes can be superimposed with an rmsd of 0.4 Å. When additional active site residues are included in the superposition the rmsd is 0.68 Å. In the I-*Ppo*I 1A73 structure, one of the waters found in the coordination sphere of the *Serratia* structure (water 5), is replaced by the 3' deoxyribose oxygen from the cleaved product.

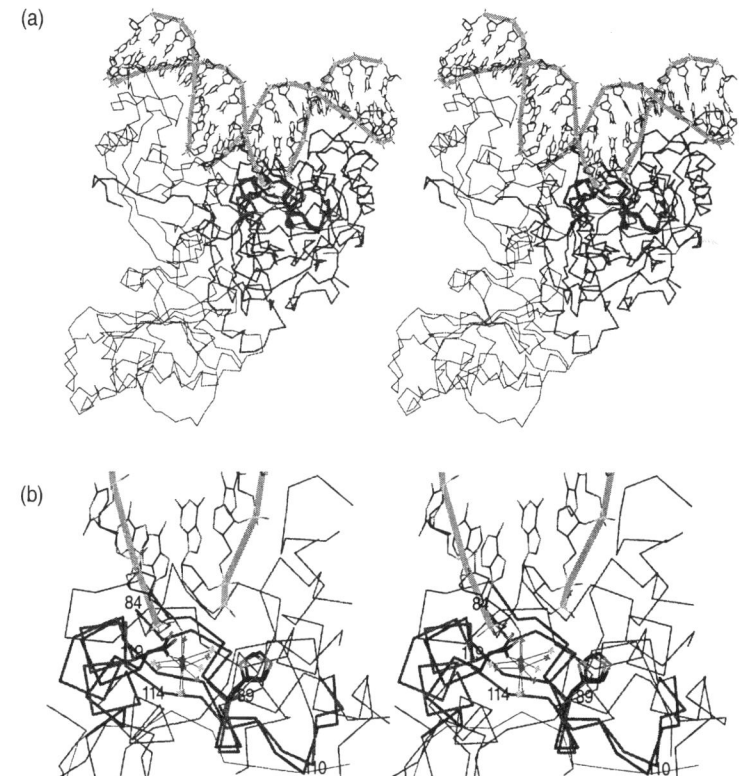

Figure 11. Superposition of the Serratia endonuclease with I-PpoI endonuclease (a) Overall view showing both dimers and the I-PpoI cleaved DNA oligomer. (b) Stereo view of the active site including the bound Mg^{2+} – water cluster and essential histidine and asparagine residues. (Adapted from reference 21)

Discussion

These two enzymes are the first nucleases described with a catalytic geometry centered around a magnesium bound water cluster adjacent to a catalytically essential histidine. In this newly described type of cleavage "engine", both enzymes bind their catalytic metal ion with a single asparagine protein ligand (Asp 119 in both). The arrangement of the catalytic histidines, the asparagines and the metal–water clusters is virtually identical in both proteins. It is, therefore, likely that they share the same catalytic mechanism with the catalytically essential histidines acting as the general base.

Properties of the cleavage engine

Since the residues that carry out an enzyme's catalyic function reside in the active site, and since *Serratia* nuclease and I-*Ppo*I nuclease share much the same active site makeup, it would be reasonable to conclude that they would share many functional properties in common. As noted, they do share the same sissicile bond, the 3' O–P bond, and the same preference for cleaving A form DNA as indicated by a preference for cleavage in GC rich regions by *Serratia* nuclease. On the other hand they have many differences as well. One nuclease cleaves both DNA and RNA in double and single stranded forms; the other cleaves only a very large (15 bp) and specific DNA site. The *Serratia* enzyme is extremely fast, 15-fold faster than DNase I, while I-*Ppo*I has a modest k_{cat} value.

The small value for the Michaelis constant in I-*Ppo*I indicates that a very tight association is present between substrate and enzyme. Therefore, it is possible to attribute the difference in speed of these two enzymes to slow product dissociation in I-*Ppo*I. Presumably the reaction coordinate for cleavage of DNA by *Serratia* nuclease would display balanced internal states, while I-*Ppo*I would display less balance with a higher barrier in the product dissociation step. But this explanation begs the question of why nature would put such a fast cleavage engine in such a slow enzyme.

I-*Ppo*I nuclease does not require rapid cleavage of nucleic acid but it does require very high specificity in the location of cleavage. I-*Ppo*I cleaves DNA at only a very small number of sites in its host organism. Once these rare sites are located, I-*Ppo*I needs to rely on a very efficient cutting engine, in order to avoid non-productive binding. Another way then to think about its catalytic geometry is to consider that it is very efficient at cleaving nucleic acid. Perhaps, this is why a "Ferrari" engine is stuck in a "Volkswagon".

A role for shifting magnesium

Both enzymes have a single asparagine liganded to the catalytic magnesium ion and a catalytically important histidine nearby (His 89 in *Serratia* and His 98 in I-*Ppo*I). However, in the absence of DNA, we find the magnesium binding site 3.8 Å removed from this common binding site.

This is a remarkable result that could be important for I-*Ppo*I catalysis or it could be simply an artifact of the relatively weak binding between magnesium and most nucleases. Arguments in favor of this being a meaningful finding include the high overall quality of the I-*Ppo*I structure determination, and the chemical reasonableness of the shifted metal binding site. Also, it is known that I-*Ppo*I does not require Mg^{2+} to bind specifically to its cognate DNA substrate *(39)*.

In some ways the secondary magnesium binding site strongly resembles the normal catalytic site. That is, it is composed of a histidine and an asparagine residue. In fact, this auxillary site might bind metal more tightly than the catalytic site because the metal is chelated by two protein residues, as opposed to only one. In the absence of bound DNA, therefore, the metal is tucked away in this alternative binding site. When cognate DNA is bound to I-*Ppo*I, the magnesium is drawn into position for catalysis, perhaps by electrostatic forces. This "lock-and-load" magnesium feature of the I-*Ppo*I active site could represent a mechanism to improve the specificity of cleavage in I-*Ppo*I, by preventing DNA cleavage of non-cognate DNA which might be transiently bind near the active site. In fact, if I-*Ppo*I endonuclease maintained its magnesium in the catalytic position at all times it would be hard to understand why more non-specific cleavage does not take place. For example, the *Serratia* nuclease which has similar active site geometry and a metal bond in the catalytic position at all times rapidly cleaves DNA and RNA with little sequence selectivity *(21)*.

Acknowledgments

We would like to thank Dr. Ron Raines for supplying the I-*Ppo*I protein, Drs. G. N. Phillips, Jr. and R. O. Fox for data collection time, Dr. M. A. White for assistance with the CCD data collection and Mr. Bart M. DeLatte for assistance in I-*Ppo*I crystal growth and structure solution. This work has been supported by The Robert A. Welch Foundation, the National Institutes of Health, the State of Texas, The Methodist Hospital Foundation, and the W. M. Keck Foundation.

References

1. Muscarella, D. E.; Ellison, E. L.; Ruoff, B. M.; Vogt, V. M., Characterization of I-*Ppo*, an intron-encoded endonuclease that mediates homing of a group I intron in the ribosomal DNA of *Physarum polycephalum. Mol Cell Biol* 1990, 10, 3386-96.
2. *Nucleases*. Linn, S. M.; Lloyd, R. S.; Roberts, R. J., Eds. 2nd ed. Cold Spring Harbor Monograph Series. Cold Spring Harbor Laboratory: Cold Spring Harbor; 1993.
3. Westheimer, F. H., Why nature chose phosphates. *Science* 1987, 235, 1173-1178.
4. Chin, J.; Banaszcyk, F.; Jubian, V.; Zou, X., Co(III) complex promoted hydrolysis of phosphate diesters: Comparison in reactivity of rigid *cis*-diaquotetraazacoboalt(III) complexes. *J Am Chem Soc* 1989, 111, 186-190.

5. Serpersu, E. H.; Shortle, D.; Mildvan, A. S., Kenetic and magnetic resonance studies of the active-site mutants of staphylococcal nuclease: Factors contributing to catalysis. *Biochemistry* 1987, 26, 1289-1300.
6. Gerlt, J. A., *Mechanistic principles of enzyme-catalyzed cleavage of phosphodiester bonds*, in *Nucleases*, S.M. Linn, R.S. Lloyd, and R.J. Roberts, Eds. Cold Spring Harbor Laboratory: Cold Spring Harbor, 1993; pp. 1-34.
7. Guthrie, J. P., Hydration and dehydration of phosphoric acid derivatives: Free enerfies of formation of the pentacoordinate intermediates for phosphate ester hydrolysis and of monomeric metaphosphate. *J. Am. Chem. Soc.* 1977, 3991-4001.
8. Eisenstein, B., *Enterobacteriaceae*, in *Principles and Practice of Infectious Diseases*, G.L. Mandell, J.E. Bennett, and R. Dolin, Eds. Churchill Livingstone: New York, 1995; pp. 1964-1980.
9. Acar, J. F., *Serratia marcescens* infections. *Infect. Control.* 1986, 7, 273-8.
10. Saito, H.; Elting, L.; Bodey, G. P.; Berkey, P., *Serratia* Bacteremia: review of 118 cases. *Rev Infect Dis* 1989, 11, 912-20.
11. Ball, T. K.; Saurugger, P. N.; Benedik, M. J., The extracellular nuclease gene of *Serratia marcescens* and its secretion from *Escherichia coli*. *Gene* 1987, 57, 183-192.
12. Molla, A.; Matsumura, Y.; Yamamoto, T.; Okamura, R.; Maeda, H., Pathogenic capacity of proteases from *Serratia marcescens* and *Pseudomonas aeruginosa* and their suppression by chicken egg white ovomacroglobulin. *Infect Immun* 1987, 55, 2509-17.
13. Heller, K., Lipolytic Activity Copurified with the Outer Membrane of *Serratia marcescens*. *J Bacteriol* 1979, 140, 1120-1122.
14. Givskov, M.; Olsen, L.; Molin, S., Cloning and Expression in *Escherichia coli* of the Gene for Extracellular Phospholipase A1 from *Serratia liquefaciens*. *J Bacteriol* 1988, 170, 5855-5862.
15. Jones, J. D. G.; Grady, K. L.; Suslow, T. V.; Bedbrook, J. R., Isolation and characterization of genes encoding two chitinase enzymes from *Serratia marcescens*. *EMBO J* 1986, 5, 496-473.
16. Fraser, M. J.; Low, R. L., *Fungal and mitochondrial nucleases*, in *Nucleases*, S.M. Linn, R.S. Lloyd, and R.J. Roberts, Eds. Cold Spring Harbor Laboratory: Cold Spring Harbor, 1993; pp. 171-207.
17. Meiss, G.; Friedhoff, P.; Hahn, M.; Gimadutdinow, O.; Pingoud, A., Sequence preferences in cleavage of dsDNA and ssDNA by the extracellular *Serratia marcescens* endonuclease. *Biochemistry* 1995, 34, 11979-11988.
18. Côté, J.; Renaud, J.; Ruiz-Carrillo, A., Recognition of $(dG)_n \cdot (dC)_n$ sequences by endonuclease G: Characterization of the calf thymus nuclease. *J. Biol. Chem.* 1989, 264, 3301-3310.
19. Tiranti, V.; Rossi, E.; Ruiz-Carrillo, A.; Rossi, G.; Rocchi, M.; DiDonato, S.; Zuffardi, O.; Zeviani, M., Chromosomal localization of mitochondrial transcription factor A (TCF6), single-stranded DNA-binding protein

(SSBP), and endonuclease G (ENDOG), three human housekeeping genes involved in mitochondrial biogenesis. *Genomics* 1995, 25, 559-64.
20. Prats, E.; Noel, M.; Letourneau, J.; Tiranti, V.; Vaque, J.; Debon, R.; Zeviani, M.; Cornudella, L.; Ruiz-Carrillo, A., Characterization and expression of the mouse endonuclease G gene. *DNA Cell Biol* 1997, 16, 1111-22.
21. Miller, M. D.; Cai, J.; Krause, K. L., The active site of *Serratia* endonuclease contains a conserved magnesium- water cluster. *J. Mol. Biol.* 1999, 288, 975-87.
22. Eaves, G. N.; Jeffries, C. D., Isolation and Properties of an Exocellular Nuclease of *Serratia marcescens. J. Bacteriol.* 1963, 85, 273-278.
23. Nestle, M.; Roberts, W. K., An extracellular nuclease from *Serratia marcescens*. II. Specificity of the enzyme. *J Biol Chem* 1969, 244, 5219-25.
24. Nestle, M.; Roberts, W. K., An extracellular nuclease from *Serratia marcescens*. I. Purification and some properties of the enzyme. *J Biol Chem* 1969, 244, 5213-8.
25. Friedhoff, P.; Meiss, G.; Kolmes, B.; Pieper, U.; Gimadutdinow, O.; Urbanke, C.; Pingoud, A., Kinetic analysis of the cleavage of natural and synthetic substrates by the *Serratia nuclease. Eur J Biochem* 1996, 241, 572-80.
26. Filimonova, M. N.; Balaban, N. P.; Sharipova, F. P.; Leshchinskaia, I. B., Isolation and physico-chemical properties of homogenous nuclease from *Serratia marcescens. Biokhimiia* 1980, 45, 2096-104.
27. Benzon, *Product Specifications for Benzonase, The first industrial endonuclease.* Benzon Pharma A/S, Helseholmen 1, P.O. Box 1185, DK-2650 Hvidovre Denmark. 1993.
28. Miller, M. D.; Tanner, J.; Alpaugh, M.; Benedik, M. J.; Krause, K. L., 2.1 Å structure of *Serratia* endonuclease suggests a mechanism for binding to double-stranded DNA. *Nat Struct Biol* 1994, 1, 461-8.
29. Murzin, A. G.; Brenner, S. E.; Hubbard, T.; Chothia, C., SCOP: a structural classification of proteins database for the investigation of sequences and structures. *J. Mol. Biol.* 1995, 247, 536-540.
30. Orengo, C. A.; Michie, A. D.; Jones, S.; Jones, D. T.; Swindells, M. B.; Thornton, J. M., CATH -- a hierarchic classification of protein domain structures. *Structure* 1997, 5, 1093-1108.
31. Miller, M.; Krause, K., Identification of the *Serratia* endonuclease dimer: structural basis and implications for catalysis. *Protein Sci* 1996, 5, 24-33.
32. Nicholls, A.; Sharp, K. A.; Honig, B., Protein folding and association: insights from the interfacial and thermodynamic properties of hydrocarbons. *Proteins* 1991, 11, 281-296.
33. Friedhoff, P.; Kolmes, B.; Gimadutdinow, O.; Wende, W.; Krause, K. L.; Pingoud, A., Analysis of the mechanism of the *Serratia* nuclease using site-directed mutagenesis. *Nucleic Acids Res* 1996, 24, 2632-9.

34. Friedhoff, P.; Gimadutdinow, O.; Pingoud, A., Identification of catalytically relevant amino acids of the extracellular *Serratia marcescens* endonuclease by alignment-guided mutagenesis. *Nucleic Acids Res* 1994, 22, 3280-7.
35. Lin, J.; Vogt, V. M., I-*Ppo*I, the endonuclease encoded by the group I intron PpLSU3, is expressed from an RNA polymerase I transcript. *Mol Cell Biol* 1998, 18, 5809-17.
36. Belfort, M.; Roberts, R. J., Homing endonucleases: keeping the house in order. *Nucleic Acids Res* 1997, 25, 3379-3388.
37. Flick, K. E.; McHugh, D.; Heath, J. D.; Stephens, K. M.; Monnat, R. J., Jr.; Stoddard, B. L., Crystallization and preliminary X-ray studies of I-*Ppo*I: a nuclear, intron-encoded homing endonuclease from *Physarum polycephalum*. *Protein Sci* 1997, 6, 2677-2680.
38. Wittmayer, P. K.; McKenzie, J. L.; Raines, R. T., Degenerate DNA recognition by I-*Ppo*I endonuclease. *Gene* 1998, 206, 11-21.
39. Wittmayer, P. K.; Raines, R. T., Substrate binding and turnover by the highly specific I-*Ppo*I endonuclease. *Biochemistry* 1996, 35, 1076-83.
40. Flick, K. E.; Jurica, M. S.; Monnat, R. J., Jr.; Stoddard, B. L., DNA binding and cleavage by the nuclear intron-encoded homing endonuclease I-*Ppo*I. *Nature* 1998, 394, 96-101.
41. Galburt, E. A.; Chadsey, M. S.; Jurica, M. S.; Chevalier, B. S.; Erho, D.; Tang, W.; Monnat, R. J., Jr.; Stoddard, B. L., Conformational changes and cleavage by the homing endonuclease I-*Ppo*I: a critical role for a leucine residue in the active site. *J Mol Bio* 2000, 300, 877-887.
42. Otwinowski, Z.; Minor, W., Processing of X-ray diffraction data collected in oscillation mode. *Meth Enzymol* 1997, 276, 307-326.
43. Bruker-AXS, *SMART/SAINT Software Program Manual*: Madison, WI. 1996.
44. Navaza, J., AMoRe: an automated package for molecular replacement. *Acta Crystallog.* 1994, A50, 157-163.
45. Bernstein, F. C.; Koetzle, T. F.; Williams, G. J. B.; Meyer, E. F.; Brice, M. D.; Rodgers, J. R.; Kennard, O.; Shimanouchi, T.; Tasumi, M., The protein data bank: a computer-based archival file for macromolecular structures. *J. Mol. Biol.* 1977, 112, 535-542.
46. Jones, T. A.; Zou, J.-Y.; Cowan, S. W., Improved methods for building protein models in electron density maps and the location of errors in these models. *Acta Crystallographica* 1991, A47, 110-119.
47. Brünger, A. T.; Krukowski, A.; Erickson, J. W., Slow-cooling protocols for crystallographic refinement by simulated annealing. *Acta Crystallogr.* 1990, A46, 585-593.
48. Rice, L. M.; Brünger, A. T., Torsion angle dynamics: reduced variable conformational sampling enhances crystallographic structure refinement. *Proteins* 1994, 19, 277-90.
49. Brünger, A. T., *X-PLOR version 3.1 A system for x-ray crystallography and NMR*. New Haven: Yale University Press; 1992.

50. Hendickson, W. A.; Konnert, J. H., in *Computing in Crystallography*, R. Diamond, S. Ramaseshan, and K. Venkatesan, Eds. Indian Institute of Science: Bangalore, 1980; pp. 13.01-13.23.
51. Engh, R. A.; Huber, R., Accurate bond and angle parameters for X-ray protein structure refinement. *Acta Crystallogr.* 1991, A47, 392-400.
52. Brünger, A. T., The Free R value: A novel statistical quantity for the assessing the accuracy of crystal structures. *Nature* 1992, 355, 472-474.
53. Perrakis, A.; Morris, R.; Lamzin, V. S., Automated protein model building combined with iterative structure refinement. *Nat Struct Biol* 1999, 6, 458-463.
54. Lamzin, V. S.; Wilson, K. S., Automated refinement for protein crystallography. *Meth Enzymol.* 1997, 277, 269-305.
55. Murshudov, G. N.; Vagin, A. A.; Dodson, E. J., Refinement of macromolecular structures by the maximum-liklihood method. *Acta Crystallogr.* 1997, D53, 240-255.
56. Laskowski, R. A.; MacArthur, M. W.; Moss, D. S.; Thornton, J. M., PROCHECK: a program to check the stereochemical quality of protein structures. *J. Appl. Cryst.* 1993, 26, 283-291.
57. Hooft, R. W. W.; Vriend, G.; Sander, C.; Abola, E. E., Errors in protein structures. *Nature* 1996, 381, 272.
58. Kleywegt, G. J.; Jones, T. A., Efficient rebuilding of protein structures. *Acta Crystallogr.* 1996, D52, 829-832.
59. Kleywegt, G. J.; Jones, T. A., Detecting folding motifs and similarities in protein structures. *Meth Enzymol* 1997, 277, 525-545.
60. Nadassy, K.; Wodak, S. J.; Janin, J., Structural features of protein-nucleic acid recognition sites. *Biochemistry* 1999, 38, 1999-2017.
61. Friedhoff, P.; Franke, I.; Meiss, G.; Wende, W.; Krause, K. L.; Pingoud, A., A similar active site for non-specific and specific endonucleases. *Nat Struct Biol* 1999, 6, 112-113.
62. Friedhoff, P.; Franke, I.; Krause, K. L.; Pingoud, A., Cleavage experiments with deoxythymidine 3'-5'-bis-(p-nitrophenyl phosphate) suggest that the homing endonuclease I-*Ppo*I follows the same mechanism of phosphodiester bond hydrolysis as the non-specific *Serratia* nuclease. *FEBS Lett* 1999, 443, 209-214.

Chapter 18

Activation of Hematopoiesis and Vasculogenesis in the Mouse Embryo: Induction and Reprogramming of Ectodermal Cell Fate by Signals from Primitive Endoderm

Margaret H. Baron

Departments of Medicine, Biochemistry and Molecular Biology, Ruttenberg Cancer Center, and Institute for Gene Therapy and Molecular Medicine, Mount Sinai School of Medicine, New York, NY 10029

During vertebrate gastrulation, the anterior-posterior axis of the embryo becomes morphologically evident, mesoderm is induced from ectoderm, and the basic body plan of the animal is established. Nascent mesoderm arises from the primitive streak, a structure which forms at the posterior pole of the embryo. The first mesodermal cell types to form are primitive erythroblasts and vascular endothelial cells, said to represent "posterior" cell fates. Using a novel transgenic explant culture system, we showed that development of these lineages is not mesoderm-autonomous but requires signals secreted from the adjacent primitive endoderm. Remarkably, these signals can also reprogram anterior embryonic ectoderm, a tissue that ordinarily would form neural structures, to form blood and endothelial cells. Therefore, primitive endoderm is a source of instructive signals for the activation of hematopoietic and vascular cell lineages and also plays a key role in anterior-posterior patterning of the mouse embryo.

The determination of cell fate during development is governed by instructive and permissive signals emitted and received between different cell populations of the embryo. This phenomenon is referred to as "embryonic induction." Among the first differentiated cell types to form in the developing vertebrate embryo in response to induction of mesoderm (reviewed in ref. *1,2*) are blood and vascular endothelial cells. The first site of hematopoiesis and vasculogenesis in nearly all vertebrate embryos is the yolk sac. In the mouse, the yolk sac is an extraembryonic, two-layered tissue in which an epithelium of primitive (visceral) endoderm is closely apposed with a layer of mesoderm. The extraembryonic mesoderm of the yolk sac gives rise to blood (largely primitive erythroid) and endothelial cells, which begin to form "blood islands" by around 7.5 days of development, late in gastrulation (for a review, see ref. *3*). Until recently, the contribution -- if any -- of primitive endoderm to hematopoiesis and vasculogenesis in the mouse embryo was unknown. Work carried out in our laboratory has demonstrated clearly that inductive signals from this secretory epithelial tissue are essential for the formation of the first hematopoietic and endothelial cells during mouse embryogenesis (*4*).

I begin this chapter by reviewing the major events of mouse gastrulation, the process in which mesoderm is generated and the final (trilaminar) body plan of the developing embryo is established. Next follow brief reviews of induction of hematopoietic mesoderm and the role of epithelial-mesenchymal interactions during embryonic development. Finally, I describe a novel explant culture assay devised in our laboratory for transgenic mouse embryos which allowed us to examine the potential role of endodermal signals in specifying hematopoietic and vascular endothelial (posterior) cell fates. On the basis of this work, we concluded that primitive endoderm signaling is a critical early determinant of hematopoietic and vascular development and plays a decisive role in anterior-posterior patterning during mouse embryogenesis.

Gastrulation and Induction of Mesoderm

The first known inductive events in vertebrates occur during gastrulation, a process in which cells of the primitive ectoderm (epiblast) delaminate on the posterior aspect of the embryo and ingress through the primitive streak, resulting in the formation of the definitive embryonic germ layers and the establishment of the basic body plan (reviewed in refs. *1,2*). As gastrulation progresses and more epiblast cells ingress, the primitive streak extends towards the distal end of the embryo. Fate mapping studies have shown that distinct types of mesoderm arise from different positions along the primitive streak (*5*). Thus, extraembryonic mesoderm is the first type of mesoderm derived from the posterior streak and it is followed by lateral, paraxial and finally axial mesoderm which arise from progressively more anterior parts. Epiblast cells that do not ingress through the streak expand into the area previously occupied by mesodermal and endodermal precursors and eventually give rise to embryonic ectoderm derivatives such as neurectoderm and surface ectoderm (5).

In nearly all vertebrate animals, embryonic blood development begins during gastrulation and results from the induction of extraembryonic mesoderm to form hematopoietic tissue. In the mouse, these events are initiated at around 6.5 days post coitum (p.c.) and lead to the formation of the yolk sac, a bilaminar membrane composed of adjacent mesodermal and primitive endodermal cell layers (reviewed in ref. 3). The yolk sac is an extraembryonic tissue which surrounds the entire embryo. Though it will not contribute cells directly to the fully formed animal, its function is essential to normal development (reviewed in refs. 3,6).

Induction of Hematopoietic Mesoderm

Primitive hematopoiesis, the formation of embryonic blood cells, and vasculogenesis, the de novo formation of blood vessels from endothelial cells, begin during gastrulation and occur essentially concurrently during early development. Induction of mesoderm and embryonic hematopoiesis has been most extensively studied in the frog Xenopus. Animal cap cells, which are equivalent to the epiblast of mouse embryos, differentiate into a variety of mesodermal derivatives when recombined in vitro with vegetal pole cells. Treatment of animal cap cells with various purified proteins also results in the induction of mesoderm in vitro (for a review, see ref. 7). These in vitro assays have identified numerous members of the transforming growth factor-β (TGF-β) and fibroblast growth factor (FGF) families as potential mesoderm inducers. Although experiments involving the injection of dominant negative receptors into frog embryos have provided additional support for the role of these molecules in mesoderm induction (8,9), interpretation of such experiments is complicated by the fact that dominant negative receptors can potentially inhibit signaling of other closely related receptors. Thus, the identity of endogenous mesoderm inducing molecules remains to be demonstrated definitively for the frog embryo.

In contrast to Xenopus, where embryological experiments indicate that the vegetal pole is the source of mesoderm inducing signals, it has been difficult to perform similar tissue recombination experiments in the mouse to determine which tissue is responsible for mesoderm induction. This is partly due to the small size of mouse tissues as well as their more demanding growth requirements. However, a major advantage of the mouse system has been the availability of naturally-occuring mutants as well as the development of gene targeting techniques to specifically abolish the function of a given gene, permitting more definitive conclusions to be drawn about the function of the gene in question.

A member of the TGF-β superfamily of extracellular signaling molecules, *Bone Morphogenetic Protein-4* (Bmp-4, see ref. *10*) is required for blood island formation in the mouse embryo. Although targeted mutagenesis of this gene compromises both blood and endothelial cell development (*10*), it is not yet known whether the signaling protein it encodes is involved solely in the induction of mesodermal progenitors of primitive hematopoietic and vascular endothelial cells or whether it plays a later role in lineage specification in the yolk sac. It is now clear that, at least in some developmental processes, BMP-4 is a morphogen (reviewed in ref. *11*), so it is possible that this protein induces mesoderm at one threshold concentration and has distinct activities at other concentrations (for a review, see ref. *12*). Not surprisingly, the phenotype of a null mutation in the type I receptor which binds BMP-4 (*13*) is very similar to that of the *Bmp-4* knockout.

Smad2, an intracellular effector in the TGFβ/activin signaling pathway, plays a key role in specification of the anterior aspect of the embryo (*14*). This gene is widely expressed and regulates an extraembryonic signal, though the tissue of origin -- primitive endoderm or extraembryonic ectoderm or both -- has not yet been identified. Interestingly, in the absence of Smad2, the embryo proper fails to develop and the resulting structure resembles a giant yolk sac, adopting cell fates associated with posterior mesoderm (*14*). In these *Smad2* mutant embryos, which lack any clear evidence of proximal/distal or antero-posterior polarity, Smad1 signaling is unopposed and presumably results in the formation of mesoderm and associated posterior cell types (see below). It is worth noting that *Bmp-4* expression early in gastrulation is restricted to distal extraembryonic ectoderm (derived from the trophectoderm lineage) abutting the proximal rim of the epiblast (*14*). Later it is expressed in posterior mesoderm (*10*) and in the mesodermal layer of the yolk sac (*15*). The signals responsible for activating expression of *Bmp-4* early in embryonic development are unknown.

Epithelial-mesenchymal interactions in early post-implantation development of the mouse

Morphogenesis and patterning of the vertebrate embryo is dependent upon finely orchestrated interactions between neighboring tissues. In the mouse, the embryonic ectoderm is situated adjacent to an outer layer of primitive (visceral) endoderm, a secretory epithelium which plays a number of critical regulatory roles during early postimplantation development (reviewed by *6,16*). For example, cavitation (*17*), growth and survival (*18-20*) of the ectoderm require primitive endoderm signals. Patterning of anterior ectoderm, the region of the embryo that will form the central nervous system, is initiated by spatially regulated gene expression in the overlying visceral endoderm (for a review, see

21). In virtually all mammalian embryos, the anterior visceral endoderm (AVE) can be distinguished only at the molecular level. Patterned expression of a variety of genes, including *Otx2, Lim1, goosecoid, cerberus-like, Hesx1, nodal, Hex,* and *Mrg1* is detected in the AVE well before formation of the primitive streak, which marks the site of mesoderm formation at the most posterior aspect of the embryo (reviewed recently in ref. *16*). *Smad2*, an intracellular effector of TGFβ/activin signaling, has also been implicated in anterior specification of the embryo by regulating an extraembryonic signal (*14*). However, the provenance of that signal (visceral endoderm or extraembryonic ectoderm or both) is not yet known.

Patterning of the posterior ectoderm is only poorly understood. During gastrulation, when the body plan of the embryo is established (reviewed by *22*), anterior movement of distal VE cells and posterior movement of proximal epiblast cells creates an embryonic "rotation" (*5*). Expression of *Brachyury (T)* (*23*), *nodal* (*24*) and *Eomesodermin* (*25*) becomes confined to posterior ectoderm, marking the prospective mesoderm prior to formation of the primitive streak. In addition to its role in maintaining the prospective anterior region of the embryo, the Smad2 pathway appears to restrict the site of primitive streak formation (*14*).

Epithelial-Mesenchymal Interactions in Hematopoiesis and Vasculogenesis

Classic tissue recombination studies with chick embryos indicated that blood formation by yolk sac mesoderm requires diffusible signal(s) from the extraembryonic endoderm (hypoblast) (reviewed in ref. *4*). Although it was later shown that a bFGF-like signal can substitute for the chick hypoblast (analogous to primitive endoderm in the mouse) in stimulating the formation of hemoglobinized tissue in ectodermal explants, the endogenous signal has not been identified (see ref. *4*).

Whether primitive endoderm is required for blood island formation in the mouse embryo was, until recently, controversial. Some evidence existed to support a role for visceral endoderm in the development of the embryonic hematopoietic and vascular endothelial lineages (reviewed in ref. *4*). Mouse embryonic stem (ES) cells deficient for the transcription factor GATA-4 do not develop an outer layer of visceral endoderm when induced to form embryoid bodies and lack any recognizable blood islands. In chimeric mice, blood islands formed normally where mutant ES cells were juxtaposed with normal visceral endoderm. Gene targeting studies have shown, however, that GATA-4 expression in the primitive endoderm is not essential for embryonic hematopoiesis and vasculogenesis. Other studies led to the conclusion that visceral endoderm signaling is not required for hematopoiesis (see below, and discussion in ref. 4).

Transgenic Embryo Explant Culture Assay for Induction of Hematopoiesis by Non-Mesodermal Signals

We were struck by the fact that nascent mesoderm, as it moves into the extraembryonic region of the embryo, comes in contact with three different lineages (embryonic and extraembryonic ectoderm and primitive endoderm), yet only the mesoderm cells adjacent to primitive endoderm will form endothelial cells ("angioblasts") and the first hematopoietic cells of the embryo. To determine whether the primitive endoderm lineage is required for embryonic hematopoiesis, we devised an explant culture system in which ectoderm dissected from pre- or early-gastrulation mouse embryos (6.0-6.25 days post coitum, dpc; see Figure 1A) was cultured in the presence or absence of primitive (visceral) endoderm and analyzed for activation of a primitive erythroid reporter transgene or of endogenous hematopoietic marker genes. For this assay (4), here termed the "induction assay" (Figure 1A), we took advantage of transgenic mouse lines generated in our laboratory in which a *β-galactosidase* (*lacZ*) reporter gene is expressed exclusively in primitive erythroid cells under the control of upstream regulatory regions (26) of the human embryonic β-like *globin* gene (27). Embryos from these animals served as a source of genetically marked ectodermal tissue. Transgenic embryos were harvested at the pre- to early gastrulation stage, prior to the formation of morphologically detectable blood cells or their molecular markers. Ectoderm and visceral endoderm (VE) layers were separated following brief enzymatic treatment of the embryos and were subsequently assayed for *lacZ* transgene or endogenous embryonic *globin* gene expression. Analysis of the expression of endoderm- and ectoderm-specific genesby a semi-quantitative reverse transcription-polymerase chain reaction (RT-PCR) (15) confirmed that there was no cross-contamination of the separated tissues.

Embryonic Hematopoiesis is not Autonomous to Mesoderm

Male transgenic mice bred to homozygosity were mated with non-transgenic females and embryos were harvested at 6.0-6.25 days. Whole embryos or ectoderms stripped of VE (15) were cultured individually for 48-72 hours, then fixed and stained with X-gal to monitor the generation of primitive erythroblasts. Cultured whole embryos formed β-galactosidase-positive blood islands but β-gal staining was not detected in cultured ectoderms separated from the VE layer. These results suggested that embryonic hematopoiesis is not autonomous to ectoderm (or intrinsic to the mesodermal cells arising from the

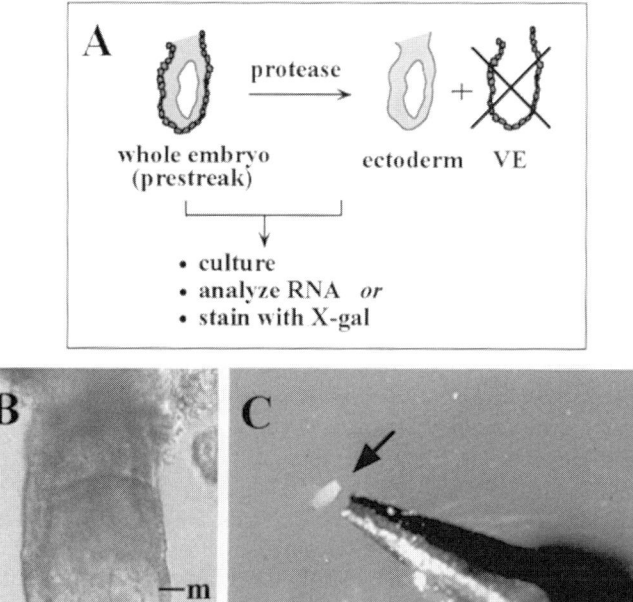

Figure 1. Explant culture assays for gastrulation stage mouse embryos. (A) Photograph of mid-streak embryo. a, anterior; p, posterior; m, mesoderm; ps, region of primitive streak. (B) Photograph of mid-streak embryo (arrow) at tip of pencil to provide indication of size. (C) Diagram of induction assay. (D) Diagram indicating anterior ectoderm region of mid-streak embryo dissected for reprogramming assay.

primitive streak) but requires contact with or signals released from visceral endoderm.

Visceral Endoderm Provides a Signal(s) Required for Embryonic Hematopoiesis

To establish more directly that visceral endoderm is required for induction of hematopoiesis in the gastrulating embryo, tissue recombination experiments were performed. A semi-quantitative RT-PCR protocol was used to assay for activation of the endogenous mouse embryonic β-like globin genes (*15*). Ectoderm and VE layers from individual embryos were cultured separately or in combination for two to four days. RNA was prepared from individual explants and analyzed for embryonic *globin* gene activation using the RT-PCR assay.

Embryonic β-like *globin* gene expression was not detected in newly dissected 6.0-6.25 dpc whole embryos or isolated ectodermal layers. After 72 hours in culture, the mouse embryonic β-like globin genes were activated in whole embryos but in isolated ectoderm little or no *globin* transcription could be detected. In contrast, recombination of ectoderm with VE from same stage embryos resulted in activation of embryonic *globin* to levels comparable to those observed with whole embryos. These observations established that, around the onset of gastrulation, induction of primitive hematopoiesis is not autonomous to ectoderm (more specifically, nascent mesoderm cells arising from the primitive streak) and requires the presence of visceral endoderm (*4*). Using analogous approaches, we established that the requirement for visceral endoderm signaling is restricted to a relatively narrow window of time (*4*).

Respecification of Anterior Ectoderm to Hematopoietic and Endothelial Cell Fates by Visceral Endoderm

Lineage tracing experiments have shown that hematopoietic mesoderm arises from the posterior primitive streak (posterior mesoderm, reviewed in ref. *2*). The explants used in the experiments described above contained posterior ectoderm, which gives rise to blood cells and other derivatives of extraembryonic mesoderm. To determine whether visceral endoderm signaling can respecify ectoderm that does not contain mesoderm and is not normally fated to express a posterior embryonic developmental program, we modified the explant culture assay (*4*). In place of pre- or early-gastrulation stage embryonic ectoderm, we used anterior ectoderm from midgastrulation stage embryos (~6.75 dpc). At this time during development, the anterior epiblast does not yet contain mesoderm (*28*) and is fated to give rise to neurectoderm. It is therefore not expected to produce hematopoietic or vascular tissue in culture. Anterior and posterior aspects of the midstreak stage embryo are easily distinguished by the presence of prominent mesodermal wings in the posterior region (Figure 1B, see area marked "m") and by the primitive streak (Figure 1B, "ps"), which marks the posterior pole of the anterior-posterior axis (*29*). However, these embryos are tiny, only about the size of the tip of a pencil (see Figure 1C), and the dissections therefore require considerable practice and skill.

Embryonic ectoderms stripped of VE were dissected into mesoderm-free anterior pieces and posterior pieces with their associated mesodermal wings. Analysis of expression of *Brachyury*, an early mesodermal marker, was carried out to ensure that the anterior ectoderm pieces were free of contaminating mesoderm. RT-PCR analysis confirmed that dissected anterior and posterior ectodermal explants differ in their capacity to activate expression of embryonic *globin* RNA during culture (*4*), as predicted from lineage tracing studies (*22*).

Anterior epiblast pieces dissected from 6.75 dpc transgenic embryos were recombined with non-transgenic VE in collagen droplets and cultured for 3-4 days (this assay is referred to here as the "reprogramming assay"). Transgenic whole embryos or posterior pieces contained large numbers of β-galactosidase positive, round erythroid cells after four days. In contrast, β-gal positive cells were not detected in cultured transgenic anterior pieces, indicating the absence of blood cells. However, when transgenic anterior ectoderms were recombined with non-transgenic VE, large numbers of β-galactosidase positive blood cells were detected in the recombinants. This experiment confirmed that the erythroid cells present in cultured recombinants were derived from anterior ectoderm (transgenic) and not from VE (non-transgenic). Sectioning of stained recombinants confirmed that X-gal staining was confined to hematopoietic cells. Furthermore, β-galactosidase positive cells remained localized to an area immediately adjacent to the VE tissue, suggesting either that primitive endoderm signaling to underlying ectoderm requires cell-cell contact or that the signal(s) are diffusible but act within short range.

Analysis by RT-PCR of expression of a panel of hematopoietic markers indicated that visceral endoderm signaling results in activation not only of genes characteristic of differentiated erythroid cells (*globin*) but also of early hematopoietic genes such as *GATA-1* and *CD-34*. The presence of VE in recombinants did not result in promiscuous activation of mesoderm differentiation markers: cardiac myosin, which is expected to be expressed only in cardiac tissue (and therefore only at a later developmental stage), was not detected in anterior or posterior epiblast pieces or in recombinants during the first 4 days in culture, but was detected in an older (10.5 dpc) embryonic control.

Early in gastrulation, the homeobox gene *Otx2* is expressed throughout the ectoderm but by the headfold stage becomes restricted to the anterior of the embryo (reviewed in ref. *4*). *Otx2* was expressed at significant levels in both posterior and anterior pieces at the time of their dissection from 6.75 dpc embryos but during culture was dramatically reduced or absent from posterior explants and recombinants. In contrast, *Otx2* expression continued in isolated anterior explants. These results suggest that signals from primitive (visceral) endoderm can respecify tissue (anterior ectoderm) that does not contain mesoderm and has no intrinsic potential to activate expression of markers of a posterior (hematopoietic) cell fate. The reduction of *Otx2* expression and concomitant induction of hematopoietic marker genes in anterior ectoderm/VE recombinants suggest that anterior ectoderm can be respecified by the visceral endoderm to adopt a posterior fate (*4*).

The common origin of blood and endothelial cells from extraembryonic mesoderm and the close temporal association of their development in the yolk sac blood islands has led to the hypothesis that these two cell lineages share a common progenitor, the "hemangioblast" (for a review, see ref. *30*). For example, targeted mutagenesis of a number of mouse genes prevents the formation of both embryonic blood and endothelial cells in the yolk sac (discussed in ref. *31*). Recombination of anterior ectoderm with VE resulted not only in formation of primitive erythroblasts and activation of hematopoietic markers but also in the activation of three endothelial cell genes, *flk-1*, *Vezf-1* and *PECAM*-1 (*4*).

Our results demonstrate that VE can reprogram anterior ectoderm to form cells of both the hematopoietic and vascular endothelial lineages, perhaps by acting on a target hemangioblast. However, the evidence in support of the existence of this hypothetical cell remains circumstantial and awaits more definitive confirmation through lineage tracing studies. The induction of blood cells and markers of hematopoiesis and vasculogenesis, together with downregulation of *Otx2* in anterior ectoderm/VE recombinants, suggest that the inductive signals from VE are instructive (*4*). Whether induction of hematopoiesis and vasculogenesis by VE signaling is direct or indirect (a primary or secondary event) remains to be determined (see below).

Activation of Hematopoiesis by Primitive Endoderm is Mediated by Soluble Molecules

To determine whether cell-cell contact is required for respecification of anterior ectoderm, we carried out two types of experiment. In the first, medium conditioned by a visceral endoderm-like cell line was tested for its ability to substitute for VE tissue in the induction assay. Indeed, activation of an embryonic β-like globin gene was detected using RT-PCR (our unpublished results), indicating that at least some functions of the visceral endoderm can be mediated by soluble molecules. In the second type of experiment, transgenic anterior ectoderm and non-transgenic VE were cultured on opposite sides of a

Nucleopore filter of pore size (0.1 µm) too small to permit the passage of individual cells. β-galactosidase positive erythroblasts were detected only on the side of the filter containing anterior ectoderm and were absent from the VE side. This observation was reinforced by sectioning through an X-gal stained ectoderm-filter-VE sandwich. Therefore, the reprogramming of cells in anterior ectoderm to a posterior fate is mediated by diffusible signaling molecule(s) from VE (*4*).

The explant culture system described here provides an assay that can be used to identify signal(s) involved in activation of hematopoiesis and vasculogenesis. Presently we are testing candidate molecules by using recombinant forms of proteins that we or others have shown to be expressed in visceral endoderm early in development. While Bmp-4 is required for embryonic hematopoiesis and vasculogenesis (*10*), it is not expressed in visceral endoderm (unpublished data and ref. *15*). Another Bmp which is expressed in VE around the onset of gastrulation can, however, as recombinant protein substitute for VE in both the induction and reprogramming assays and is currently under intense investigation in our laboratory. In addition, we are attempting to simplify the reprogramming assay for use in a cDNA expression cloning approach for identifying VE signals.

Evolutionary Conservation of Primitive Endoderm Signaling in Hematopoietic Induction

Embryonic blood and endothelial cells of the yolk sac blood islands arise from extraembryonic mesoderm by around 7.5 days of development in the mouse. The work summarized above was initiated by two simple observations. First, blood islands form in a ring at the level of the exocoelomic cavity of the late-gastrulation stage embryo. This is the only region of the embryo at this stage where mesodermal cells are in direct contact with visceral (primitive) endoderm cells. Second, examination of the structure of the yolk sac reveals that its two tissue layers -- mesoderm and endoderm -- are in very close apposition. In view of the well described secretory function of the visceral endoderm (reviewed in refs. *3,6*), the possibility of epithelial-mesenchymal signaling interactions between the two tissues immediately becomes pertinent. Experiments carried out in the chick more than 30 years ago pointed to a role for primitive endoderm signaling in embryonic hematopoiesis and vasculogenesis (reviewed in ref. *4*). On the basis of our experiments, which establish a critical role for visceral endoderm signaling in embryonic hematopoiesis and vasculogenesis in the mouse, we conclude that these processes are at least partially conserved between mouse and chick.

Anterior-Posterior Axis Determination and Specification of Cell Fate

The primitive hematopoietic and endothelial cell lineages arise from nascent mesoderm from the posterior primitive streak and are considered posterior cell fates. This point is highlighted by recent gene targeting experiments in which anterior-posterior (A-P) axis determination is disrupted in a way that results in the formation of an embryo containing only posterior cell types. Embryos deficient in the transcription factor Smad2, a downstream effector of TGFβ signaling, develop normal yolk sacs in the absence of any embryonic tissues (*14*). The Smad2 pathway evidently serves to restrict the site of primitive streak formation by specifying the anterior pole at the onset of gastrulation.

Smad2 is genetically intact in the embryos used to create the recombinants in our experiments, yet its anteriorizing activity is apparently overcome when visceral endoderm is cultured with anterior ectoderm. The primitive endoderm thus is a source of potent signals that antagonize the Smad2 pathway. Moreover, posterior fate in the epiblast is evidently not a default state but is regulated by visceral endoderm signal(s).

During normal development, the anterior ectoderm is in direct contact with primitive endoderm. It now seems clear that prior to and during gastrulation the anterior visceral endoderm constitutes a signaling center that functions in maintenance of anterior character and head induction (reviewed in ref. *16*). Yet, when cultured in isolation with visceral endoderm, anterior ectoderm can be reprogrammed to posterior (hematopoietic and vascular) cell lineages (*4*).

One explanation for this apparent paradox is that the signal(s) involved in induction of hematopoiesis and vasculogenesis early in gastrulation may normally be regionalized within the posterior visceral endoderm. Our failure to detect blood formation in recombinants containing later stage visceral endoderm (*4*) could therefore have resulted from inadvertent use of tissue taken from the wrong part of the embryo. Though there is clear evidence that some regulatory molecules (e.g. the secreted proteins Cer-1 and nodal and the transcription factors lim-1, goosecoid, and HNF-3β) are restricted in their expression at some point during development to the anterior primitive endoderm and are likely to be involved in the patterning of the underlying anterior ectoderm (for recent review see ref. *16*), specific expression in posterior primitive endoderm has not yet been reported for any molecule. Identification of the visceral endoderm signal and analysis of its developmental expression pattern may help to clarify this question.

Concluding Remarks

We have demonstrated (4) that primitive (visceral) endoderm secretes diffusible signal(s) that are required for the initiation of hematopoiesis and vasculogenesis during embryogenesis. Induction of mesoderm precedes the cellular and molecular events reflected in the explant cultures (our unpublished results). A number of important questions remain unresolved:

1. Might the molecules responsible for respecification of cell fate observed in the reprogramming assay be distinct from those involved in activation of hematopoiesis from nascent posterior mesoderm *in vivo*? The induction (epiblast explant) assay presumably reflects what occurs *in vivo*, while the reprogramming (anterior ectoderm) assay reflects the potential activity of visceral endoderm signals. Anterior ectoderm is normally fated to form neurectoderm, not blood or endothelial cells.
2. Does VE signaling induce hematopoiesis and vasculogenesis indirectly -- through a mesodermal intermediate -- or directly? What, if any, role does BMP-4 play in respecification of anterior ectoderm by VE signaling? BMP-4 is not expressed in VE or in midgastrulation stage anterior ectoderm but is expressed in the posterior primitive streak and then later in nascent mesoderm. *Bmp-4* expression could be the initial event triggered by VE signaling in the reprogramming assay, followed by mesoderm induction and then formation of blood and endothelial cells.
3. How many distinct signaling molecules are involved in these pathways? What are the identities of these molecules? Do they function by stimulating proliferation, survival, or differentiation?
4. Do similar or identical signals play a role in formation of definitive blood and endothelial cells at later stages of development? During gastrulation, nascent mesodermal cells are displaced from the posterior primitive streak to form both extraembryonic and intraembryonic mesoderm. These cells may share a common lineage and their development could certainly be regulated by some of the same molecules.
5. Do any of the signaling molecules produced by visceral endoderm play a role in adult hematopoiesis? While primitive (visceral) endoderm cells are not thought to contribute directly to tissues within the developing animal (those endodermal derivatives form from a different, definitive lineage), they express many of the same genes as endodermal cells of the liver or gut (3,6).

Whether or not the same signals are normally required for hematopoiesis in adult bone marrow, they might nevertheless have the potential to stimulate stem cell proliferation or survival. The possibility that such embryonic inducers of hematopoiesis might lead to advances in stem cell transplantation is an exciting one.

References

1. Conlon, F.; Beddington, R. *Seminars in Dev. Biol.* **1995**, *6*, 249-256.
2. Tam, P. L.; Behringer, R. R. *Mech. Dev.* **1997**, *68*, 3-25.
3. Farrington, S. M., Ph.D. Thesis, Harvard University, Cambridge, MA, 1996.
4. Belaoussoff, M.; Farrington, S. M.; Baron, M. H. *Development* **1998**, *125*, 5009-5018.
5. Lawson, K. A.; Meneses, J. J.; Pedersen, R. A. *Development* **1991**, *113*, 891-911.
6. Rossant, J. *Semin. Dev. Biol.* **1995**, *6*, 237-247.
7. Kessler, D. S.; Melton, D. A. *Science* **1994**, *266*, 596-604.
8. Graff, J. M.; Thies, R. S.; Song, J. J.; Celeste, A. J.; Melton, D. A. *Cell* **1994**, *79*, 169-179.
9. Harland, R. M. *Proc. Natl. Acad. Sci. U.S.A.* **1994**, *91*, 10243-10246.
10. Winnier, G.; Blessing, M.; Labosky, P. A.; Hogan, B. L. M. *Genes & Develop.* **1995**, *9*, 2105-2116.
11. Lemaire, P.; Yasuo, H. *Current Biol.* **1998**, *8*, R228-231.
12. Dale, L. *Curr. Biol.* **1997**, *7*, R698-R700.
13. Mishina, Y.; Suzuki, A.; Ueno, N.; Behringer, R. R. *Genes & Develop.* **1995**, *9*, 3027-3037.
14. Waldrip, W. R.; Bikoff, E. K.; Hoodless, P. A.; Wrana, J. L.; Robertson, E. J. *Cell* **1998**, *92*, 797-808.
15. Farrington, S. M.; Belaoussoff, M.; Baron, M. H. *Mech. Dev.* **1997**, *62*, 197-211.
16. Beddington, R. S. P.; Robertson, E. J. *Cell* **1999**, *96*, 195-209.
17. Coucouvanis, E.; Martin, G. R. *Cell* **1995**, *83*, 279-287.
18. Chen, W. S.; Manova, K.; Weinstein, D. C.; Duncan, S. A.; Plump, A. S.; Prezioso, V. R.; Bachvarova, R. F.; J.E. Darnell, J. *Genes & Develop.* **1994**, *8*, 2466-2477.
19. Spyropoulos, D. D.; Capecchi, M. R. *Genes & Develop.* **1994**, *8*, 1949-1961.
20. Sirard, C.; de la Pompa, J. L.; Elia, A.; Ities, A.; Mirtsos, C.; Cheung, A.; Hahn, S.; Wakeham, A.; Schwartz, L.; Kern, S. E.; Rossant, J.; Mak, T. W. *Genes & Develop.* **1998**, *12*, 107-119.
21. Beddington, R. S. P.; Robertson, E. J. *Trends Genet.* **1998**, *14*, 277-284.
22. Tam, P. P. L.; Beddington, R. S. P. *Development* **1987**, *99*, 109-126.
23. Herrmann, B. G. *Develop.* **1991**, *113*, 913-917.
24. Conlon, F. L.; Lyons, K. M.; Takaesu, N.; Barth, K. S.; Kispert, A.; Herrmann, B.; Robertson, E. J. *Development* **1994**, *120*, 1919-1928.
25. Russ, A. P.; Wattler, S.; Colledge, W. H.; Aparicio, S. A.; Carlton, M. B.; Pearce, J. J.; Barton, S. C.; Surani, M. A.; Ryan, K.; Nehls, M. C.; Wilson, V.; Evans, M. J. *Nature* **2000**, *404*, 95-99.
26. Trepicchio, W. L.; Dyer, M. A.; Baron, M. H. *Mol. Cell. Biol.* **1993**, *13*, 7457-7468.

27. Belaoussoff, M., Ph.D. Thesis, Harvard University, Cambridge, MA, 1998.
28. Beddington, R. S. P. *J. Embryol. Exp. Morphol.* **1982**, *69*, 265-285.
29. Downs, K. M.; Davies, T. *Development* **1993**, *118*, 1255-1266.
30. Dieterlen-Lievre, F. *Current Biol.* **1998**, *8*, R727-730.
31. Shalaby, F.; Ho, J.; Stanford, W. L.; Fischer, K.-D.; Schuh, A. C.; Schwartz, L.; Bernstein, A.; Rossant, J. *Cell* **1997**, *89*, 981-990.

Chapter 19

Scoring Functions Sensitive to Alignment Error Have a More Difficult Search: A Paradox for Threading

Jeffrey Chang, Michelle Whirl Carrillo, Allison Waugh, Liping Wei, and Russ B. Altman*

Stanford Medical Informatics, Stanford University Medical Center, MSOB X–215, Stanford, CA 94305–5479
({jchang, whirl, waugh, wei, altman}@smi.stanford.edu)

The task of recognizing protein sequence-structure compatibility (or threading) holds promise for detecting similarities that can not be seen with sequence analysis alone. The development of a method that can recognize a fold as well as the correct correspondence (or alignment) to the sequence remains a major challenge. We have created a new threading evaluation function that (1) uses a full atomic representation of structure (rapidly instantiating a plausible set of sidechains), (2) reflects the idiosyncrasies of particular environments within a fold family, and (3) is very sensitive to shifts away from the correct alignment. We illustrate these characteristics in the context of the globin family. We created a statistical model of globin environments using a structural alignment of 13 globin (and globin-like) molecules. We tested these with cross-validation, and found that our method reliably recognizes the correct alignments with sequence identities in the range of 11% to 18% identity, as long as the backbone upon which the sequence is threaded (the template backbone) is within 2.9 Å RMSD of the actual backbone. When the backbone is more than 3.0 Å from the template backbone, we are unable to detect the correct alignment. Our method shows sensitivity to both systematic and random shifts in the alignment—even with shifts of a single residue. These results suggest that threading methods using atomic detail and statistical descriptions of structural environments can be effective. Such methods, however, make the search for sequence-structure compatibility more difficult, since they create narrow, steep optima.

Introduction

The database of protein sequences is growing at an exponential rate. Analyses of these sequences alone do not capture the physical and chemical properties of the 3D structure of the folded protein. Knowing the structure of a protein gives insight into its biochemical function and therefore its role in the molecular processes in the cell. Unfortunately, experimental methods to elucidate protein structures are time-consuming and can not keep up with the rate at which sequences are generated. Thus, there is considerable interest in computational methods to predict the protein structure from its sequence accurately

Despite the success of sequence alignment methods in recognizing similar proteins when sequence identity is above 25 to 30%, there are many sequences with similar folds that have much lower sequence identity (1). In particular, it is difficult to distinguish structurally homologous sequences from unrelated sequences when the sequence identity is in the "twilight zone" region below 20-25% sequence identity (2).

Threading methods have received considerable attention recently as partial solutions to the problem of recognizing structural similarity in the context of distant sequences (3, 4). Threading methods take advantage of the growing library of protein structures to determine whether a new sequence adopts a structure similar to one already observed. They generally require: 1) a library of known structural backbones, 2) a method for mounting a new sequence on a template structural backbone, and 3) an objective function to evaluate the compatibility of the mounted sequence with the template backbone. A prediction is made when a new sequence scores highly when aligned in a particular manner with the template structural backbone. Most threading potentials are pairwise potentials that evaluate the compatibility between amino acids that are in proximity once the sequence is placed upon the template backbone (5). Alternative alignments can be generated randomly, heuristically, or in some cases by exhaustive search (6). Recent evaluations of threading methods have shown increasing success in recognizing some correct folds; however, the most distantly related ones are still difficult to detect (7), and in particular, the quality of the alignments decrease with more distant homology (8). We hypothesize that correct alignments can be detected by using full atomic representations of amino acid sidechains (and not single pseudo-atoms), and by creating environmental descriptions of sites within proteins that reflect the detailed geometry and biochemical/biophysical features of the environments around particular alpha carbons.

Previously, we developed a computer program that can generate a statistical description of the 3D biochemical environments within proteins (9, 10, 11). The FEATURE system calculates the spatial distribution of a comprehensive set of physico-chemical properties. Given a set of *sites* that share some structural or functional role and a set of control sites (or *nonsites*) without such a role, FEATURE compares the distribution of values for each property and reports those regions of space containing significant differences in the abundance of particular properties. In an extension of this work, we have developed an algorithm for using the statistical description produced by FEATURE to compute the likelihood that a new query environment is a site of interest (12).

One can view a protein structure as a series of microenvironments surrounding the alpha carbons within the sequence. These environments are represented as concentric radial shells, each containing detailed information about the shell's biochemical and physical properties.

Given a set of aligned structures from a fold family, we can generate a statistical description of the environment surrounding each set of corresponding alpha carbons. We first take a template backbone and mount the sidechains of a new sequence upon the template. Then we can compute a likelihood score for each alpha carbon environment (created by the sidechains as they hang off the template backbone) and sum these to get an overall measure of the compatibility of this sequence to the template backbone.

In this paper, we studied diverse globin-like structures with sequence identities ranging from 9 to 25 percent. The set contains proteins that are closely related in terms of function, as well as two that are functionally diverse. We studied the ability of our scoring function to distinguish the correct alignments from shifted alignments.

Methods

We obtained the 1HLB structural alignment file from the FSSP database (*13*). We selected structures with less than 25% sequence identity yielding thirteen PDB proteins: 1HLB, 1MBC, 1FLP, 2FAL, 1ITH-A, 2HBG, 1ASH, 3SDH-A, 2GDM, 1BAB-A, 1COL-A, 1CPC-A, and 1CPC-B. These proteins range in length from 141 to 197 residues, but share 104 common "core" amino acid positions. We performed threading using leave-one-out cross-validation, in which we selected a test protein and used the remaining 12 proteins to create an average backbone template. The template was obtained by computing an unbiased average of the twelve backbones (*14, 15*) and then choosing, from among those twelve, the backbone closest to the computed average. Our test threading targets (representing a range of structures within the family) were 1HLB, 1BAB-A, 2GDM, 1COL-A and 1CPC-A. These proteins have a range of RMS distances from their template backbones of 1.4 to 3.4 Å.

To extract the statistical environmental descriptions, FEATURE was applied to the region surrounding each of the core 104 alpha carbons within the 12 training proteins. For the nonsite controls, we randomly selected alpha carbons from a set of twenty non-homologous PDB proteins (1ACX, 1AVR, 1F3G, 1FKF, 1GAL, 1IPD, 1PAF, 1PDA, 1PYP, 1REC, 1RHD, 1SNC, 1TEN, 2ACT, 2BP2, 2BPA, 2HIP, 3GLY, 4TMS, 9INS; *16*). Using the algorithm as previously described, FEATURE computes the key biochemical and biophysical properties of the microenvironments around the aligned globin alpha carbons, as compared with the randomly chosen alpha carbons from nonhomologous proteins (*9, 10*; the analysis was centered at the alpha carbon of the sites and nonsites and a threshold of $p = 0.01$ was used to determine if the differences in the environments for given properties at specific radii were statistically significant). The FEATURE program is implemented in C++ and compiles on Windows or Unix-type environments.

Generating Sample Alignments

Since our 104 core positions occur within 13 ungapped segments, the task of generating an alignment is equivalent to generating loop lengths between the 13 segments. For each query sequence (a sequence not used in training the model), we generated possible alignments with the core in two ways. First, we took the correct alignment and systematically shifted it along the query sequence. Second, we generated random sample alignments using information about the average loop lengths gathered from the multiple-alignment file of the FSSP family. Using the multiple sequence alignment, we compute the mean length of each loop, its variance, and its covariance with other loop lengths. Assuming a multivariate, correlated normal distribution of loop lengths, we generated a set of correlated sample gap lengths using a Cholesky factorization of the resulting covariance matrix (17; to check our sampling procedure, we calculated the means and variances of the sampled loop lengths to ensure convergence to the means and variances of the observed loop lengths).

Generating and Scoring Actual and Sample Structures

Given a test alignment, our program requires that full atomic sidechains be mounted upon the template backbone. The SCWRL (Side Chain placement With a Rotamer Library) program uses a backbone-dependent rotamer library to place sidechains on a protein backbone (18). We supplied SCWRL with the template backbone and the alignment of the query sequence with the backbone. SCWRL positioned the side chains onto the template backbone and performed optimizations to minimize steric clashes. After instantiating the side chains onto the template backbone, we used the scoring method described in (19, 20) to determine the compatibility of the newly created environments with those observed in real globins. The scoring method can be summarized here: for each property within corresponding volumes, the system places the observed range of site and nonsite values into k bins. Given a new observation, it uses Bayes' Rule to compute the log odds that the observed value is drawn from the site versus nonsite distributions. The resulting score has a positive value if the evidence favors a site and a negative value if the evidence favors a nonsite. The sum of scores, assuming independence of evidence, is computed and normalized by the number of residues considered.

For comparison with the sampled alignments, we evaluated the scores of shuffled globin sequences and non-globin sequences (including all alpha and other fold families).

Results

The scores for perfect alignments (using the FSSP multiple alignment as a gold standard) for the target proteins were 13.4 for 1HLB, 5.2 for 1BAB, 3.8 for 2GDM, -9.5 for 1COL-A, and -12.9 for 1CPC-A. The background distribution of scores for shuffled homologous and non-homologous proteins ranged from -28 to 0. Thus, 1COL-A and 1CPC-A could not be distinguished from background.

For 1HLB, we generated roughly 4800 random sample alignments to test on the template backbone. Shifting the correct alignment even one position to the left or to the right results in a dramatic decrease in the score, all below zero (shown in Figure 1). The randomly generated alignments score in the same range, with only five positive scores for the sample alignments, the highest being 3.7, thus establishing an absolute distance of almost 10 between correct alignment and highest scoring incorrect alignment. Interestingly, the score of the alignment increases slightly with a periodicity of three to four residues, but never above zero (as shown in Figure 2).

For 1BAB-A and 2GDM, the highest scoring alignment for each protein is the correct alignment, and shifts from the correct alignment also lead to a decrease in scores (as shown in Figure 3). Again, the scores of shifted alignments show a periodicity of three to four (Figure 2). The highest scores received by sampled alignments were -2.1 for 1BAB-A and -4.9 for 2GDM, thus showing a separation from the correct alignment of 5.9 and 8.7, respectively.

We see significantly different results for the globin-like structures 1CPC-A and 1COL-A. The correct alignment for 1COL-A protein sequence scores only -9.5. This score is, in fact, *improved* to 2.6 by shifting the alignment as many as 68 residues toward the amino terminus. Similarly, several randomly generated alignments also score higher than the correct alignment. The same pattern of scoring is seen with the 1CPC-A protein sequence. The score for the 1CPC-A correct alignment is -12.9. The highest score for this protein is also achieved by shifting it towards the amino terminus.

Discussion

Our results for 1HLB, 1BAB-A, and 2GDM suggest that our scoring function is very sensitive to alignment errors. If the loop lengths between core segments are incorrect, the structure created by mounting the sequence on the template backbone does not score well, as demonstrated by the scores for shifted and sampled alignments. In addition, the RMSD of the protein backbone to the template backbone seems to be very important. The proteins 1HLB, 1BAB-A, and 2GDM all have RMSDs to the average backbone less than 2.8Å. However, 1COL-A and 1CPC-A have RMSDs to the average backbone greater than 3.3Å. Our method was unable effectively to identify the correct alignment of the 1COL-A and 1CPC-A proteins to the average backbone.

Our sampling for these five target proteins is not absolutely exhaustive, and so it is possible that some untested alignments would score more closely to the correct alignment. The expense of our threading technique (which requires that

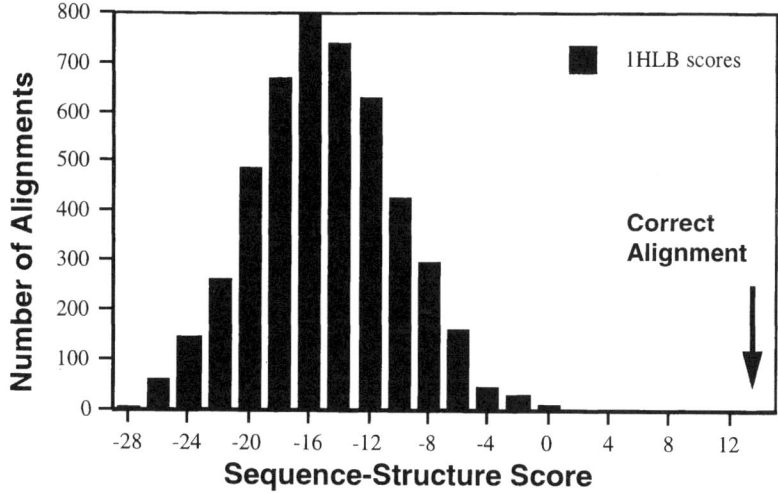

Figure 1: Scores for sampled alignments of 1HLB, and value of correct alignment shown with arrow.

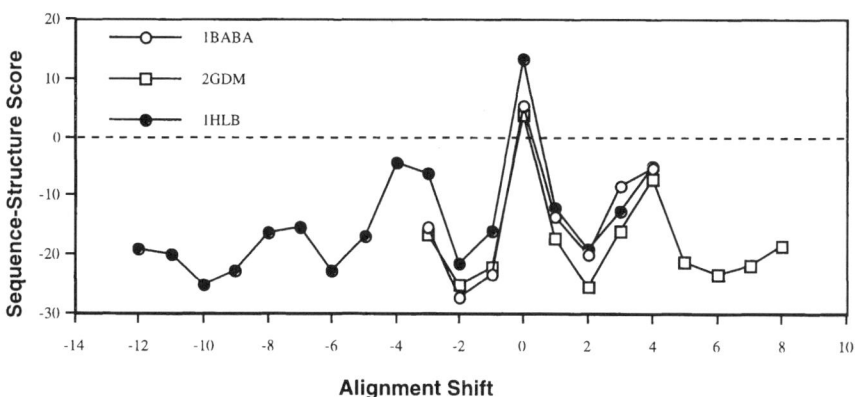

Figure 2. Scores for shifted alignments of 1BAB-A, 2GDM and 1HLB. Positive score (above dashed line) indicates weight of evidence favors assignment as globin. In all three cases, the highest score (and only positive score) is achieved with the correct alignment, and even small shifts bring the score to baseline. A periodicity in score is seen for shifts of three to four residues, as discussed in the text.

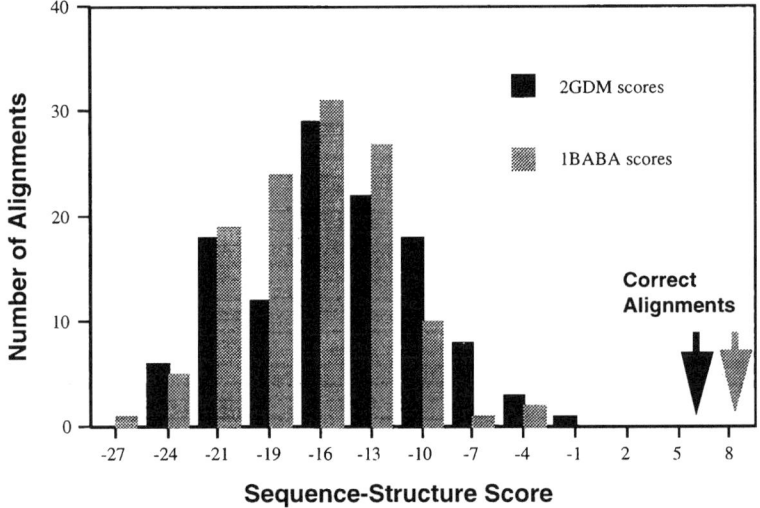

Figure 3. Scores for sampled alignments of 2GDM and 1BAB-A, along with corresponding correct alignments shown with arrows.

we instantiate sidechains for each tested alignment) prohibits exhaustive sampling. It is important to note, however, that the alignments we generated were *not* random. Instead, we generated a very globin-like model of loop lengths from our training set in an attempt to generate a relatively rich collection of plausible alignments. In addition, we tested shifts of the correct alignment to test if we were merely detecting similarities in secondary structure or amino acid composition.

Our results seem to demonstrate that the environments of 1COL-A and 1CPC-A are different than the environments of the other globin family members. In fact, these two structures are not globins, but are globin-like, and share no more than 16% sequence identity with the globins. However, the percent of sequence identity does not explain our results entirely. 1BAB, 2GDM and 1HLB all have relatively low sequence similarity to other members of the globin family (no more than 20% for 1HLB, 21% for 1GDM and 25% for 1BAB). Phycocyanins (1CPC-A and 1CPC-B) are classified in SCOP (Structural Classification of Proteins) (*21*) to have a globin-like fold, but are not truly globins as they contain 2 more helices at the N terminus than the globins and bind a chromophore instead of heme. Similarly, 1COL-A (colicin) is in a completely different SCOP class and is described as having a globin-like fold, with additional helices at the N- and C-termini. It is classified as a membrane and cell surface protein. Colicin is rich in alpha helices, and is a toxin found in the cell membrane. Since this protein is located in the membrane, it is not surprising that its structural environments are different from the other proteins in our test set. In order to understand why the alignments of 1COL-A and 1CPC-A received the best scores when they were shifted toward the N-termini of both proteins, we ran the PhD secondary structural prediction program on both sequences (*22, 23*). PhD predicts strong helical regions in the beginning of each sequence. Most of the higher scoring alignments for both 1CPC-A and 1COL-A include the beginning helical regions. It appears that our scoring function is detecting isolated helical environments, but is not detecting any globin-like interactions between these helices.

The SCWRL program for placing sidechains is able to reproduce crystal structure sidechains to about 2 Å RMSD. In early (unpublished) work, we used average side chain rotamers, not dependent on backbone torsion angles, and the resulting scores were much lower than the SCWRL scores. In fact, the template backbones with SCWRL-placed sidechains score as well as template backbones with sidechains taken from the known crystal structure of the query sequence.

Our scoring method is sensitive to the distance of the template backbone from the actual structure. Up to about 2.9 Å, we are able to place sidechains such that the main features of the alpha carbon environments remain recognizable by the program. However, above 3 Å, the sidechains are simply too far from one another to create the interactions found in actual structures, and the environments become distorted and difficult to recognize. These observations suggest that threading methods that are sensitive to atomic detail and alignment accuracy (such as ours) will require template backbones that are within 2.8 Å of the unknown query structure. Unfortunately, even with such template backbones, the search for the correct alignment is quite difficult because the minima tend to be steep compared to the surrounding scores, with very little indication from alignments that are "close" (for example, with small shift errors) that the search is in the correct neighborhood.

We observe that the peaks in the globin alignment scores do show a periodicity of 3 to 4 amino acids. This corresponds to the number of amino acids in one turn of the helix (3.6), and appears chiefly because globins have a high percentage of residues in a helical conformation. As the alignment moves three to four residues, the faces of the helices coincide with the correct alignment, and the environments created by these interactions score marginally better than for other shifts. It is important to observe, however, that these shifts do *not* bring the scores very close to the peak score of the correct alignment. In addition, such local clues about the location of the correct alignment may not arise in structures that have a mixture of alpha helices and beta structure, and so the peak at 3 to 4 residue shift is not likely to be as exaggerated.

Finally, we should point out that our results are dependent on the quality of the structural alignment in the FSSP resource. Our gold standard for correct alignments is the multiple alignment provided by FSSP. Any inaccuracies in this alignment might affect our results. In fact, improvements in the alignment would most likely lead to a more pronounced difference in the score between the correct alignment and incorrect alignments, so the results here should be considered a lower bound. It is possible that a consideration of environmental match can be used in the future to improve multiple structural alignments that are created using geometric criteria alone.

Conclusion

We have shown that the biochemical microenvironments of the residues in a protein can be used to evaluate structure-sequence alignments. The sensitivity of our scoring function to alignment errors has features that in retrospect may be obvious. First, when using full side chain instantiations, the template backbone must be relatively close to the actual backbone of the query sequence. It is unreasonable to expect that with backbone RMS errors more than 3.0 Å, scoring functions sensitive to alignment will be able to detect similar environments. Second, although the ability to detect the correct alignment is gratifying, it comes at a cost. The computational complexity of the search for a correct alignment is NP-hard (6). Heuristic methods that depend on the gradient of the scoring function (or other local clues that the alignment is improving) will not work in cases where there is a single, deep minima for the scoring function only at the correct alignment. Unless the correct alignment is sampled directly, the scoring function will give little indication that the search is "close." Thus, we are left with a paradox for threading: we want to develop threading functions that clearly distinguish correct alignments from incorrect ones, but we need functions that give us some indication that we are in the correct neighborhood—an indication that comes at the cost of decreased resolution between correct and incorrect alignments. It is possible that a combination of methods with different operating characteristics will provide the best solution: a relatively smooth threading function for identifying the overall compatability of a sequence with a template structure and identifying the general location of an alignment, followed by an atomic-level threading function which can sensitively perform a local search for the correct alignment.

Acknowledgments

This work is supported in part under grants NIH-LM05652, LM-06244 and NSF-BIR9600637, and IBM. Computing facilities were provided under NIH LM-05305. MW is supported by the NIH Molecular Biophysics Training Program. We thank Scott Schmidler for his advice on sampling alignments and various readers for thoughtful comments. JC developed the original code for protein threading. LW developed the scoring function. AW developed the sampling strategy for generating alignments. AW and MW ran additional tests on the globins and created the first draft of this paper.

References

1. Chothia, C.; Lesk, A. M. *EMBO J.* **1986**, *5*, 823-826.
2. Vogt, G.; Etzold, T.; Argos, P. *J. Mol. Biol.* **1995**, *249*, 816-831.
3. Bowie, J. U.; Luthy, R.; Eisenberg, D. *Science.* **1991**, *253*, 164-170.
4. Christian, M.; Lemer, R.; Marianne, J. R.; Shoshana, J. W. *Proteins: Structure, Function, and Genetics.* **1995**, *23*, 337-355.
5. Sippl, M. J. *J. Mol. Biol.* **1990**, *213*, 859-83.
6. Lathrop, R. *Protein Eng.* Sep, 7, 1994, pp 1059-1068.
7. Marchler-Bauer, A.; Bryant, S. H. *Proteins* **1999**, *37:S3*, 218-225.
8. Venclovas, C.; Zemla, A.; Fidelis K.; Moult J. *Proteins* **1999**, *37:S3*, 231-237.

9. Bagley, S. C.; Altman, R. B. *Protein Sci.* **1995**, *4*, 622-635.
10. Wei, L., Altman, R. B. & Chang, J. T. In *Proceedings of Pacific Symposium on Biocomputing 1997*; Altman R. B.; Dunker A. K.; Hunter L.; Klein T. E., Eds.; World Scientific Publishing Co. Pte. Ltd.:River, NJ, 1997; pp 465-476.
11. Waugh, A.; Williams, G. A.; Wei, L.; Altman, R. B. In *Proceedings of Pacific Symposium on Biocomputing 2001*; Altman R. B.; Dunker A. K.; Hunter L.; Lauderdale K.; Klein T. E., Eds.; World Scientific Publishing Co. Pte. Ltd.:River, NJ, 2001; pp 360-371.
12. Wei, L.; Huang, E.; Altman, R. B. *Structure Fold Des.* **1999**, *7*, 643-650.
13. Holm, L.; Sander, C. *Nucleic Acids Research* **1994**, *22*, 3600-3609.
14. Gerstein, M.; Altman, R. B. *CABIOS* **1995**, *11*, 633-644.
15. Schmidt, R.; Gerstein, M.; Altman, R. B. *Protein Sci.* **1997**, *6*, 246-248.
16. Bernstein, R. C.; Koetzle, T. F.; Williams, G. J. B.; Meyer, E. F. Jr.; Brice, M. C.; Rodgers, J. R.; Kennard, O.; Shimanouchi, T.; Tasumi, M. *J. Mol. Biol.* **1977**, *112*, 535-542.
17. Press, W. H.; Teulosky, S. A.; Vetterling, W. T.; Flannery, B. P. *Numerical Recipes in FORTRAN: The Art of Scientific Computing*; Cambridge University Press, Cambridge, UK., 1992.
18. Bower, M. J.; Cohen, F. E.; Dunbrack R. L. Jr. *J. Mol. Biol.* **1997**, *267*, 1268-1282.
19. Wei L.; & Altman R. B. In *Proceedings of Pacific Symposium on Biocomputing 1998*; Altman R. B.; Dunker A. K.; Hunter L.; Klein T. E., Eds.; World Scientific Publishing Co. Pte. Ltd.:River, NJ, 1998; pp 497-508.
20. Wei L.; Chang J.T.; Altman RB. In *Computational Methods in Molecular Biology*; Salzberg S. L.; Searls D. B.; & Kasif S., Eds.; Elsevier: New York, NY, 1998; Volume 32, pp 207-226.
21. Murzin, A. G.; Brenner, S. E.; Hubbard, T.; Chothia, C. *J. Mol. Biol.* **1995**, *247*, 536-540.
22. Rost, B.; Sander, C. *J. Mol. Biol.* **1993**, *232*, 584-599.
23. Rost, B.; Sander, C. *Proteins.* **1994**, *19*, 55-77.

Chapter 20

Electron Paramagnetic Resonance Techniques for Measuring Distances in Proteins

Sandra S. Eaton and Gareth R. Eaton

Department of Chemistry and Biochemistry, University of Denver, Denver, CO 80208

Analysis of the effects of electron-electron dipolar interactions on electron paramagnetic resonance (EPR) spectra of samples with interacting paramagnetic centers can be used to determine the distance between a pair of interacting spins. The technique required to measure the dipolar interaction depends upon the distance between the spins and on the electron spin relaxation times. Continuous wave spectra can be analyzed to obtain distances up to about 20 Å, and longer distances can be measured by pulsed methods. Changes in relaxation times can be used to determine the distance between a slowly relaxing spin and a more rapidly relaxing spin.

Deciphering Biological Function and Structure at a Molecular Level

A central goal of science is to explain biological function at a molecular level, which therefore requires knowledge of the structure of the biological system. As genome research rapidly increases the number of proteins known, the requirement for obtaining structural information increases dramatically (1).

Estimates of distances between parts of proteins or assemblies of proteins with other species can be made by exploiting diffraction, energy transfer, and dipolar interactions. The most common of these methods have been X-ray diffraction, fluorescence energy transfer, NMR, and EPR. Each physical method provides a different perspective, and multiple methods are needed to achieve a full view (2). For each method, key questions include for what species is it appropriate, what range of distances can be measured, how sensitive is the method to distributions in distance or in orientation, and what is the accuracy and precision of the method. Furthermore, one wants to know whether a method is sensitive to dynamics, and whether this sensitivity blurs the distance measurement or is an opportunity to measure dynamics of distances.

X-ray diffraction can yield high-resolution interatomic distances for the form of the protein that is stable in a crystal. When the crystal is cryocooled the resolution for a protein may be as good as 1.2 Å. However, two things that X-ray diffraction does not reveal are (a) the structure of the protein in solution or other physiological environment, and (b) the electronic configuration of a metal in a metalloprotein (2). Even in the small number of cases in which crystals have been formed of proteins that function in a membrane (ca. 20 so far), some parts have remained ill-defined by the single-crystal X-ray diffraction. Thus, membrane-bound proteins are a significant challenge and opportunity for other methods of structure determination.

A renaissance in electron microscopy, especially using freeze-trapped conformations and image-processing is providing structures with 20 Å resolution, and in some cases near 3 Å resolution (3). Single particles of large complexes can be studied, and when atomic-resolution structures of component parts are available from other techniques, such as X-ray diffraction or NMR, it is possible to put detailed structural information into functional context (3).

Many short distances (<5-6 Å), semiquantitatively evaluated from NMR NOEs, yield a dynamically-averaged structure of the form of the protein that is stable under specified solution conditions. Current work on enhancing NMR techniques include efforts to achieve partial orientation of the protein, via magnetic anisotropy, to observe residual dipolar couplings that define longer-range order in the molecule, isotopic labeling and new pulse sequences, and placing proteins in micelles in low-viscosity solvents (4, 5). The use of a few longer distances is becoming increasingly important in predicting 3-D structure, deciphering multi-protein aggregates, and defining relative positions of parts of large proteins (2). For ^{13}C-^{13}C distances longer than ca. 4 Å (particularly useful for structure determination) the couplings are less than ca. 120 Hz, and the standard deviation in the derived distances are about an order of magnitude larger for powder samples than for single crystals (6). A new pulse sequence provides distances between methyl groups via selective NOE data (7). Since many methyl groups are located in the hydrophobic core of a protein, inter-methyl constraints help define the structure.

In many cases, key insights about function may be obtainable from a relatively small number of distances that are larger than those routinely available from NMR. Electron-electron dipolar interactions can be observed directly by EPR, independent of other features of the environment of the spins and measure longer distances than do nuclear-nuclear interactions. In some cases, even a qualitative readout of the distance as short, intermediate, or long, suffices to test a model of biological organization. Sometimes only one or a few such distances provide the crucial information to permit constructing a 3-dimensional picture of the system, and if the readout of distance is also sensitive to a substrate or cofactor or change in ions, or can reveal motion, then function as well as structure is revealed.

Fluorescence energy transfer also can measure relatively long distances (8). However, a key problem in using this method is that one needs to know all of the angles that define the relative orientation of the molecular axes of the donor and the acceptor. Assumptions of average values for the orientation-dependent parameters can lead to substantial errors.

Protein Dynamics

Structure provides a vantage point from which to envision dynamics that facilitate function. However, for some systems the detailed protein structures obtained by single-crystal X-ray diffraction are not compatible with the ingress and egress of substates unless there is substantial motion of the protein. Dynamics also complicates distance and structure determination because structures are fluctuating. Proteins undergo significant motion, even at very low temperatures (9, 10). For example, spectral broadening of holes burnt in the optical spectrum of hemes at 1.7 K is governed by a power law in time that is attributed to conformational dynamics (11). Many proteins exhibit large amplitude motions above ca. 200 K. Numerous modes are known in the tens to hundreds of cm^{-1} range. Myoglobin and lysozyme exhibit modes below 10 cm^{-1} in the dry state (12). Neutron scattering measurements show that there is a lot of motion in proteins. Neutron diffraction of a hen egg white lysozyme crystal (2x2x1.5 mm) in which H has been replaced by D, provided 2 Å resolution after 7 days of data collection (13).

Increasingly, molecular mechanics and molecular dynamics calculations are being extended to biological molecules and assemblies. In combination with physical methods of estimating structure and motion, computational methods are helping researchers to understand the data acquired. For example, recently Buck and Karplus (14) used molecular dynamics calculations to estimate that the N-H bond vectors in lysozyme undergo "rapid in-plane and out-of-plane fluctuations with average amplitudes of 4.7° ± 0.1° and 7.4° ± 0.4° respectively." A fundamental problem in comparing computations with experiment is how the solvent is included in the calculations (*15, 16*). Tarek and Tobias (*17*) performed molecular dynamics calculations of the mobility of water in the presence of ribonuclease A in a crystal, in powders with low and high water content, and in a "cluster" of 399 waters around one protein. On the scale of picoseconds, water mobility was similar in the crystal and powders, but significantly higher in the cluster. In molecular dynamics calculations, anharmonic motion of proteins, and especially of side-chains, is enhanced in the presence of water (*18*). The root-mean-squared deviations of the positions of the α-carbons from the crystal structure were 1.3 and 1.4 Å in the molecular dynamics simulations of ribonuclease A in a crystal, but 2.4 and 2.8 Å in the simulations of the powder samples, with the larger deviation for the more hydrated sample (*17*). It was concluded that neutron scattering experiments on powders are representative of proteins in crystals.

EPR Methods for Measuring Distances

Special features of EPR that provide unique power for biological structure determination are that (1) the method detects only unpaired electrons, and thus can literally see the needle in the haystack; (2) EPR can see unpaired electrons in any phase and over a wide range of temperatures; and (3) the large electron magnetic dipole results in very-long-range effects on EPR line shapes and relaxation times. EPR spectra can be observed for proteins in membranes and even in whole cells and tissues, environments that can confound other physical methods. EPR can be used to study proteins that cannot be crystallized for X-ray diffraction study, proteins that are too large for study by NMR in solution, and proteins in a much more natural environment than is possible for either NMR or X-ray crystallography, including in whole cells.

In addition to being selective and long-distance, the EPR method is also applicable to a vast range of biological species. Electron-electron spin-spin interactions are common in biochemistry (>16 different types of biological systems are listed by Reed and Orosz, (*19*)). The pairs of spins can be naturally-occurring, or one or both spins can be introduced to make the distance measurement. As of the end of 1999, over 10,700 metal binding sites in proteins had been characterized by X-ray crystallography. In addition, large numbers of metalloenzymes involve free radicals (*20, 21*). Site-directed mutagenesis permits placing a cysteine almost anywhere in a protein, and then a nitroxyl spin label can be attached to that cysteine (*22, 23*). The same method of substituting amino acids can be used to engineer metal binding sites (*24-33*). For example, molecular genetics was used to introduce both a metal-binding site and a nitroxide spin label intoT4 lysozyme, with three metal-nitroxyl distances in the range 10-25 Å (*24*). The same methodology was used with distances of 8 - >23 Å in the membrane-bound protein lactose permease (*25*). The helix packing model proposed for lactose permease was also tested by introducing histidines at selected locations in the helices and measuring the extent to which the site created bound Mn^{2+} (*26-28*). The extensive literature of engineered metal sites has been reviewed frequently (*29-32*). It is possible to change the relative affinity of sites for various metals, and to insert metal sites in proteins that do not have them. For example, a Cu(II) site can be introduced into almost any protein by attaching a GlyGlyHis sequence to the N terminus of the existing protein (*29*). Ribonuclease S' has been shown to respond to Cu^{2+} binding to an unnatural amino acid with an iminodiacetic acid moiety as the metal binding site (*33*). These developments in site-directed creation of paramagnetic centers, and the profusion of new proteins that will come as result of the genome project, create many opportunities for applications of EPR methodology.

Studies of electron-nuclear distances by electron spin echo envelope modulation (ESEEM) or by electron nuclear dipolar interaction (ENDOR) are powerful tools for determining the fine details of the geometry around an unpaired electron. ESEEM has been reviewed recently by Dikanov and Tsvetkov (*34*) and by McCracken (*35*). ENDOR has been reviewed by Grupp and Mehring (*36*) and by Huttermann (*37*). ENDOR data were reviewed by Kispert and coworkers (*38, 39*). Originally, it appeared that ESEEM and ENDOR were applicable to different electron-nuclear coupling regimes. However, recently, Hoffman and coworkers have demonstrated the ability to measure very weak couplings to distant nuclei, extending the range of ENDOR, so that while ENDOR and ESEEM remain complementary techniques, their coupling regimes now overlap.

Distance measurements based on electron-electron interactions have been reviewed by Eaton and Eaton (*40-42*) and by Maret (*43*), and our book titled "Distance Measurements in Biological Systems by EPR" (*44*) will be in press in 2000. A variety of techniques have been developed for various types of samples and magnitudes of dipolar interaction as outlined in the following paragraphs. The appropriate methodology depends on the relative magnitudes of electron spin relaxation times, dipolar couplings, and molecular dynamics.

Distance Between Two Slowly-Relaxing Spins in Immobilized Sample

In this context slowly-relaxing means that the electron spin relaxation rates are slow relative to the magnitude of the dipolar coupling, expressed in hz. This case includes nitroxyl spin labels at room temperature and below and some transition metals at lower temperatures.

In addition to the dipolar interaction between two unpaired electrons there can also be a through-bond exchange interaction. Unless there is a direct bonding pathway between the two unpaired electrons the exchange interaction usually is assumed to be negligible relative to the dipolar interaction at distances longer than 10 to 12 Å. At distances shorter than about 12 Å it is necessary to take both contributions into consideration.

Half-Field Transitions

In a rigid lattice a pair of electron spins yields, in addition to the "allowed" transitions near g = 2, a "forbidden" transition at half the magnetic field. The ratio of the integrated intensity of the half-field transition to the integrated intensity of the allowed transitions is a function of interspin distance, r, and independent of the exchange interaction, J, and hence is a direct measure of the distance between the spins (*45-48*). Note that the relative intensity is inversely proportional to the square of the microwave frequency. Thus, longer distances can be measured at lower microwave frequencies (*45*). The method is useful for distances from about 4-5 Å to about 12 Å (*42, 45, 46*). When the interspin distance is less than about 4 Å the position of the half-field transition depends upon the dipolar interaction and that shift can be analyzed to determine the interspin distance (*49*). At distances less than about 5 Å it is necessary to consider the impact of anisotropic exchange, and the point dipole approximation may not be appropriate. The relative intensity of the half-field transition has been used to determine the structure of copper porphyrin dimers (*50*) and to analyze the structure of the trichogin GA IV peptide (*47*).

Analysis of Dipolar Splitting that is Significant Relative to Linewidths

When the dipolar splitting is significant relative to linewidths in the continuous wave (CW) EPR spectra, the analysis of the lineshape broadening can be used to determine the magnitude of the dipolar interaction. A qualitative indication of spin-spin interaction can be obtained by measuring the relative amplitudes of characteristic peaks in a nitroxyl spectrum (the ratio d_1/d) (*51*).

Both exchange and dipolar contributions to lineshape changes can be included in computer simulations of the spectral lineshapes. Since simulated spectral lineshapes depend upon the degree of ordering of the relative orientations of the magnetic axes, information can be obtained concerning the range of conformations present in the sample. The simulations involve multiple parameters, so the uniqueness of solutions depends strongly upon the resolution of the spectra and can be substantially enhanced by fitting data at multiple microwave frequencies (*52*). Lineshape simulations have been used to determine distances in spin-labeled metal complexes (*53-57*). Simulations of the $S_2Y_Z^{\cdot}$ state of acetate-inhibited photosystem II demonstrated a point-dipole distance of 7.7 Å between the tyrosyl radical and the Mn_4 cluster and an exchange coupling of - 280×10^{-4} cm^{-1} (*58, 59*). This distance is an important clue to understanding the mechanism of action of the oxygen-evolving center. Beth and coworkers have determined the structural changes that accompany function of glyceraldehyde-3-phosphate dehydrogenase (60). The 0.6±0.1 G increase in Cu^{2+} line width due to interaction with Fe^{3+} observed by Zweier yielded a Cu-Fe distance in transferrin of 41.6±2.8 Å, in good agreement with the Fe-Fe distance of 42 Å by X-ray in human lactoferrin (61). Hustedt and Beth and coworkers are extending CW simulation of nitroxyl-nitroxyl interaction spectra, seeking to identify the distance and five angles of the interaction, and to include the effects of dynamics (52, 62).

If the dipolar splittings are not well resolved, it may be difficult to obtain a unique simulation of the CW lineshape. Commonly, the flexibility of the linkage between the nitroxyl and the protein results in a distribution of conformations, and the dipolar splittings are not well resolved. To determine the dipolar contributions to the CW line shapes for such cases, Rabinstein and Shin (63) used Fourier deconvolution for distances between 10 and 20 Å, and Steinhoff and coworkers used convolution of the non-interacting line shape with a Gaussian broadening function for distances between 8 and 25 Å (48). These methods assume that there is an essentially random distribution of the orientation of the interspin vector relative to the magnetic axes of the paramagnetic centers. In both approaches it is necessary to have the spectra of the corresponding monoradicals in order to analyze the spectrum of the diradical. The upper limit on the distance that can be obtained is determined by the inherent inhomogeneous linewidths in the CW spectra, and distances become more uncertain the longer the distance. Examples of applications of these methods include measurement of conformational changes in the aspartate chemoreceptor (64) and study of the tetrameric structure of the PcsA potassium channel from *Streptomyces lividans* (65).

Pulsed Techniques

Smaller dipolar interactions can be measured by pulsed techniques than by CW techniques because the distance scale for the pulse techniques is limited by the spin-packet width (proportional to the rate constant for echo dephasing), which usually is much smaller than the inhomogeneously broadened CW linewidth. The pulse techniques however face another challenge. Some of the methods require excitation of most of the spins in order to define the full dipolar broadening function. With current technology, it is difficult to achieve a bandwidth that is broad enough to excite the full spectrum of an immobilized nitroxyl. This challenge becomes even greater as the spectrum is broadened by increasingly strong dipolar interaction. Thus, the pulse techniques are more useful at longer distances than are accessible by CW techniques. Calibrations often can be used to account for the partial excitation of the spectrum. The first of these pulse techniques was the "2+1" pulse sequence developed by Milov et al. (*66*) and subsequently applied by Raitsimring and others, in which all three pulses are at the same microwave frequency (*67-71*). In the pulsed electron-electron double resonance (ELDOR or PELDOR or DEER) technique (*72, 73*) one of the pulses in a three-pulse sequence is at a second microwave frequency. This technique and examples of its applications have been described by Tsetkov and co-workers (*74-76*) and is reviewed in Milov et al. (*77*). In the 2+1 pulse sequence and in 3-pulse ELDOR experiments there is an inherent experimental deadtime that limits the magnitude of the dipolar interaction that can be characterized. By using a 4-pulse experiment that deadtime can be eliminated (*78, 79*). Comparing dipolar interactions to spin packet widths, Mims and Peisach (*80*) and Spiess and coworkers (*78*) estimate that the distance that can be measured by electron spin echo (ESE) techniques is about 70-80 Å. Distances >30 Å have been measured (*70, 71*), but most of the effort is very recent and has been applied to small test molecules with spin-spin distances of ca. 15 – 20 Å. Double-quantum coherence in a 2-D pulsed EPR spectrum can be generated with 3-pulse or 5-pulse sequences. Phase cycling in the 5-pulse sequence removes unwanted contributions. The dipolar interaction can be read directly from one dimension of the 2-D plot (*81, 82*).

Examples of the application of pulsed techniques include the distance between spin-labels attached to β-93 cysteine in hemoglobin (*70*), the distance between the Mn_4 cluster and the dark stable tyrosyl radical in photosystem II (*83*), and distances between spin labels in model compounds and polypeptides (*72, 73, 75, 76, 78, 79*). Several papers have examined the effect of distributions of distances and orientations on the "2+1" and DEER ESE results (*48, 67, 76*).

Analysis of the dephasing of the out-of-phase echo for radical pairs produced by photo-excitation can give the interspin distance. This technique has been applied predominantly in studies related to photosynthesis. Following creation of the radical pair, a two-pulse spin echo is created. Crucial to the interpretation is that there is phase coherence between eigenstates present in the photo-induced spin-correlated radical pair (*84*). The spin-spin interaction causes deep modulation in the out-of-phase echo, whereas the normal, in-phase, echo vanishes (*84*). Distances in the range of 25 – 34 Å have been measured to define distances between donor and acceptors in the photosystem (*84-96*).

Echo detection of selectively-burned holes in photosynthetic systems led to estimates of distances between 25 and 50 Å. If the contributions to spectral diffusion from motion, nuclear spin flip-flops, and instantaneous diffusion are smaller than the contribution from dipolar interaction between unpaired electrons, the spectral diffusion can be used to determine the interspin distance. (*97-100*).

An example of the special insights obtainable with EPR are the magnitude and orientation of the vector connecting the radical pair $P_{700}^{+}Q_K^{-}$ within the crystal axis system. This quinone radical in photosystem I (*85*) was defined by

EPR before the group was even detected by X-ray diffraction. It was subsequently located in an improved 4-Å resolution electron density map (*101*). Other applications of distance measurements by EPR include studies of protein folding, studies of conformational changes during protein function, geometry of assembly of subunits in multicomponent systems, arrangement of proteins in membranes, and structure of proteins in intact biological systems.

Distance Between a Rapidly-Relaxing Spin and a Slow-Relaxing Spin in an Immobilized Sample

The impact on the CW spectrum of a slowly-relaxing spin that is dipolar-coupled to a more rapidly-relaxing spin is temperature dependent (Table I). When the relaxation rate of the more-rapidly relaxing spin is slow compared with the dipolar splitting, broadening or splitting of the CW spectrum is observed provided that the splittings are sufficiently large compared to the linewidth of the EPR signal. As temperature is increased and the relaxation rate of the rapidly-relaxing center increases, the CW spectrum of the slowly-relaxing center broadens. Further increases in the relaxation rate cause collapse of the splitting due to the electron-electron coupling and then narrowing to an unbroadened spectrum. Examples of this temperature dependence are shown in Fielding et al. (*102*) and Rakowsky et al. (*103*).

Table I. Temperature Dependence of Effects of Rapidly Relaxing Spin on Slowly Relaxing Spin

temperature/ relaxation rate	*low/slow*	*intermediate*	*high/fast*
impact of dipolar interaction on CW spectra	potentially resolvable splitting	broadened	splitting averaged by the rapid relaxation
impact of dipolar interaction on echo dephasing	T_m determined by other factors	T_m decreases, goes through minimum, then increases	T_m determined by other factors

Impact on Amplitude of CW Spectrum

In a rigid lattice, and in the temperature interval where the rapid relaxation rate is fast enough to collapse the dipolar splitting, but not fast enough to result in full narrowing, the broadening of a nitroxide spin label CW spectrum can be interpreted in terms of the distance between the spins (*104*). The Leigh method has been one of the most commonly applied EPR methods for estimating distances in biological system, but in many applications the assumptions of the method have not been met (*40, 42, 105-111*).

Impact on Spin Echo Dephasing of Slowly Relaxing Spin

The effect of a rapidly-relaxing metal on the 2-pulse electron spin echo decay for a spin label in its proximity is analogous to a two-site dynamic exchange process, where the dynamic process is the electron spin relaxation of the rapidly relaxing spin (Table I). In the temperature range where the relaxation of the rapidly-relaxing spin causes broadening and partial collapse of the CW spectrum of the slowly-relaxing spin, the spin echo dephasing rate ($1/T_m$) of the slowly-relaxing spin is enhanced (Table I). The maximum impact on T_m occurs when the rate of relaxation of the faster-relaxing spin is comparable to the dipolar splitting between the two paramagnetic centers. The effect on T_m in this "intermediate" region depends upon r^{-3} and thus can be used to determine the interspin distance. The enhancement of the echo dephasing can be seen clearly even in cases where the interspin distance is so long that there is little impact on the CW lineshape. Since the dipolar interaction is orientation dependent, the experimental spin echo decay curves are the superposition of the dynamic effects on a powder pattern of dipolar splittings. Using this phenomenon, we developed a method for obtaining distances from the effect of a rapidly-relaxing spin on the electron spin echo of a slowly-relaxing spin (*103, 112-115*).

FepA is an 81 kDa ligand-gated channel that mediates iron transport across the outer membrane of *Escherichia coli*. A recent X-ray crystal structure at 2.4Å resolution showed a 22-stranded β-barrel with large extracellular loops and an N-terminal globular domain that folds into the barrel, inhibiting access to the periplasm (*116*). From the X-ray data it was not possible to determine the conformations of key extracellular loops or to identify specific residues that interact with FeEnt. FeEnt bound to FepA had a dramatic impact on spin-echo dephasing for a spin label attached at V338C (*114*). The estimated distance between the iron and the nitroxyl was 20 – 30 Å. CW EPR studies performed by Prof. Feix showed that there was a large change in mobility of the spin label attached to V338C when FepA was bound. The combined results suggested that V338C occupied a hinge region connecting a ligand-binding loop to the β-barrel and provided the strongest evidence available at the time of an *in vitro* conformational change of FepA due to FeEnt binding (*114*).

Impact of Rapidly Relaxing Spin on Spin-Lattice Relaxation for a More Slowly Relaxing Spin

A rapidly–relaxing spin also increases the spin-lattice relaxation rate of a slowly-relaxing spin to which it is coupled. The maximal effect occurs when either of two criteria are met. If the relaxation rate, T_1^{-1}, of the rapidly-relaxing spin is equal to the Larmor frequency, $2\pi\nu$, of the slowly-relaxing spin, or if the T_2^{-1} of the rapidly-relaxing spin is equal to the difference between the Larmor frequencies for the two paramagnetic centers, the effect of the rapidly-relaxing spin on the slowly-relaxing spin will be greater than for a slower or faster relaxation of the faster relaxing center. These effects are well known in NMR.

The effects of a fast-relaxing spin on a slower-relaxing spin depend on distance. Applications to biological systems started with interpretations of changes in CW EPR power saturation curves (*40, 41, 106-111, 115*). More recently effects on the spin-lattice relaxation time, T_1, were measured directly by saturation recovery. Interpretation of the effect on T_1 is made in terms of an expression originally derived by Bloembergen about 50 years ago in an NMR context (*117-119*) to express the effect of a rapidly relaxing nuclear spin on the T_1 of an interacting slowly relaxing spin. Early attempts to apply this equation to EPR spectra of interacting spin systems attempted to use a single value of T_1 to describe the relaxation of the slowly relaxing center (*120-125*). However, the anisotropy of the dipolar interaction (*120*) results in a distribution of T_1 values, which is poorly modeled by a single exponential. The distribution of T_1 values is reflected more accurately in the shape of a saturation recovery curve than by a single value of T_1. Simplified forms of Bloembergen's equation, that do not require knowledge of the relaxation times for the fast relaxing paramagnetic partner, have been used to analyze saturation recovery data and thereby estimate distances between paramagnetic centers in photosystem II and between the diferric center and tyrosine radicals in ribonucleotide reductase (*125-130*). We have examined systems in which it is possible to directly measure the relaxation rates for the rapidly relaxing partner. Analysis of the saturation recovery curves for interacting spin labels permitted distance determinations for spin-labeled iron porphyrins, methemoglobin, and metmyoglobin (*112, 113, 115, 131*).

Distance between Spins in a Slowly-Tumbling Macromolecule

Current methods for rigorous distance measurements require immobilization of the sample. However, very useful qualitative information can be obtained from the effects of dipolar interactions in macromolecules in solution. In fluid solution the tumbling rate of macromolecules is too slow to average the anisotropic dipolar interaction, so dipolar interactions cause broadening of the nitroxyl EPR spectra that are significant out to distances of about 15 Å (*132-*

136). There may be substantial flexibility of the linkage between a protein and the spin label that can result in partial motional averaging of the anisotropic g and A values. To distinguish broadening due to dipolar interaction from motional effects, spectra of singly-labeled mutants can be compared with spectra of doubly-labeled mutants (*132*). The dipolar broadening causes a decrease in signal amplitude as interspin distance decreases, so relative signal amplitude is a convenient qualitative indicator of interspin distance (*137*). Electron-electron exchange also contributes to broadening of the spectra at short interspin distances, but can be viewed as simply an additional indication of short interspin distances. These techniques have been used to characterize conformational changes in T4 lysozyme (*133*), structural rearrangements in K^+ channel activation gating (*137*), ligand-induced conformational changes in the aspartate chemoreceptor (*64*), packing of ß-sheets in αA-crystallin (*135*), subunit interactions in αA-crystallin (*138*), and the conformation of the diphtheria toxin T domain in membranes (*139*).

Accessibility of Slowly Relaxing Spin to Paramagnetic Species in Solution

Collisions between rapidly-relaxing paramagnets and slower-relaxing spins causes broadening of the CW EPR spectrum of the slower-relaxing spins (*22, 111, 132, 135, 140-144*). Likhtenshtein (*145*) pioneered metal-broadening of nitroxyl spin label CW spectra to estimate the distance from the solvent-exposed surface to a spin label. The relative broadening effectiveness of numerous metal complexes has been quantified in our laboratory (*111, 141-144*). Hubbell and coworkers compared the effects of metal complexes and oxygen, taking advantage of differential partitioning between aqueous and membrane environments, to map label locations in transmembrane proteins (*22, 140*), and structure of heat-shock proteins (*132, 135*). A series of these qualitative estimates of distance can provide a detailed understanding of the structure of the assembly including distinguishing between α-helices and β-sheets. For example, Hubbell and coworkers have mapped out the 3-dimensional arrangement of bacteriorhodopsin (*22, 23, 140, 146*) and of lactose permease (*25-28, 147*). Hubbell, Mchaourab and coworkers identified the 3-dimensional arrangement of T4 lysozyme (*24, 132, 133, 148*).

The Future

We expect significant advances in both CW and pulsed methods for determining distances between spins. Limitations of computer power available to most labs have inhibited using exact simulations of CW EPR spectra. There are simply too many terms, and too many interactions and rates of similar magnitudes. Furthermore, experimental spectra are broad enough that many interactions are not resolved. Consequently, unique fits of simulated spectra with experimental spectra are rare. As more information accumulates on the effects of various solvent and protein environments on CW EPR spectra of common labels, the range of parameters that should be used in simulations will be better defined. As readily-available computers become more powerful, less approximate simulations will be more feasible. Improvements in spectrometer signal-to-noise will improve the definition of the experimental spectra, so that one can better state what is a significant disagreement between theory and experiment. Pulsed EPR is only beginning to exploit sophisticated pulse sequences (e.g., *149-151*), as the available instrumentation makes it possible (e.g., Bruker Elexsys spectrometers). The relaxation times in EPR are so much shorter than in NMR that some of the pulse sequences that have proven useful in NMR will usually not be feasible in EPR. However, the recent demonstrations of HYSCORE, pulsed ELDOR, 3-pulse and 4-pulse DEER, and other sequences will stimulate additional pulse sequences designed to query particular aspects of the spin systems.

Acknowledgments

Research in our laboratory on developing methods to measure distances between electron spins has been supported by NIH grant GM21156.

References

1. Read, R. J.; Wemmer, D. E. *Curr. Opin. Struct. Biol.* **1999**, *9*, 591-593.
2. *Protein Folds. A Distance-Based Approach.* Bohr, H.; Brunak, S., Eds., 1996, CRC Press, Boca Raton, FL, p. 34, 85.
3. Stowell, M. H. B.; Miyazawa, A.; Unwin, N. *Curr. Opin. Struct. Biol.* **1998**, *8*, 595-600.
4. Banci, L.; Bertini, I.; Huber, J. G.; Luchinat, C.; Rosato, A. *J. Am. Chem. Soc.* **1998**, *120*, 12903-12909.
5. Wider, G.; Wüthrich, K. *Curr. Opin. Struct. Biol.* **1999**, *9*, 594-601.
6. Hodgkinson, P.; Emsley, L. *J. Magn. Reson.* **1999**, *139*, 46-59.
7. Zwahlen, C.; Gardner, K. H.; Sarma, S. P.; Horita, D. A.; Byrd, R. A.; Kay, L. E. *J. Am. Chem. Soc.* **1998**, *120*, 7617-7625.
8. Dewey, T. G. *Biophysical and Biochemical Aspects of Fluorescence Spectroscopy*, Plenum, NY, 1991.

9. *Proteins: A Theoretical Perspective of Dynamics, Structure, and Thermodynamics.* Brooks, C. L. III; Karplus, M.; Pettitt, B. M., Eds., Adv. Chem. Phys., 1988, Vol. 71.
10. Bizzarri A. R.; Cannistraro, S. *Physica A* **1999**, *267*, 257-270.
11. Schlichter, J.; Friedrich, J.; Herenyi, L.; Fidy, J. *J. Chem. Phys.* **2000**, *112*, 3045-3050.
12. Diehl, M.; Doster, W.; Petry, W.; Schober, H. *Biophys. J.* **1997**, *73*, 2726-2732.
13. Niimura, N. *Curr. Opin. Struct. Biol.* **1999**, *9*, 602-608.
14. Buck, M.; Karplus, M. *J. Am. Chem. Soc.* **1999**, *121*, 9645-9658.
15. Gans, W.; Amann, A.; Boeyens, J. C. A. *Fundamental Principles of Molecular Modeling.* Plenum Press, New York, 1996.
16. Höltje, H.-D.; Folkers, G. *Molecular Modeling. Basic Principles and Applications.* VCH, Weinheim, 1997.
17. Tarek, M.; Tobias, D. J. *J. Am. Chem. Soc.* **1999**, *121*, 9740-9741.
18. Arcangeli, C.; Bizzarri, A. R.; Cannistraro, S. *Chem. Phys. Lett.* **1998**, *291*, 7-14.
19. Reed, C. A.; Orosz, R. D. *Concepts in Bioinorganic Chemistry, Research Frontiers in Magnetochemistry* C. J. O'Connor, Ed., World Scientific, Singapore, 1993.
20. Sigel, H.; Sigel, A. Metal Ions in Biological Systems, Vol. 30, *Metalloenzymes Involving Amino Acid-Residue and Related Radicals.* Marcel Dekker, N. Y., NY, 1994.
21. Sigel, H.; Sigel, A. Metal Ions in Biological Systems, Vol. 36, *Interelations Between Free Radicals and Metal Ions in Life Processes.* Marcel Dekker, N. Y., NY, 1999.
22. Hubbell, W. L.; Fross, A.; Langren, R.; Lietzow, M. A. *Curr. Opinion Struct. Biol.* **1998**, *8*, 649-656.
23. Altenbach, C.; Flitsch, S. L.; Khorana, H. G.; Hubbell, W. L. *Biochemistry* **1989**, *28*, 7806-7812.
24. Voss, J.; Salwinski, L.; Kaback, H. R.; Hubbell, W. L. *Proc. Natl. Acad. Sci. USA* **1995**, *92*, 12295-12299.
25. Voss, J.; Hubbell, W. L.; Kaback, H. R. *Proc. Natl. Acad. Sci. USA* **1995**, *92*, 12300-12303.
26. Jung, K.; Voss, J.; He., M.; Hubbell, W. L.; Kaback, H. R. *Biochemistry* **1995**, *34*, 6272-6277.
27. He, M. M.; Voss, J.; Hubbell, W. L.; Kaback, H. R. *Biochemistry* **1995**, *34*, 15661-15666.
28. He, M. M.; Voss, J.; Hubbell, W. L.; Kaback, H. R. *Biochemistry* **1995**, *34*, 15667-15670.
29. Regan, L. *Ann. Rev. Biophys. Biomol. Struct.* **1993**, *22*, 257-281.
30. Hellinga, H. W. In *Protein Engineering, Principles and Practice,* Cleland, J. L.; Craik, C. S. Eds.; Wiley-Liss, 1996, pp. 369-398.

31. Lu, Y.; Valentine, J. S. *Curr. Opin. Struct. Biol.* **1997**, *7*, 495-500.
32. Klemba, M. W.; Munson, M.; Regan, L. In *Proteins: Analysis and Design;* R. H. Angeletti, Ed., Academic Press, San Diego, CA, 1998, pp. 313-353.
33. Hamachi, I.; Yamada, Y.; Matsugi, T.; Shinkai, S. *Chem. Eur. J.* **1999**, *5*, 1503-1511.
34. Dikanov, S. A.; Tsvetkov, Yu. D. *Electron Spin Echo Envelope Modulation Spectroscopy*, CRC Press, Boca Raton, FL, 1992.
35. McCracken, J. In *Handbook of Electron Spin Resonance*, C. P. Poole, Jr.; H. A. Farach, Eds., AIP Press, Springer-Verlag, N. Y., NY, 1999, Vol. 2, pp. 69-84.
36. Grupp, A.; Mehring, M. In *Modern Pulsed and Continuous-Wave Electron Spin Resonance.* L. Kevan; M. K. Bowman, Eds., Wiley, N. Y., NY, 1990, pp. 195-229.
37. Huttermann, J. *Biol. Magn. Reson.* **1993**, *13*, 219-252.
38. Piekara-Sady, L.; Kispert, L. D. In *Handbook of Electron Spin Resonance*, C. P. Poole, Jr.; H. A. Farach, Eds., American Institute of Physics, N. Y., NY, 1994, Vol. 1, pp. 312-356.
39. Goslar, J.; Piekara-Sady, L.; Kispert, L. D. In *Handbook of Electron Spin Resonance*, C. P. Poole, Jr.; H. A. Farach, Eds., American Institute of Physics, N. Y., NY, 1994, Vol. 1, pp. 360-651.
40. Eaton, S. S.; Eaton, G. R. *Coord. Chem. Rev.* **1978**, *26*, 207-262.
41. Eaton, S. S.; Eaton, G. R. *Coord. Chem. Rev.* **1988**, *83*, 29-72.
42. Eaton, G. R; Eaton, S. S. *Biol. Magn. Reson.* **1989**, *8*, 339-397.
43. Maret, W. *Meth. Enzymol.* **1993**, *226*, 594-618.
44. *Distance Measurements in Biological Systems by EPR*, Eaton, G. R.; Eaton, S. S.; Berliner, L. J., Eds., Biol. Magn. Reson.; Kluwer, N.Y., NY, 2000, Vol. 19.
45. Eaton, S. S.; More, K. M.; Sawant, B. M.; Eaton, G. R. *J. Am. Chem. Soc.* **1983**, *105*, 6560-6567.
46. Coffman, R. E; Pezeshk, A. *J. Magn. Reson.* **1986**, *70*, 21-33.
47. Anderson, D. J.; Hanson, P.; McNulty, J.; Millhauser, G.; Monaco, V.; Formaggio, F.; Crisma, M.; Toniolo, C. *J. Am. Chem. Soc.* **1999**, *121*, 6919-6927.
48. Steinhoff, H.-J.; Radzwill, N.; Thevis, W.; Lenz, V.; Brandenburg, D.; Antson, A.; Dodson, G.; Wollmer, A. *Biophys. J.* **1997**, 73, 3287-3298.
49. Thomson, C. *Q. Rev. Phys. Soc.* **1968**, *22*, 45-74.
50. Eaton, S. S.; Eaton, G. R.; Chang, C. K. *J. Am. Chem. Soc.* **1985**, *107*, 3177-3184.
51. Kokorin, A. I.; Zamarayev, K. I.; Grigoryan, G. L.; Ivanov, V. P.; Rozantsev, E. G. *Biofizika* **1972**, *17*, 34-41 (p. 31-39 in transl.)
52. Hustedt, E. J.; Smirnov, A. I.; Laub, C. F.; Cobb, C. E.; Beth, A. H. *Biophys. J.* **1997**, *72*, 1861-1877.

53. Damoder, R.; More, K. M.; Eaton, G. R.; Eaton, S. S. *J. Am. Chem. Soc.* **1983**, *105*, 2147-2154.
54. Damoder, R.; More, K. M.; Eaton, G. R.; Eaton, S. S. *Inorg. Chem.* **1983**, *22*, 2836-2841.
55. Damoder, R.; More, K. M.; Eaton, G. R.; Eaton, S. S. *Inorg. Chem.* **1983**, *22*, 3738-3744.
56. Damoder, R.; More, K. M.; Eaton, G. R.; Eaton, S. S. *Inorg. Chem.* **1984**, *23*, 1320-1326.
57. Damoder, R.; More, K. M.; Eaton, G. R.; Eaton, S. S. *Inorg. Chem.* **1984**, *23*, 1326-1330.
58. Lakshmi, K. V.; Eaton, S. S.; Eaton, G. R.; Frank, H. A.; Brudvig, G. W. *J. Phys. Chem. B* **1998**, *102*, 8327-8355.
59. Lakshmi, K. V.; Eaton, S. S.; Eaton, G. R.; Brudvig, G. W. *Biochemistry* **1999**, *38*, 12758-12767.
60. Beth, A. H.; Robinson, B. R.; Cobb, C. E.; Dalton, L. R.; Trommer, W. E.; Birktoft, J. J.; Park, J. H. *J. Biol. Chem.* **1984**, *259*, 9717-9728.
61. Zweier, J. L. *J. Biol. Chem.* **1983**, *258*, 13759-13760.
62. Hustedt, E. J.; Beth, A. H. *Annu. Rev. Biophys. Biomol. Struct.* **1999**, *28*, 129-153.
63. Rabenstein, M. D.; Shin, Y.-K. *Proc. Natl. Acad. Sci. USA* **1995**, *92*, 8239-8243.
64. Ottemann, K. M.; Thorgeirsson, T. E.; Kolodziej, A. F.; Shin, Y.-K.; Koshland, D. E., Jr. *Biochemistry* **1997**, *37*, 7062-7069.
65. Gross, A.; Columbus, L.; Hideg, K.; Altenbach, C.; Hubbell, W. L. *Biochemistry* **1998**, *38*, 10324-10335.
66. Milov, A. D.; Salikhov, K. M.; Shchirov, M. D. *Sov. Phys. Solid State* **1981**, *23*, 565-569.
67. Astashkin, A. V.; Hara, H.; Kawamori, A. *J. Chem. Phys.* **1998**, *108*, 3805-3812.
68. Kurshev, V. V.; Raitsimring, A. M.; Salikhov, K. M. *Sov. Phys. Solid State* **1988**, *30*, 239-242.
69. Kurshev, V. M.; Raitsimring, A. M.; Tsvetkov, Yu. D. *J. Magn. Reson.* **1989**, *81*, 441-454.
70. Raitsimring, A.; Peisach, J.; Lee, H. C.; Chen, X. *J. Phys. Chem.* **1992**, *96*, 3526-3531.
71. Shigemori, K.; Hara, H.; Kawamori, A.; Akabori, K. *Biochim. Biophys. Acta* **1998**, *1363*, 187-198.
72. Larsen, R. G.; Singel, D. J. *J. Chem. Phys.* **1993**, *98*, 5134-5146.
73. Pfannenbecker, V.; Klos, H.; Hubrich, M.; Volkmer, T.; Heuer, A.; Wiesner, U.; Spiess, H. W. *J. Phys. Chem.* **1996**, *100*, 13428-13432.
74. Milov A. D.; Tsvetkov, Yu. D. *Appl. Magn. Reson.* **1997**, *12*, 495-504.
75. Maryasov, A. G.; Tsvetkov, Y. D.; Raap, J. *Appl. Magn. Reson.* **1998**, *14*, 101-113.

76. Milov, A. D.; Maryasov, A. G.; Tvestkov, Yu. D.; Raap, J. *Chem. Phys. Lett.* **1999**, *303*, 135-143.
77. Milov, A. D.; Maryasov, A. G.; Tsvetkov, Yu. D. *Appl. Magn. Reson.* **1998**, *15*, 107-143.
78. Martin, R. E.; Pannier, M.; Diederich, F.; Gramlich, V.; Hubrich, M.; Spiess, H. W. *Angew. Chem.* **1998**, *37*, 2834-2837.
79. Pannier, M.; Veit, S.; Godt, A.; Jeschke, G.; Spiess, H. W. *J. Magn. Reson.* **2000**, *142*, 331-340.
80. Mims, W. B.; Peisach, J. *Biol. Magn. Reson.* **1981**, *3*, 213-263.
81. Saxena, S.; Freed, J. H. *Chem. Phys. Lett.* **1996,** *251*, 102-110.
82. Saxena, S.; Freed, J. H. *J. Chem. Phys.* **1997**, *107*, 1317-1340.
83. Hara, H.; Dzuba, S. A.; Kawamori, A.; Akabori, K.; Tomo, T.; Satoh, K.; Iwaki, M.; Itoh, S. *Biochim. Biophys. Acta* **1997**, *1322*, 77-85.
84. Tang, J.; Thurnauer, M. C.; Norris, J. R. *Chem. Phys. Lett.* **1994**, *219*, 283-290.
85. Bittl, R.; Zech, S. G.; Fromme, P.; Witt, H. T.; Lubitz, W. *Biochemistry* **1997**, *36*, 12001-12004.
86. Dzuba, S. A.; Gast, P.; Hoff, A. J. *Chem. Phys. Lett.* **1995**, *236*, 595-602.
87. Dzuba, S. A.; Gast, P.; Hoff, A. J. *Chem. Phys. Lett.* **1997**, *268*, 273-279.
88. Dzuba, S. A.; Hara, H.; Kawamori, A.; Iwaki, M.; Itoh, S.; Tsvetkov, Yu. D. *Chem. Phys. Lett.* **1997**, *264*, 238-244.
89. Dzuba, S. A. *Chem. Phys. Lett.* **1997**, *278*, 333-340.
90. Hoff, A. J.; Gast, P.; Dzuba, S. A.; Timmel, C. R.; Fursman, C. E.; Hore, P. J. *Spectrochim. Acta A* **1998**, *54*, 2283-2293.
91. Zech, S. G.; van der Est, A. J.; Bittl, R. *Biochem.* **1997**, *36*, 9774-9779.
92. Zech, S. G.; Kurreck, J.; Renger, G.; Lubitz, W.; Bittl, R. *FEBS Lett.* **1999**, *442*, 79-82.
93. Bittl, R.; Zech, S. G. *J. Phys. Chem. B* **1997**, *101*, 1429-1436.
94. Borovykh, I. V.; Dzuba, S. A.; Proskuryakov, I. I.; Gast., P.; Hoff, A. J. *Biochim. Biophys. Acta* **1998**, *1363*, 182-186.
95. Iwaki, M.; Itoh, S.; Hara, H.; Kawamori, A. *J. Phys. Chem. B* **1998**, *102*, 10440-10445.
96. Fursman, C. E; Hore, P. J. *Chem. Phys. Lett.* **1999**, *303*, 593-600.
97. Dzuba, S. A.; Kodera, Y.; Hara, H.; Kawamori, A. *J. Magn. Reson. A* **1993**, *102*, 257-260.
98. Kodera, Y.; Dzuba, S. A.; Hara, H.; Kawamori, A. *Biochim. Biophys. Acta* **1994**, *1186*, 91-99.
99. Dzuba, S. A.; Kawamori, A. *Concepts in Magn. Reson.* **1996**, *8*, 49-61.
100. Hara, H.; Kawamori, A. *Appl. Magn. Reson.* **1997**, *13*, 241-257.
101. Klukas, O.; Schubert, W.-D.; Jordan, P.; Krauss, N.; Fromme, P.; Witt, H. T.; Saenger, W. *J. Biol. Chem.* **1999**, *274*, 7361-7367.
102. Fielding, L.; More, K. M.; Eaton, G. R.; Eaton, S. S. *J. Am. Chem. Soc.* **1986**, *108*, 8194-8196.

103. Rakowsky, M. H.; More, K. M.; Kulikov, A. V.; Eaton, G. R.; Eaton, S. S. *J. Am. Chem. Soc.* **1995**, *117*, 2049-2057.
104. Leigh, J. S., Jr. *J. Chem. Phys.* **1970**, *52*, 2608-2612.
105. Case, G. D; Leigh, J. S., Jr. *Biochem. J.* **1976**, *160*, 769-783.
106. Morris, A. T.; Dwek, R. A. *Quart. Rev. Biophys.* **1977**, *10*, 421-484.
107. Eaton, S. S.; Law, M. L.; Peterson, J.; Eaton, G. R.; Greenslade, D. J. *J. Magn. Reson.* **1979**, *33*, 135-141.
108. Hyde, J. S.; Swartz, H. M.; Antholine, W. E. in *Spin Labeling II: Theory and Applications*; L. J. Berliner, Ed., Academic Press, NY, NY, 1979, ch. 2.
109. Kuo, L. C.; Fukuyama, J. M.; Makinen, M. W. *J. Mol. Biol.* **1983**, *163*, 63-105.
110. Makinen, M. W.; Kuo, L. C. *Magn. Reson. Biol.* **1983**, *2*, 53-94.
111. More, K. M.; Eaton, G. R.; Eaton, S. S. *Inorg. Chem.* **1985**, *24*, 3820-3823.
112. Rakowsky, M. H.; Zecevic, A.; Eaton, G. R.; Eaton, S. S. *J. Magn. Reson.* **1998**, *131*, 97-110.
113. Seiter, M.; Budker, V.; Du, J.-L.; Eaton, G. R.; Eaton, S. S. *Inorg. Chim. Acta* **1998**, *273*, 354-366.
114. Klug, C. S.; Eaton, S. S.; Eaton, G. R.; Feix, J. B. *Biochemistry* **1998**, *37*, 9016-9023.
115. Budker, V.; Du, J.-L.; Seiter, M.; Eaton, G. R.; Eaton, S. S. *Biophys. J.* **1995**, *68*, 2531-2542.
116. Buchanan, S. K.; Smith, B. S.; Venatramani, L.; Xia, D.; Esser, L.; Palnitkar, M.; Chakraborty, R.; Van Der Helm, D.; Deisenhofer, J. *Nat. Struct. Biol.* **1999**, *6*, 56-63.
117. Bloembergen, N. *Physica* **1949**, *15*, 386-426.
118. Bloembergen, N.; Purcell, E. M.; Pound, R. V. *Phys. Rev.* **1948**, *73*, 679-712.
119. Bloembergen, N.; Shapiro, S.; Pershan, P. S.; Artman, J. O. *Phys. Rev.* **1959**, *114*, 445-459.
120. Poole, C. P., Jr.; Farach, H. A. *Relaxation in Magnetic Resonance*, Academic Press, N. Y., NY, 1971, ch. 6 and 13.
121. Goodman, A.; Leigh, J. S., Jr. *Biochem*istry **1985**, *24*, 2310-2317.
122. Makinen, M. W.; Wells, G. B. *Metal Ions Biol. Syst.* **1987**, *22*, 129-206.
123. Scholes, C. P.; Janakiraman, R.; Taylor, H. *Biophys. J.* **1984**, *45*, 1027-1030.
124. Ohnishi, T.; LoBrutto, R.; Salerno, J. C.; Bruckner, R. C.; Frey, T. G. *J. Biol. Chem.* **1982**, *257*, 14821-14825.
125. Brudvig, G. W.; Blair, D. F.; Chan, S. I. *J. Biol. Chem.* **1984**, *259*, 11001-11009.
126. Hirsh, D. J.; Beck, W. F.; Innes, J. B.; Brudvig, G. W. *Biochem.* **1992**, *31*, 532-541.
127. Hirsh, D. J.; Beck, W. F.; Lynch, J. B.; Que, L., Jr.; Brudvig, G. W. *J. Am. Chem. Soc.* **1992**, *114*, 7475-7481.

128. Koulougliotis, D.; Innes, J. B.; Brudvig, G. W. *Biochem*istry **1994**, *94*, 11814-11822.
129. Koulougliotis, D.; Tang, X.-S.; Diner, B. A.; Brudvig, G. W. *Biochemistry* **1995**, *34*, 2850-2856.
130. Hirsch, D. J.; Brudvig, G. W. *J. Phys. Chem.* **1993**, *97*, 13216-13222.
131. Zhou, Y.; Bowler, B. E.; Lynch, K.; Eaton, S. S.; Eaton, G. R. *Biophys. J.* **2000**, in press.
132. Mchaourab, H. S.; Berengian, A. R.; Koteiche, H. A. *Biochemistry* **1997**, *36*, 14627-14634.
133. Mchaourab, H. S.; Oh, K. J.; Fang, C. J.; Hubbell, W. L. *Biochemistry* **1997**, *36*, 307-316.
134. Mchaourab, H. S.; Kálai, T.; Hideg, K.; Hubbell, W. L. *Biochemistry* **1999**, *38*, 2947-2955.
135. Koteiche, H. A.; Berengian, A. R.; Mchaourab, H. S. *Biochemistry* **1998**, *37*, 12681-12688.
136. Hanson, P.; Martinez, G.; Millhauser, G.; Formaggio, F.; Crisma, M.; Toniolo, C.; Vita, C. *J. Am. Chem. Soc.* **1999**, *118*, 271-272.
137. Perozo, E.; Cortes, D. M.; Cuello, L. G. *Science* **1999**, *285*, 73-78.
138. Berengian, A. R.; Parfenova, M.; Mchaourab, H. S. *J. Biol. Chem.* **1999**, *274*, 6305-6314.
139. Oh, K. J.; Zhan, H.; Cui, C.; Altenbach, C.; Hubbell, W. L.; Collier, R. J. *Biochemistry* **1999**, *38*, 10336-10343.
140. Sun, J.; Voss, J.; Hubbell, W. L.; Kaback, H. R. *Biochem.* **1999**, *38*, 3100-3105.
141. Yager, T. D.; Eaton, G. R.; Eaton, S. S. *J. C. S. Chem. Commun.* **1978**, 944-945.
142. Dalal, D. P.; Damoder, R.; Eaton, G. R.; Eaton, S. S. *J. Magn. Reson.* **1985**, *63*, 327-332.
143. Dalal, D. P.; Damoder, R.; Benner, C.; Eaton, G. R.; Eaton, S. S. *J. Magn. Reson.* **1985**, *63*, 125-132.
144. Burchfield, J.; Telehowski, P.; Rosenberg, R. C.; Eaton, S. S.; Eaton, G. R. *J. Magn. Reson. B* **1994**, *104*, 69-72.
145. Likhtenshtein, G. I. *Spin Labeling Methods in Molecular Biology*; Wiley, N.Y., NY, 1976.
146. Greenhalgh, D. A.; Altenbach, C.; Hubbell, W. L.; Khorana, H. G. *Proc. Natl. Acad. Sci. USA* **1991**, *88*, 8626-8630.
147. Voss, J.; Hubbell, W. L.; Kaback, H. R. *Biochemistry* **1998**, *37*, 211-216.
148. Mchaourab, H. S.; Lietzow, M. A.; Hideg, K.; Hubbell, W. L., *Biochemistry* **1996**, *35*, 7692-7704.
149. Forrer, J.; Pfenninger, S.; Wagner, B. Weiland, T. *Pure & Appl. Chem.* **1992**, *64*, 865-872.
150. Schweiger, A. *Angew. Chem. Int. Ed. Engl.* **1991**, *30*, 265-292.
151. Schweiger, A. *Pure & Appl. Chem.* **1992**, *64*, 809-814.

Chapter 21

Allostery and Induced Fit: NMR and Molecular Modeling Study of the trp Repressor–mtr DNA Complex

Luciano Brocchieri[1], Guo-Ping Zhou[2,3], and Oleg Jardetzky[2,*]

Departments of [1]Mathematics and [2]Molecular Pharmacology, Stanford University, Stanford, CA 94305
[3]Current address: Department of Structural Biology, The Burnham Institute, La Jolla, CA 92037

The complex of the trp repressor from *E.coli* with the mtr operator DNA (one of five to which it is known to bind) has been studied by NMR and molecular modeling and compared to the repressor - consensus operator complex, whose structure has been determined previously (*1*). In contrast to the consensus operator complex, which is symmetric in all respects, the complex with the asymmetric mtr operator is asymmetric in both its structure and its dynamics: a significantly larger number of NOE contacts to the repressor is observed for one half of the operator, as compared to the other. Loss of intensity and broadening of the NOE cross peaks from this region can be attributed to rapid exchange between a bound and a free state. Presence of rapid exchange decreases the number of NOE data that can be obtained, with the result that the available data set is sufficient to define a structure of moderate, but not one of high precision. The structures,

© 2002 American Chemical Society

calculated using restrained molecular dynamics and sequential simulated annealing with 3982 NOE and other experimental constraints have RMSD values of 1.50±0.45 Å for the protein backbone atoms and 2.19±0.80 Å for the mtr DNA. Nevertheless some important features of the structure and its dynamic behavior can be deduced. The affinity of the repressor helix-turn-helix DNA binding domain for one of the two sequences in the binding regions of the mtr DNA appears to be greater, resulting in a difference in the dynamics of the two regions: during the lifetime of the more stable half of the complex, the DNA binding domain forming the less stable complex undergoes repeated exchange between the free and the bound state. Among the best defined features of the contact region are the contacts between the methyl groups of thymines and hydrophobic side chains of the repressor. Their possible role in sequence specific recognition is discussed.

Understanding the catalytic function of proteins requires knowledge of the structure of the active site and an elementary understanding of amino acid side chain motions. Understanding their more complex functions - allosteric control, membrane transport, information transfer, mechanical work - requires in addition to the knowledge of the entire structure a detailed quantitative description of its dynamics. While there has been phenomenal progress in our knowledge of protein structures over the past two decades - thousands of structures have by now been reported as a result of the rapid development of x-ray crystallography and to a lesser extent of Nuclear Magnetic Resonance (NMR) as structural methods - our understanding of protein dynamics has not kept pace. This is due to the scarcity of suitable methods for the study of dynamics. Molecular Dynamics (MD) - the solution of equations of motion for each atom in the structure - could *in principle* give a complete description of all motions within each protein structure, but in practice it is subject to severe limitations, both of a practical and theoretical nature. The practical limitation that MD simulations on proteins can only be carried out on time scales several orders of magnitude shorter (nanoseconds to microseconds) than the time scales of biologically significant motions within the structure (milliseconds or longer) might eventually be overcome by the development of faster computers. More significant - and more crippling - is the limitation of principle, that binary interatomic potentials (between pairs of atoms) have to be used in the computations to make them tractable. This neglects the higher order interactions which occur in the system, and therefore requires the introduction of corrections, so that an accurate result could only be reached by successive approximation. To ascertain that such a result is indeed accurate one would need to test it against an experimental finding. NMR is the only experimental method that can provide detailed information on protein dynamics at atomic resolution, but the information is somewhat indirect - describing the motion of internuclear vectors - and requires the construction of models - best derived from MD concepts - to be translated into the motions of individual atoms. The success in devising such models has thus far been rather modest. As a result our knowledge of protein dynamics is at this point mostly qualitative. Nevertheless, important features of the dynamic processes which can occur in proteins and protein complexes have become apparent even from an examination of qualitative results.

Several years ago we undertook a comprehensive NMR study of the structure and dynamics of the simplest allosteric system we could find - the tryptophan repressor from E.coli, and its complexes with operator DNAs. Among the most significant results of this study have been the findings that the DNA binding helix-turn-helix domain is highly unstable and stabilized by the interaction with operator DNA, and that the selectivity of the repressor molecule for different DNA sequences is correlated with the degree of instability - or flexibility - of the DNA binding domain. Mutations which increase the rigidity of the domain convert the molecule to a super-repressor for some operators, but not for others. The functional significance of the flexibility appears to be to allow induced fit when binding to different DNA sequences. These findings have been published in a series of reports and reviewed on several occasions, as cited below.

In this paper we report some of the more recent results on the binding of the repressor to an asymmetric operator DNA sequence and discuss them in the context of the earlier findings.

The System

The trp repressor from *Escherichia coli* is a DNA binding protein which in the presence of the amino acid tryptophan inhibits the transcription of a least five operons: trpEDCBA, trpR, aroH, aroL and mtr (2-6). Structures of the ligand-free form (aporepressor) and the tryptophan containing form (holorepressor) have been determined by both x-ray crystallography (7-9) and high resolution NMR (10). The crystal and solution structures are identical in all essential features and show that the trp repressor is an intertwined dimer; each monomer is comprised of six alpha helices with helices A, B, C, F forming the hydrophobic central core and helices D and E extending from the central core to form a helix-turn-helix DNA binding motif.

The mutation studies of Bass et al. (11) have demonstrated that the CTAG block in the consensus sequence is the most sensitive to mutation and so are the flanking adenine and thymine bases. Both solution studies (12-14) and dynamic studies led to the conclusion that the DNA structure is mostly B-form (14, 15) and that it is not a rigid structure but is capable of conformational adjustments.

Two crystal structures (16, 17) and a family of NMR solution structures (1) of the trp repressor - operator DNA (ODNA) complex have also been reported.

The structure of the repressor in the complex is similar to that found in the isolated repressor structures (*10*). The overall topology observed in the solution structure of the complex agrees with that of the crystal structure (*16*). The structure of the hydrophobic core is essentially identical in different forms of the repressor, although the N-terminal segments were found to be highly disordered in both the crystal (*16*) and NMR (*1*) structures. Helices D and E were found in a different conformation in the complex than in the isolated apo- or holorepressor, with helix E making contacts with the major groove of the operator DNA.

The structures of the D and E helices are better defined in the solution structure of the complex than in those of the holo- and aporepressors (*1*). This indicates that these segments are stabilized by binding to operator DNA. The DNA is distorted upon binding (*1, 16*), suggesting that there is an "induced fit" for both the protein and the DNA upon the formation of their complex (*18-20*). The aim of the present study was to determine whether binding to operators with different sequences involves different points of contact and a different pattern of "induced fit".

To answer this question we have undertaken a structure determination of the complex of trp repressor with one of its natural operators, mtr. Its sequence is:

```
     1  2  3  4  5  6  7  8  9 10 11 12 13 14 15 16 17 18 19 20
5'   T  G  T  A  C  T  C  G  T  G  T  A  C  T  G  G  T  A  C  A
3'   A  C  A  T  G  A  G  C  A  C  A  T  G  A  C  C  A  T  G  T
    40 39 38 37 36 35 34 33 32 31 30 29 28 27 26 25 24 23 22 21
```

This operator has two sites which are different from the consensus sequences 5'CTAG3' at positions 5-8 and 25-28. The changes in the sequence in this region have a negligible effect for one site, and a mild effect for the other on the binding of wild-type trp repressor (*21*). Therefore, the structural study had to include a careful examination of a possible asymmetry of binding.

Results

A. Proton Assignments in the Complex

(i) mtr Operator

The chemical shifts of the protons of the free mtr operator and of the bound mtr operator are given in Table I and II, respectively. Not all of the chemical shifts for the mtr operator bases could be determined, due to overlap, low intensity or large linewidths of NOE crosspeaks. The chemical shifts of the CH3 protons of T1, H2 protons of A4, A20, A32 and A40 and the NH2 protons of C13, G15, C33 and G34 could not be detected when the mtr operator was bound to the trp repressor. Comparing Tables I and II makes it apparent that the

Table I. Chemical Shifts (p.p.m.) of the Protons of the mtr Operator DNA at 45°C.

Base	H6	H8	H1'	H2'	H2"	H3'	H4'	H5'	H5"	T-Me	CH5	H2	NH$_2$	NH
T1	7.35		5.95	1.87	2.27	4.66	4.06	3.68		1.56				
G2		8.05	6.05	2.74	2.88	4.97	4.40	4.06						
T3	7.30		5.77	2.12	2.48	4.90	4.23	4.18		1.50				
A4		8.28	6.24	2.75	2.88	5.04	4.44	4.22	4.13			7.49		
C5	7.29		5.78	1.96	2.46	4.62	4.22	4.18			5.25		7.89/6.53	
T6	7.37		6.01	2.12	2.49	4.85	4.18	4.08		1.50				
C7	7.44		5.64	2.10	2.40	4.84	4.13	4.09			5.61		8.33/6.64	
G8		7.84	5.92	2.58	2.74	4.95	4.35	4.13						12.61
T9	7.10		5.78	2.05	2.42	4.84	4.35	4.16	4.15	1.42				
G10		7.78	5.90	2.55	2.72	4.92	4.33	4.12					6.45	12.40
T11	7.16		5.70	2.00	2.42	4.85	4.18	4.12		1.42				
A12		8.23	6.18	2.69	2.84	5.01	4.39	4.13	4.09					
C13	7.24		5.76	1.90	2.40	4.65	4.16	4.14			5.21		7.90/6.44	
T14	7.21		5.75	1.99	2.40	4.84	4.15	4.04		1.53				
G15		7.79	5.68	2.65	2.73	4.96	4.34						6.68	12.62
G16		7.59	5.91	2.49	2.72	4.86	4.35	4.20					6.52/6.68	12.66
T17	7.20		5.74	1.99	2.40	4.86	4.17	4.15		1.39				
A18		8.23	6.18	2.62	2.79	5.01	4.39	4.16	4.09			7.62		
C19	7.27		5.81	1.92	2.28	4.75	4.13				5.39			
A20		8.20	6.31	2.48	2.62	4.66	4.18	4.13	4.12			7.73		
T21	7.35		5.95	1.87	2.27	4.66	4.06	3.68		1.56				
G22		8.05	6.05	2.74	2.88	4.97	4.40	4.06						
T23	7.30		5.75	2.12	2.48	4.90	4.23	4.18		1.50				
A24		8.27	6.24	2.72	2.87	5.04	4.44	4.20	4.13					
C25	7.25		5.81	1.96	2.39	4.75	4.13				5.27		7.98/6.33	
C26	7.38		5.45	1.95	2.29	4.81	4.06				5.50		8.36/6.51	
A27		8.15	6.04	2.73	2.88	5.02	4.38					7.59	4.13/4.02	
G28		7.52	5.79	2.39	2.66	4.87	4.35	4.20					6.30	12.62
T29	7.11		5.69	1.96	2.39	4.84	4.16	4.12		1.32				13.66
A30		8.20	6.15	2.64	2.82	5.00	4.38					7.39		
C31	7.24		5.45	1.90	2.29	4.78	4.15				5.29		8.28/6.44	
A32		8.16	6.12	2.62	2.82	4.99	4.35					7.59		
C33	7.13		5.48	1.79	2.20	4.76	4.08						8.05/6.30	
G34		7.77	5.48	2.60	2.73	4.96	4.28	4.07					6.37	12.57
A35		8.03	6.07	2.68	2.88	5.02	4.42					7.59		
G36		7.45	5.79	2.39	2.67	4.84	4.34	4.12					6.76	12.71
T37	7.13		5.73	1.96	2.38	4.85	4.17			1.31				
A38		8.23	6.18	2.62	2.79	5.01	4.39	4.16	4.09			7.62		
C39	7.27		5.81	1.92	2.28	4.75	4.13				5.39			
A40		8.20	6.31	2.48	2.62	4.66	4.18	4.13	4.12			7.73		

NOTE: Chemical shifts are referenced to internal TSP.

Table II. Chemical Shifts (p.p.m.) of the Protons of the mtr-Operator DNA in Complex with trp Repressor at 45°C.

Base	H6	H8	H1'	H2'	H2"	H3'	H4'	H5	T-Me	CH5	H2	NH$_2$	NH
T1	7.32		6.09	1.93	2.35	4.67	4.06	3.69					
G2		7.90	6.05	2.64	2.73	4.97							
T3	7.27		5.93	2.56	2.64	4.97	4.36		1.63				
A4		8.19	6.34		2.83	5.01	4.14	4.12					
C5													
T6													
C7													
G8													
T9													
G10													12.19
T11													
A12		8.33	6.40	2.77	2.87	5.03	4.41	4.13					
C13	7.43		5.76	2.03	2.41		4.92			5.44			
T14	7.28		5.97	1.86	2.24	4.79			1.48				
G15		7.77	5.97	2.60	2.68	4.97	4.35						
G16	7.48	6.08	2.57	2.63	4.96	4.34							
T17													
A18													
C19	7.32		5.76	1.78	2.25	4.74	4.16			5.27			
A20		8.21	6.35	2.75	2.56	4.72	4.27	4.09					
T21	7.32		6.09	1.93	2.35	4.67	4.06	3.69					
G22		7.90	6.05	2.64	2.73	4.97							
T23	7.27		5.93	2.56	2.64	4.97	4.36		1.63				
A24		8.19	6.34		2.83	5.01	4.14	4.12					
C25													
C26													
A27													
G28													
T29													
A30													
C31	7.35		5.49							5.24		8.28/6.44	
A32		8.23	6.30	2.75		4.99							
C33	7.24		5.58	1.89	2.44	3.79				5.14			
G34		7.50	5.66	2.38	2.52	4.84							
A35													
G36													
T37													
A38													
C39	7.32		6.04	1.78	2.56	4.74	4.16			5.27			
A40		8.21	6.34	2.75	2.56	4.72	4.27	4.09					

NOTE: Chemical shifts are referenced to internal TSP.

chemical shifts of most protons were changed after the mtr operator docked to the trp repressor. The most significant chemical shift changes are in the stretches T1 – A4, A12 – G16, C19 – A24, and C31 – G34. These results indicate that the average conformation of the mtr ODNA changes when it interacts with the repressor, but also that the dynamic equilibrium between different conformations observed in the free DNA persists in the complex.

(ii) trp Repressor

Sequence-specific assignments for about 93% of the backbone amide protons and 88% of the side chain protons of the trp repressor in the bound mtr operator complex were obtained using NOESY-watergate, TOCSY, sensitivity enhanced HSQC and NOESY-HSQC. Figure 1 shows that the repressor in the complex has an intrachain NOE pattern almost identical to that of the holorepressor (*10, 22*) and the trp repressor in the trp repressor-consensus ODNA complex (*1*). No NOE connectivities were detected from residues 2 to 18. The only region in which more NOE connectivities were observed in the repressor-operator complex than in the free holorepressor is the helix-turn-helix DNA-binding domain formed by helices D and E. These results indicate that the core of the repressor remains essentially unchanged on complex formation, except for the stabilization of helices D and E, as previously observed in the consensus repressor-operator complex. The same conclusion was reached when the chemical shifts were analyzed using the Chemical Shift Index (*23*).

B. Long Range NOEs

(i) NOEs within the Protein Structure

The pattern of long-range NOEs within the trp repressor in its complex with the mtr operator is very similar, if not identical to that observed in the holorepressor (*10, 22*) and in the trp repressor-consensus ODNA complex (*1*). 350 NOEs previously identified as intra-chain and 536 NOEs previously identified as inter-chain have been observed in the trp repressor-mtr ODNA complex. This is a clear indication that the overall conformation of the repressor molecule is preserved in the mtr ODNA, as it is in the consensus ODNA complex, and any conformational differences are small and localized, largely to the DNA-binding domain, as discussed below.

(ii) NOEs between the Repressor and the Co-Repressor Tryptophan

There are significant differences between NOEs observed in the trp repressor-consensus ODNA and the trp repressor-mtr ODNA complexes. The NOEs

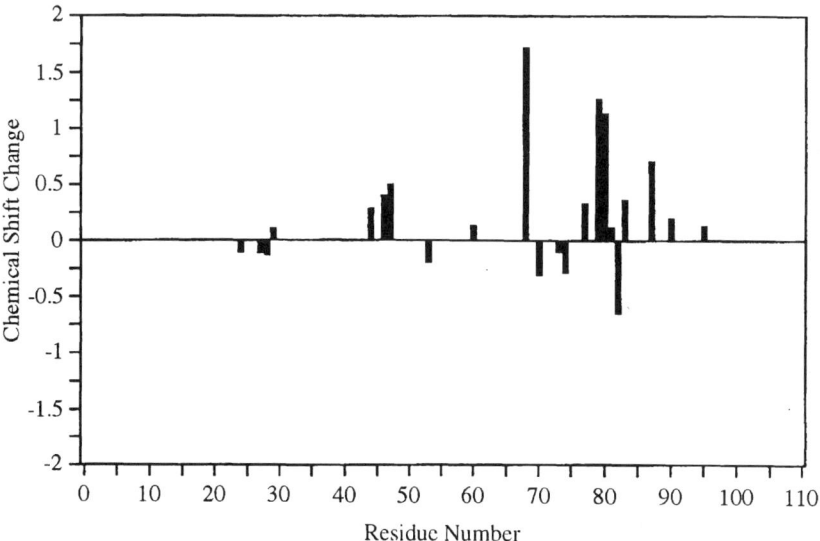

Figure 1. Amide proton chemical shift changes in the trp repressor-mtr ODNA operator upon complex formation.

between Gly52 and the tryptophan ligands and between Thr83 and the ligands seen in the complex with the consensus ODNA, are not observed in the complex with the mtr ODNA. Instead of the NOEs between the tryptophan and Met42 and Leu43 found in the consensus ODNA complex NOE connectivities between the ligands and Thr44 and Pro45 are seen in the mtr ODNA complex. A total of 38 NOEs are assigned in the repressor-mtr ODNA complex, while 36 NOEs were assigned in the repressor-consensus ODNA complex. This indicates a small but detectable conformational change in the trp-binding pocket.

(iii) Contact NOEs between the Protein and the mtr ODNA

Unlike in the trp repressor-consensus ODNA complex, the contact NOEs between the repressor and the mtr DNA are not symmetrically distributed in the mtr ODNA complex. There are 47 NOE connectivities between monomer I and the mtr DNA and 14 between monomer II and the mtr DNA (Table III) indicating that the dynamic stability and hence the binding affinity between repressor and operator is much higher for site I than for site II.

As in the consensus ODNA complex, there are many contacts between T14 of the DNA and the residues in the DE helix region. For example, there are 7 NOEs between T14 and Ala77. This is also consistent with the observation that the NH proton of Ala77 has a large downfield chemical shift change (1.25 ppm) upon complex formation (Figure 1). There are two NOEs between Lys72 and G15. However, only the central bases A32, C33, and G34 have NOE contacts with the repressor (4, 3, and 4, respectively). These results confirm that a) the helix-turn-helix is the DNA binding region, b) T14 of the mtr DNA may be a key base for recognition of the repressor and that c) compared to the consensus ODNA the transversion A:T -> C:G at positions 7 - 34 has a much more dramatic effect on the interaction with the repressor than the transition A:T -> G:C at positions 15-26.

Unlike in the case of the trp repressor-consensus ODNA complex, no NOE was detected between L-tryptophan and DNA in the trp repressor-mtr ODNA complex. This most likely indicates that the interactions between cofactors and mtr-DNA are less stable.

C. Structure of the Complex

(i) Distance and Dihedral Angle Constraints

Because the sequential NOE pattern of the repressor in the complex (Figure 2) shows clear helical connectivities for all of the six α helices, continuous stretches of 5 or more NOEs characteristic of an α helix were assigned typical dihedral angle constraints phi=-65($\pm$45)Å and psi=-35($\pm$45)Å in both the previous (*1, 10*) and present structural calculations.

Table III. Contact NOEs between trp Repressor and mtr ODNA

ODNA	Protein	ODNA	Protein
G34 H3'	Ala577 β	T14 H2'	Ile79 NH
G34 H2"	Ile579 β	T14 H2"	Ile79 β
G34 H3'	Ala580 β	T14 CH3	Ala80 β
G34 H1'	Ala580 β	T14 CH3	Thr81 α
C33 H2'	Ala550 β	G15 H8	Lys72 γ
C33 H2"	Ala550 β	G15 H2'	Lys72 γ
C33 H2"	Thr553 γ2	G15 H8	Ala77 β
A32 H2'	Thr44 γ2	G15 H8	Ile79 δ
A32 H3'	Thr44 γ2	G15 H8	Ile79 γ1
A32 H8	Thr44 γ2	A24 H8	Ala80 β
A32 H2'	Ala550 β	A24 H8	Thr83 γ2
		T23 CH3	Gln68 α
A12 H1'	Ala50 β	T23 CH3	Gln68 NH
A12 H3'	Thr81 γ2	T23 H6	Ile79 δ
A12 H1'	Thr544 γ2	T23 H6	Ile79 γ1
A12 H2'	Thr544 γ2	T23 H2'	Ile79 γ1
A12 H2"	Thr544 γ2	T23 CH3	Ile79 α
A12 H3'	Thr544 γ2	T23 CH3	Ile79 δ
A12 H4'	Thr544 γ2	T23 CH3	Ile79 γ1
A12 H8	Thr544 γ2	T23 CH3	Ile79 γ2
C13 H2"	Thr53 γ2	T23 H6	Ile79 β
C13 H5	Ala77 β	T23 CH3	Ala80 α
C13 H5	Thr81 γ2	T23 H6	Ala80 β
T14 H2'	Ala77 β	T23 CH3	Ile82 δ
T14 H2'	Ala77 NH	T23 CH3	Ile82 γ2
T14 H2"	Ala77 β	T23 H6	Ile82 δ
T14 H3'	Ala77 β	T23 H6	Ile82 γ2
T14 CH3	Ala77 α	T23 H3'	Thr83 γ2
T14 CH3	Ala77 β	T23 H6	Thr83 γ2
T14 H6	Ala77 β	G22 H3'	Ile79 δ
T14 H2'	Gly78 NH	G22 H3'	Ile82 δ

NOTE: A residue number ≥ 500 indicates that the residue belongs to monomer II.

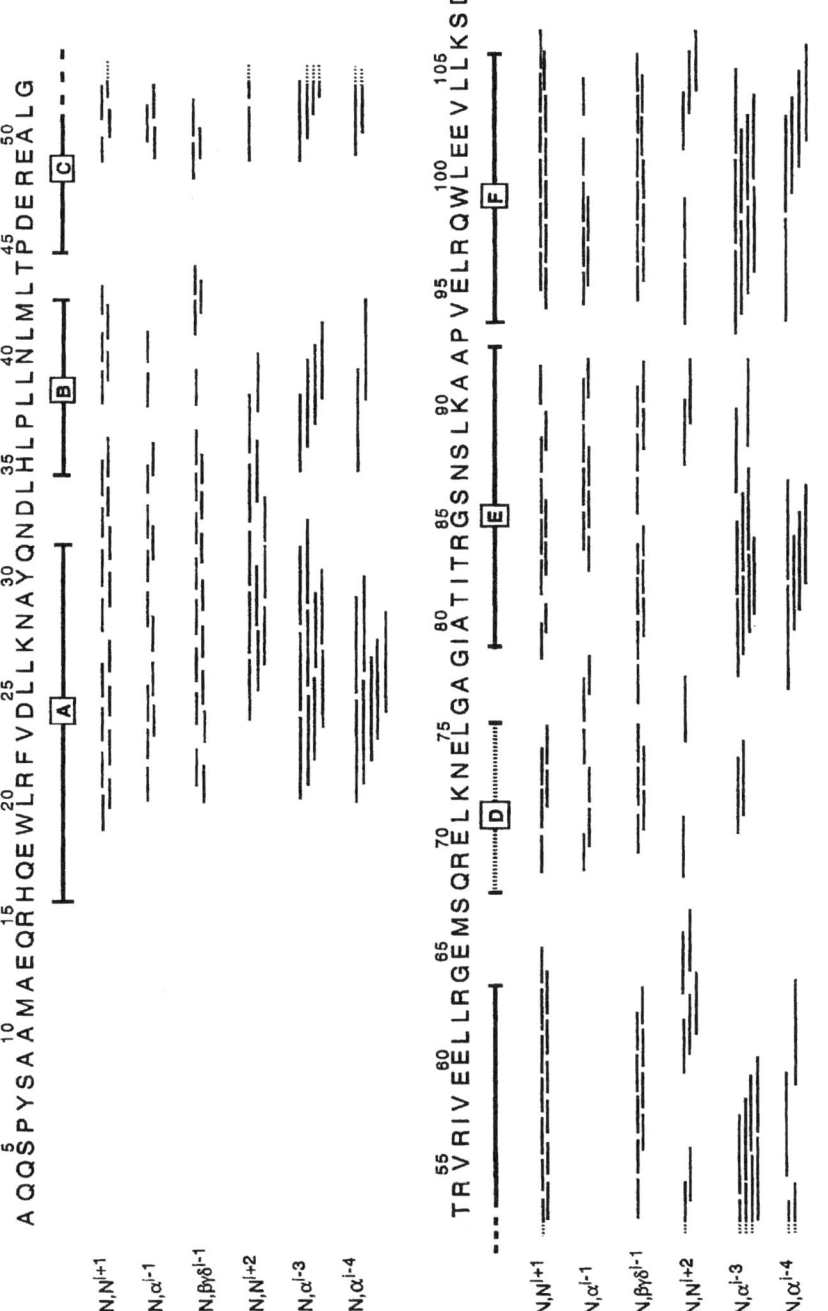

Figure 2. Sequential NOEs in the trp repressor-mtr ODNA complex.

The number of the NOE and other constraints (total constraints = 3982) obtained for the structure calculation is given in Table IV. As in the cases of the holorepressor (*10*) and the trp repressor-consensus ODNA complex (*1*), the distribution of the NOEs in the protein is not uniform, with fewer NOEs near and within helix D. There are on the average 15 NOEs per residue in the core, only 5 NOEs per residue in helix D and 12 NOEs per residue in helix E. The distribution of NOEs in the DNA is also not uniform. Comparing Table I and II, it can be seen that there are fewer NOEs in the complex than in the mtr ODNA control.

(ii) Structure Calculation

The refinement output statistics of the simulated annealing procedure used to model the trp repressor-mtr operator complex are given in Table V and Figure 3. Figures 3A, 3B and 3C represent the RMSD values of the average of the 30 structures relative to their mean structure versus amino acid number and DNA base sequence number, respectively. The average RMSD among a set of 30 structures is 1.71±0.75 Å and 1.91±0.84 Å for all side chain heavy atoms of monomer I and monomer II, respectively, 1.38±0.41 Å and 1.62±0.49 Å for the backbone ($^\alpha$C, carbonyl C, amide N in the protein) heavy atoms of the repressor monomers I and II, respectively, and 2.23±0.89 Å for the backbone atoms (O-3', C-3', C-4', C-5', O-5', P) in the DNA. The definition of the trp repressor-mtr operator complex is then analogous to that obtained for the holorepressor structure (*10*) but less precise than compared to the consensus ODNA complex (*1*).

The graphs representing the RMSD values for monomers I and II in Figures 3A and 3B have several similar features. For example, there are peaks between residues 30 and 35 and between residues 530 and 535. These indicate that the turn between helices A and B are not well defined for these two monomers due to fewer NOEs in the turns than in the remainder of the core regions. Although the average RMSD values of monomer II are very close to those of monomer I, we obtained significant differences in their DE regions, which are much less defined in monomer II (positions 568-585) than in monomer I (positions 68-85).

As in the trp repressor-consensus ODNA complex, residues 4 to 19 (monomer I) and residues 504 and 519 (monomer II) appear as a coil in our structure due to the lack of NOE constraints. All backbone atom positions in the core (helices A, B, C and F) are well defined in both monomers. The backbone atom positions in helix D are not as well defined as those in the rest of the sequence. Both the backbone and side chain atoms at the N-terminal part of helix C and the turn between helices C and D have larger than average RMSD values, as a consequence of the reduced number of constraints in these regions.

Figure 3C shows that in the mtr ODNA the largest RMSD values are distributed near the ends while both backbone and side chain atom positions of the nucleotides 12-15, 23, 24, and 32-35 have locally minimal RMSD values.

Table IV. Summary of NOE and Other Constraints

Type of Constraints	Number of Constraints
mtr DNA NOEs	54
protein NOEs (per dimer)	3445
sequential	2215
intraresidue	344
Long Range	
intra-chain	350
inter-chain	536
conzyme *trp*-protein contact NOEs	38
ODNA-protein contact NOEs	61
Total NOEs	3598
H-bond	58
Dihedral angle	252
Total other	74
Total Constraints	3982

Table V. Refinement Output Statistics for X-PLOR

Total energies (kcal/mol)	18,496±73
NOE violations >0.5 Å	1–3
NOE energies (kcal/mol)	430±29
RMS NOE violations	0.063
Bond energies (kcal/mol)	1,587±6
RMS bond length violations (Å)	0.018
RMS angle violations (deg)	1.93
Bond angle energies (kcal/mol)	4,957±19
Dihedral angle constraint violations >5°	0–1
RMS dihedral angle constraint violations (deg)	0.6±0.2
Improper energies (kcal/mol)	9,900±12
Van der Waals repulsion energies (kcal/mol)	720±4

NOTE: Based on 30 calculated structures.

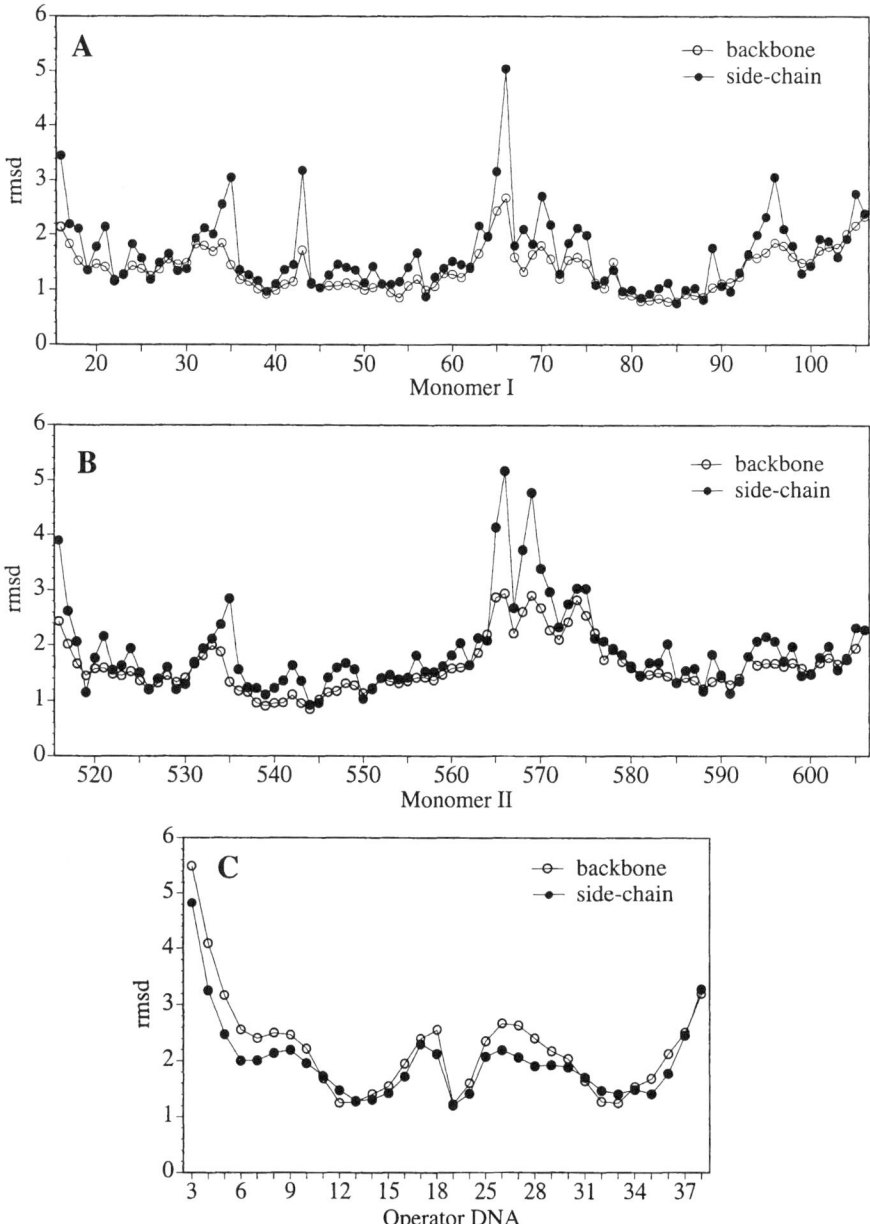

Figure 3. RMSD by residue from the average of 30 structures of trp repressor and mtr ODNA complex. (A) monomer I of the trp repressor, (B) monomer II of the trp repressor and (C) operator DNA.

The average of the calculated structure is shown in Figure 4. Figures 4A - C show the repressor backbone structure (monomer I dark gray, monomer II light gray). The conformation of the repressor in the complex is to a first approximation symmetric and the reading heads fit into the major groove. Actually, their conformations are not identical. Because more protons from monomer I (Table III) are in contact with the mtr ODNA upper half, which contains base pairs C13:G28, T14:A27, G15:C26, G16:C25, T17:A24, A18:T23, and C19:G22, this part of the DNA appears closer to the protein than the part facing monomer II. The DNA base pair planes between the two halves are bent by about 29°. It can be observed that the N terminal segment of helix E in monomer I makes contact with the DNA bases and a large portion of helix E and the turn between helices D and E are embraced by the major groove. The backbone of helix D is mostly outside of the major groove. The backbones of helices D and E in monomer II appear on the average more distant from the major groove, consistent with the fewer contact NOEs between DNA and monomer II. This could be considered as an artefact of the structure calculation, which assumes that the calculated structure is rigid. In a complex subject to rapid and frequent exchange between the free and the bound state the apparent average distance will be larger than for a complex in slower exchange. Figures 4D and E use space-filling models to show the docking structure of the trp repressor-mtr ODNA complex.

In contrast to the trp repressor-consensus ODNA complex model, we could not identify with a reasonable probability a set of hydrogen bond interactions between repressor and DNA. However, after the calculation of a family of structures, we found that the backbone positions of eight residues previously identified as hydrogen bonding with the DNA bases (1) were within a distance compatible with such interactions. Each possible protein-DNA hydrogen bond was then tested using simulated annealing to see whether it was consistent with the rest of the experimental constraints and energies. Our results show that all hydrogen bonds, except for G16.O6-Ile79.N (atom O6 of G16 with atom N of Ile79), can be simultaneously satisfied within 10% of the energy and constraint violations for both monomers. However, in the absence of adequate NOE data for monomer II, we chose to model these hydrogen bond interactions only relative to monomer I (Table VI), fully recognizing that they may also exist for monomer II, but be rendered undetectable by the relative dynamic instability of this half of the complex.

Discussion

The key question – how does a protein with a single well defined structure recognize a variety of operator DNA sequences? - has thus far received a partial answer: the instability – effectively the flexibility - of the DNA binding domain allows it to make minor adjustments in its geometry, in order to make contacts with different constellations of DNA ligand groups. This "induced fit" appears

Table VI. Potential Direct H-Bonds between trp Repressor and mtr DNA per Dimer.

Protein	mtr ODNA
Arg69 $H^{\eta 12}$	G22 O6
Arg69 $H^{\eta 22}$ and $H^{\eta 21}$	T23 O4 or G22 N7
Lys72 $H^{\xi 3}$	G16 N7
Ile79 HN	G16 N7
Ala80 HN	G15 N7
Thr83 $H^{\gamma 1}$	A24 N7
Arg84 $H^{\eta 12}$	C13 N4

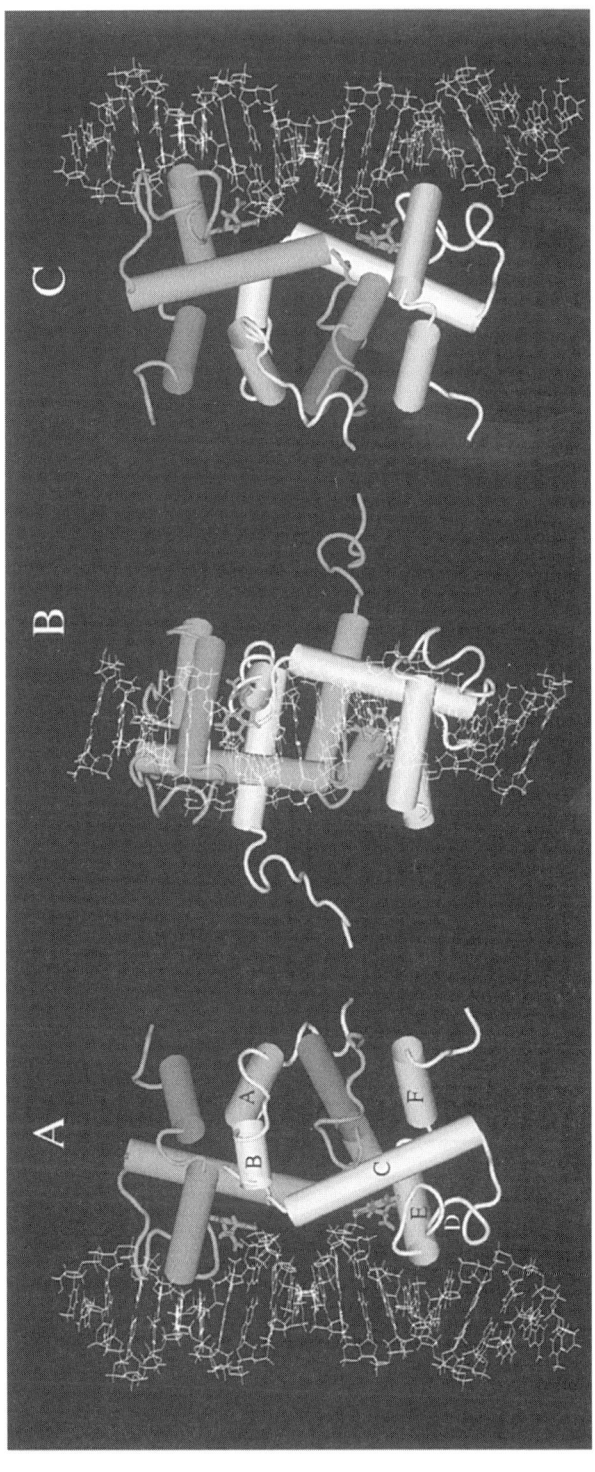

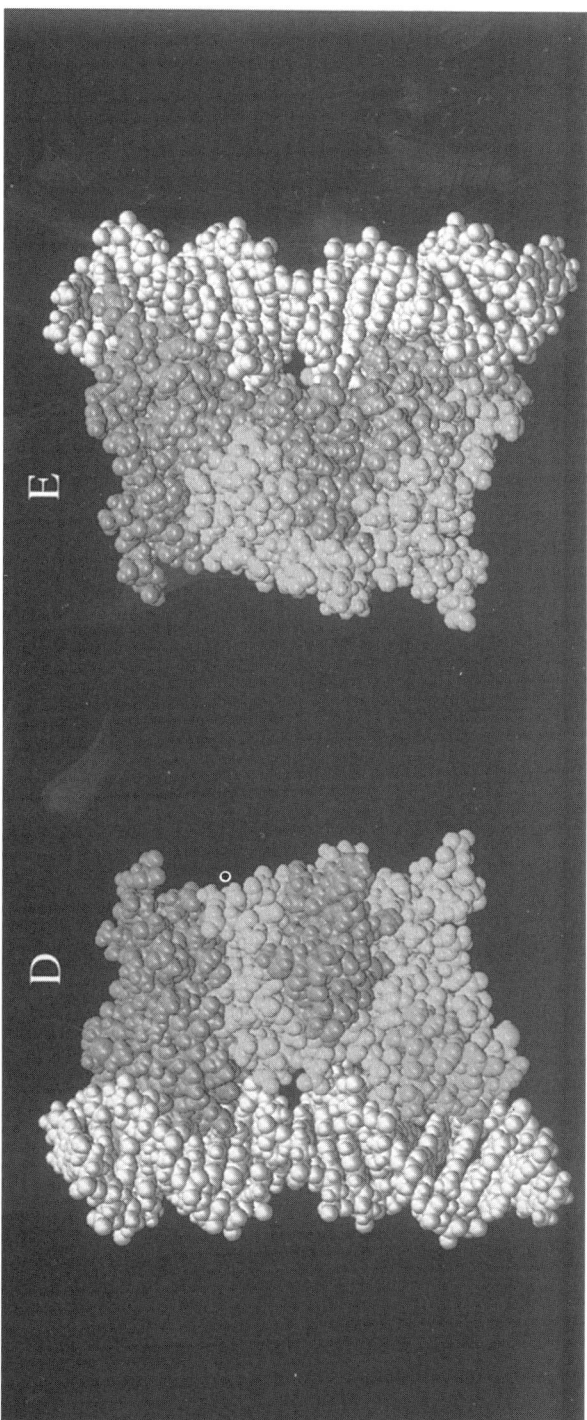

Figure 4. Models of the trp repressor-mtr ODNA solution complex mean structures: (A-C) with the repressor at backbone representation and the mtr ODNA (white) and L-trp cofactors (dark gray) at full atomic representation. Monomer I of the repressor is shown in dark gray and monomer II is shown in light gray; (D-E) space filling models with the mtr ODNA molecule shown in white and the repressor in gray. This and the following figure prepared with VMD (24).

to be reciprocal, since a considerable degree of flexibility is observed in the DNA as well. The differences in the geometry are likely to be accompanied by corresponding differences in stability of the resulting complexes. Insofar as the stability of the complexes determines the degree of repression that can be achieved, they will also be reflected in the efficiency with which each of the different operons is regulated. The more complete answer requires detailed knowledge of each of the complexes formed, their relative stability and their similarities and differences. Comparing the structures and dynamics of the mtr-DNA-repressor and the consensus ODNA-repressor complexes can provide at least a partial insight into the mechanisms at play. At the present stage of our knowledge more can be said about the nature of molecular interactions involved in determining the specific geometry of the different complexes, and their stability, than about the detailed dynamics of the complexes.

The topology of the trp repressor-mtr operator DNA (ODNA) complex structure is similar to that of the previously studied trp repressor-consensus ODNA complex structure (*1*). The structure of the ABCF core of the repressor is substantially unchanged when the repressor is bound to either operator and similar to the aporepressor or the holorepressor. Helices E in both monomers of the consensus ODNA complex and in one monomer of the mtr ODNA complex fit into the major groove of the DNA. Helices D of all monomers are mostly outside of the major groove. The first 16 N-terminal residues are disordered and do not form stable interactions in either case. In both complexes and in the apo- and holorepressors the D and E regions have larger RMSD values than the other regions and display a similar pattern of variability. The DNA-binding domain (helices DE) has more flexibility than the core regions (helices ABCF) of the repressor as in the trp repressor- consensus ODNA complex and in the apo- and holorepressors (*1, 10, 25-27*).

The most striking finding of the study was that in the mtr asymmetric operator, one half of the oligonucleotide, including positions 11-20 in one strand and positions 21-30 in the opposite strand, has a much greater number of contact NOEs with the repressor than the other half and the NOEs observed for this half are of higher intensity and narrower crossover line width. We will refer to this part of the operator/repressor complex as the "strongly-binding-half", and to the other part as the "weakly-binding-half". The strongly-binding-half is characterized by the sequence 5'-CTGG-3' (positions 13-16), while the weakly-binding-half has the sequence 5'-CGAG-3' (positions 33-36). Boldface are the mutated bases compared to the corresponding symmetric consensus operator sequence 5'-CTAG-3'. The contact NOE pattern of the region of the repressor binding to the strongly-binding-half of the mtr operator largely overlaps the pattern observed in the complex with the consensus operator (*1*). However, we observe fewer interactions between the repressor and the weakly-binding-half of the mtr operator. The NOE contacts in the weakly-binding-half are essentially reduced to interactions of the sugar atoms of A32 (with the side chains of Thr44 and Ala550), of C33 (with Ala550 and Thr553), and of the sugar of the mutated position G34 (with Ala577, Ile579, and Ala580). At the same position of the consensus operator, T34 shows interactions with the repressor residues Ala77, Gly78, Ile79, Ala80, Thr81, involving sugar atoms and also the methyl group of thymine (*1*). Why does the mutation T-34-G in the weakly-binding-half of the mtr operator have such a large effect on its affinity for the repressor? None of the potential hydrogen bonds observed in the trp repressor-consensus operator structure involve T34. We propose that the loss of the methyl group of T34 upon substitution with G34 is at the origin of the instability. This explanation is consistent with the abundance of hydrophobic side chains in the vicinity of the methyl group of T34 in the trp repressor-consensus operator structure and in the strongly-binding-half of the mtr operator. While protein-DNA interaction studies have frequently focused on hydrogen bonding and ionic interaction patterns (*28, 29*) this interpretation stresses the importance of hydrophobic contacts as a possible mechanism of sequence specific recognition. Consistent with this hypothesis, substitutions of these bases or of the contacting hydrophobic residues from the repressor produce changes in stability of the complex. Günes and Müller-Hill (*30*) determined that the mutant Arg-69-Ile changes the specificity from an operator with sequence G2 to one of sequence T2. Günes et al. (*31*) observe that the substitution of T3 with uracil, equivalent to removing the methyl group of thymine, is not tolerated. T3 interacts with Ile79 of the repressor. On the contrary, the mutation T-37-U (T37 is paired to A4) is tolerated, consistently with the lack of contacts at this position both in the mtr ODNA model and in the consensus ODNA model (*1*). The same authors show that mutations of Ala80 to Leu, Ile, or Met, all of increased hydrophobicity, broaden the specificity of the repressor, again consistently with the importance we attribute to hydrophobic contacts.

In both repressor-operator complexes low definition is observed for residue Met66 and Leu75. Met66 is positioned in the turn amino terminal to helix D; Leu75 is positioned at the C-cap of the same helix. These two residues are exposed at the surface of the molecule and their hydrophobic character may be crucial in determining the dynamic properties of helices D and E. We propose that these hydrophobic residues have a specific functional role in destabilizing the D-E region of the repressor. Such destabilization would prevent repression in the absence of free L-Trp. Upon binding of L-Trp the dynamics of the region would favor the conformation which recognizes and binds to the DNA operator sequence. This hypothesis predicts that substitutions at these positions with residues of reduced hydrophobicity would stabilize this region and induce binding to the operator. Substitutions with more hydrophobic residues on the other hand would overcome the stabilizing effect of L-Trp and prevent stable formation of the holorepressor and/or of the complex. In fact, Jin et al. (*32*) show that the mutant Leu-75-Phe reduces affinity to L-Trp. Positions 66 and 75 are indeed evolutionarily conserved among proteobacteria (data not shown).

The asymmetry in binding affinity that we observe between the strongly-binding- and the weakly-binding-half of the mtr operator are consistent with the results of Bass et al. (*11*), which show that a symmetric operator with the sequence of the weakly-binding-half of the mtr operator produces much lower efficiency of survival than a symmetric operator with the sequence 13-CTGG-16 of the strongly-binding-half. Bass et al. (*11*) however report a mild reduction of efficiency of survival also for the symmetric operator with the sequence of the strongly-binding-half, compared to the consensus. Why this is the case is not obvious. In fact, the substitution A-15-G maintains the hydrogen-bonding capabilities described in the complex with the consensus operator of atom N7 with the backbone amino groups of Ala80 or Ile79. However, careful examination of the model structures reveals some structural differences in the helix-turn-helix DE of the two complexes. In particular, in the trp repressor-mtr ODNA complex the side chain of Ala77 contacts the methyl group of T14 (Figure 5A). In the consensus ODNA complex on the other hand the side chain of Ala77 faces away from the ODNA, packing against the side chains of Thr53 and Ile57 from helix E, while the methyl group of T14 contacts instead the methyl group of Ala80 (Figure 5B). The packing reconfiguration in our models compared to the consensus ODNA structure is a consequence of the great number of contact NOEs that we observe between the side chain of Ala77 and T14 and the absence of NOEs between the side chain of Ala77 and those of Thr53 or Ile57. It is possible that the mutation A-15-G in the mtr ODNA produces unfavorable changes in the DNA structural conformation preferences associated with the transition A:T -> G:C. Besides providing substrate for hydrophobic interactions, the A:T pairs are indeed also observed to generate DNA bending towards the major groove, as opposed to bending towards the minor groove generated by G:C pairs (*28, 33*). In this respect, it is worth noting that the consensus operator tetranucleotide CTAG is conspicuously under-represented in most bacterial species (*34*). Its rarity has been tentatively related to the specific DNA structural conformation that it appears to induce. The loss of such specific structural conformation may cause the packing rearrangements that we observe between the trp repressor and the strongly-binding-half of the asymmetric mtr ODNA and explain its reduction in binding affinity. Such reconfiguration can also be expected to affect the dynamical properties of the DE helix-turn-helix region, where intramonomer hydrophobic interactions seem to compete with those provided by the DNA methyl groups in modulating recognition and affinity for different operators.

This study and its predecessors have made at least modest inroads into the largely uncharted territory of the intricate interplay between static interactions and their dynamics in determining the stability of biologically active complexes. They have also raised a host of new questions, which remain to be answered before we could claim to understand the mechanisms of allostery and induced fit. Massive collection of data will be necessary to achieve this end. A correlation between the affinity (thermodynamic stability) and dynamic stability of a complex remains to be established – whether one considers intramolecular – intra- and interchain - complexes that hold the protein together, or the intermolecular complexes of a protein with its ligands. The atomic motions reflected in such processes as nuclear relaxation and backbone proton exchange – the main NMR indicators of protein dynamics remain to be described. Last, not least the role of dynamic processes in storing and transmitting information within a protein structure remains to be established. Only a large number of studies of the kind presented here is likely to lead to answers that will have generality.

Materials and Methods

Preparation of the trp Repressor and Operator DNA and NMR Spectroscopy

The purification procedure of the trp repressor was as previously described (1, 35, 36). For the synthesis of the 20 base pair mtr ODNA procedure refer to (1). The strategy of assignments of the protons of the mtr operator and the trp repressor were as the proton assignments reported for the consensus ODNA in (26, 36) and the trp repressor assignments reported in the trp repressor-consensus ODNA complex in (1). All measurements were made on a Bruker 500Mhz spectrometer. Detailed reference to the pulse sequences has been given by (1). The pulse sequences used in the present study are identical.

First, the free mtr operator was characterized alone. A series of NOESY spectra were recorded with various mixing times at 45°C leading to a nearly complete set of assignments. The major set of resonance assignments for the complex based on NOESY-watergate, TOCSY, sensitivity enhanced HSQC and NOESY-HSQC, was obtained using a specific deuterated analog of trp repressor, PITAK, completely deuterated except for Proline, Isoleucine, Threonine, Alanine and Lysine residues and, to a lesser extent, Aspartate, Glutamate, Asparagine and Glutamine, as well as a randomly ^{15}N labeled, fully protonated repressor. This specifically deuterated analog has two major advantages. First, undesirable effects of spin diffusion are significantly decreased, resulting in a smaller line width for the remaining protonated signals. Second, the spectra are simplified by decreasing the number of signals and their potential overlapping. This particular analog was chosen because most of the residues of interest located in the DNA binding region would be protonated in it. The NOESY spectra were recorded at different mixing times, at temperatures varying from 45° to 30°C in a 10 mM NaH_2PO_4 buffer (pH 5.7) containing from 100 to 0 mM NaCl. These experiments showed that best results were obtained with a salt concentration lower than 10 mM.

Computational Methods

The solution structure of the trp repressor-mtr ODNA complex was calculated as in (1). The solution structure of the trp repressor-consensus ODNA complex was also used to generate an initial unambiguous set of NOE

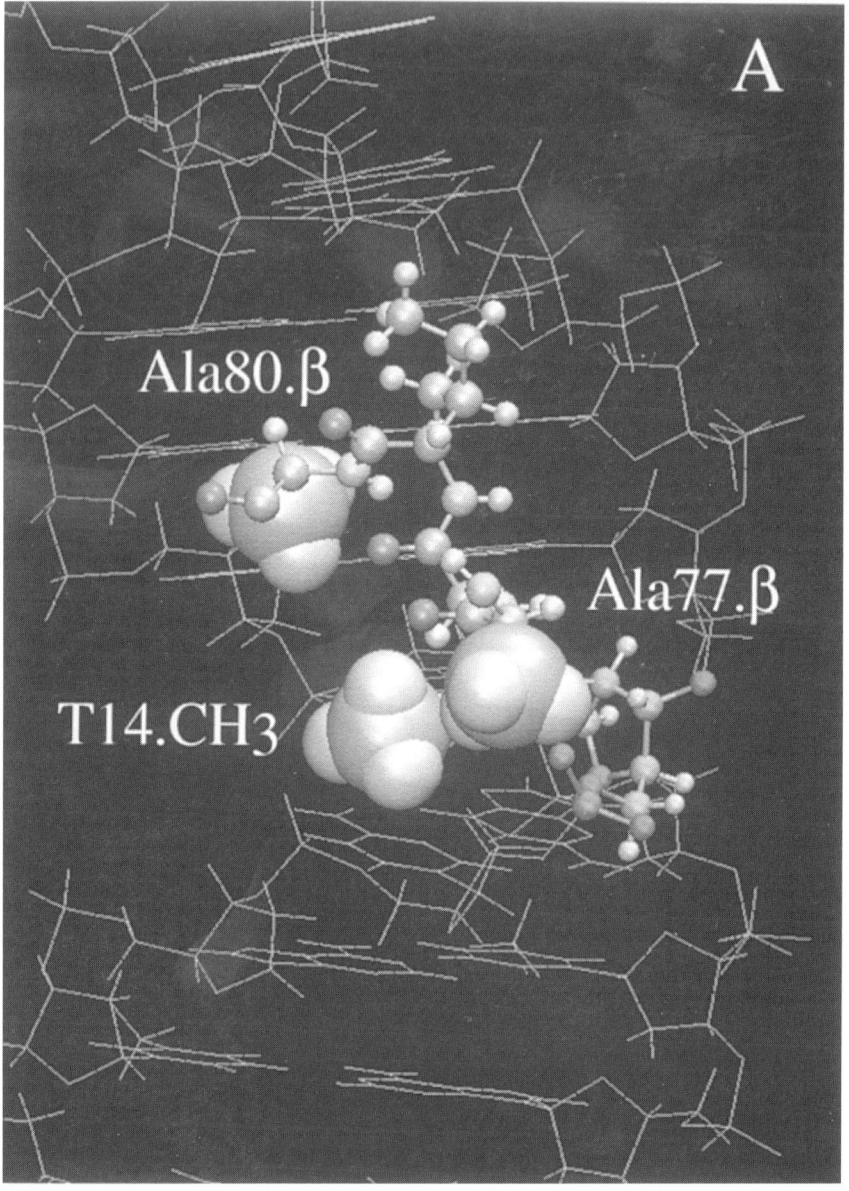

Figure 5. Representation of the relative positions of the methyl group of T14 with respect to Ala77 and Ala80, as seen in (A) the trp repressor-mtr ODNA solution structure and (B) in the trp repressor-consensus ODNA complex (1). Methyl groups are represented as full van der Waal spheres for T14, Ala77, and Ala80. Other atoms and bonds of these residues, and of residues 78 and 79, are shown as stick and ball. DNA is represented as wireframe.

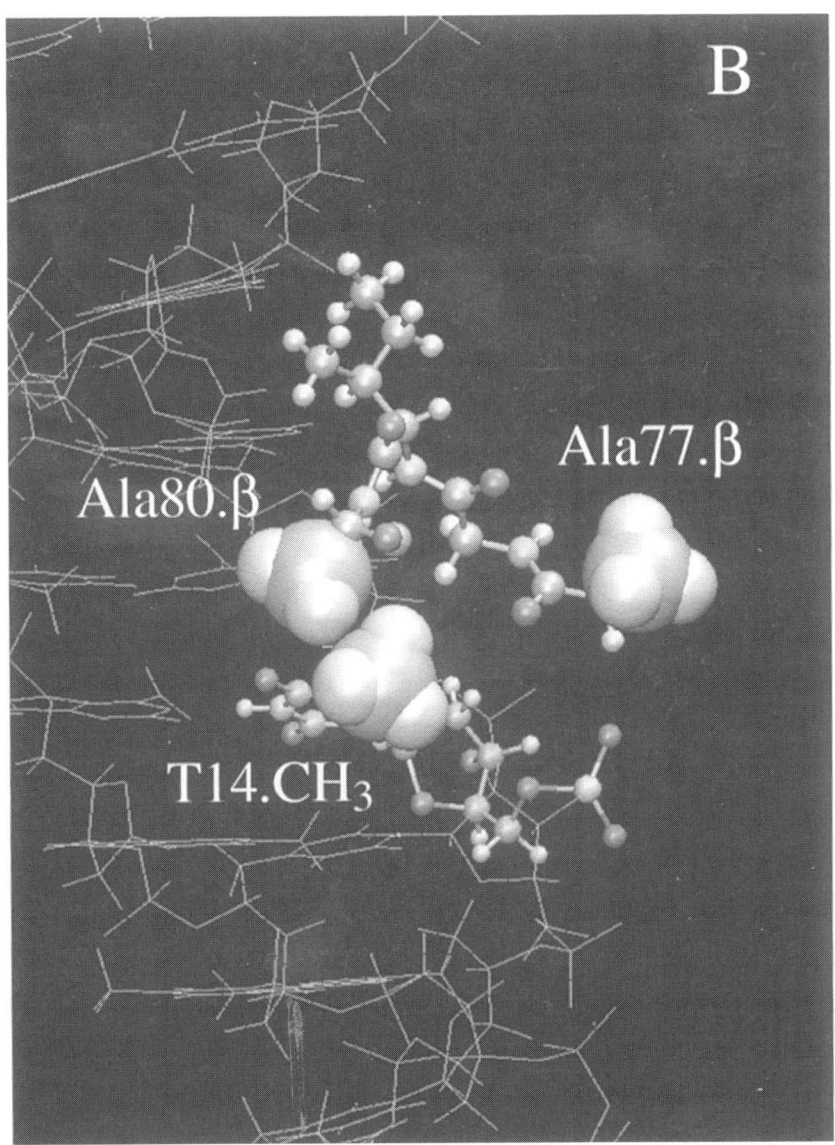

Figure 5. *Continued.*

constraints in the following way. An initial set of unambiguous NOE constraints from the spectrum was selected. Among these, constraints <u>within</u> monomers I and II or <u>between</u> monomers I and II were identified as "unambiguously within" if the distance measured in the initial structure within the same monomer was ≤ 7.0 Å and the distance measured between monomers was > 10 Å. Analogously, they were classified "between" if the distance between was ≤ 7.0 Å and the distance within was > 10.0 Å. Asymmetries of distances within each of the two monomers could not be detected from the spectrum analysis. Unambiguous NOE constraints between the asymmetric DNA oligomer end each of the two protein monomers where analogously selected. Asymmetric contacts of the two monomers with DNA could generally be assigned based on the different shift produced by the asymmetric mtr DNA. All constraints where approximated as weak (from 1.8 to 6.0 Å). This initial set of NOE constraints was used to perform sequential simulated annealing (37) at full atomic representation in vacuum. As with previous calculations (1), electrostatic empirical van der Waals attraction and the empirical hydrogen bond energies were excluded, but covalent bond, bond angles and van der Waals repulsion energies were included. Improper torsion angles (38, 39) were also used to maintain the planarity of the aromatic rings and the handedness of the chiral centers. Standard X-PLOR 3.1 (39) input parameter and topology files were used for the hydrogen force field. The parameters used for the X-PLOR simulated annealing procedure were similar to those used in the previous calculation of the structure of the trp repressor-consensus ODNA complex, as follows: the covalent bond force constant was 1000 kcal/mol per Å and bond angle force constant was 500 kcal/mol per Å^2; the improper force constant was 200 kcal/mol per Å^2; the dihedral constraint force constant was 200 kcal/mol per Å^2; the non-bonding cutoff was 4.5 Å. 20 ps of molecular dynamics was performed at 1000 K and the system was cooled down slowly, followed by final energy minimization. The final sequential simulated annealing simulation was implemented using a total of 3598 NOE constraints.

Hydrogen bond constraints were constructed based on characteristic NOEs. Dihedral phi and psi angle constraints were based on the sequential NOE pattern for α-helices and on proton and carbon chemical shifts and were assigned typical dihedral angle constraints phi = $-65(+/-45)°$ and psi = $-35(+/-45)°$. Hydrogen bonds between the Watson-Crick base pairs were also added, as well as angular constraints maintaining the planar geometry of the base pairs. The propeller rotations in the base pairs and DNA dihedral angles were not constrained.

Acknowledgments

GPZ and OJ are supported by NIH grant GM33385. LB is supported by NIH grants 5 R01 HG00335-14 and 5 R01 GM10452-36. We thank Dr. Lionel B. D. Gilet for assistance in obtaining the initial set of NMR data.

References

1. Zhang, H.; Zhao, D.; Revington, M.; Lee, W.; Jia, X.; Arrowsmith, C.; Jardetzky, O. *J. Mol. Biol.* **1994**, *238*, 592-614.
2. Zubay, G.; Morse, D. E.; Schrenk, W. J.; Miller, J. H. M. *Proc. Natl. Acad. Sci. USA* **1972**, *69*, 1100-1103.
3. Rose, J. K.; Squires, C. L.; Yanofsky, C.; Yang, H.-Y.; Zubay, G. *Nature New Biology* **1973**, *245*, 133-137.
4. Zurawski G, Gunsulas RP; Brown, K. D.; Yanofsky, C. *J. Mol. Biol.* **1981**, *145*, 47-73.
5. Heatwole, V. M.; Somerville, R. L. *J. Bacteriology* **1991**, *173*, 3601-3604.
6. Heatwole, V. M.; Somerville, R. L. *J. Bacteriology* **1992**, *174*, 331-335.
7. Schevitz, R. W.; Otwinowski, Z.; Joachimiak, A.; Lawson, C. L.; Sigler, P. B. *Nature* **1985**, *317*, 782-786.
8. Zhang, R.-G.; Joachimiak, A.; Lawson, C. L.; Schevitz, R. W.; Otwinowski, Z.; Sigler, P. B. *Nature* **1987**, *327*, 591-597.
9. Lawson, C. L.; Zhang, R.-G.; Schevitz, R. W.; Otwinowski, Z.; Joachimiak, A.; Sigler, P. B. *Proteins: Struct. Funct. Genet.* **1988**, *3*, 18-31.
10. Zhao, D.; Arrowsmith, C. H.; Jia, X.; Jardetzky, O. *J. Mol. Biol.* **1993**, *229*, 735-746.
11. Bass, S.; Sugiono, P.; Arvidson, D. N.; Gunsalus, R. P.; Youderian, P. *Genes Develop.* **1987**, *1*, 565-572.
12. Lefèvre, J.-F.; Lane, A. N.; Jardetzky, O. *FEBS Letts.* **1985**, *190*, 37-40.
13. Lefèvre, J.-F.; Lane, A. N.; Jardetzky, O. *J. Mol. Biol.* **1985**, *185*, 689-699.
14. Lefèvre, J.-F.; Lane, A. N.; Jardetzky, O. *Biochemistry* **1987**, *26*, 5076-5090.
15. Carey, C.; Lewis, D. E. A.; Lavoie, T. A.; Yang, J. *J. Biol. Chem.* **1991**, *266*, 24509-24513.
16. Otwinowski, Z.; Schevitz, R. W.; Zhang, R.-G.; Lawson, C. L.; Joachimiak, A.; Marmorstein, R. Q.; Luisi, B. F.; Sigler, P. B. *Nature* **1988**, *335*, 321-329.
17. Lawson, C. L.; Carey, J. *Nature* **1993**, *366*, 178-182.
18. Gryk, M. R.; Finucane, M. D.; Zheng, Z.; Jardetzky, O. *J. Mol. Biol.* **1995**, *246*, 618-627.
19. Finucane, M. D.; Jardetzky, O. *J. Mol. Biol.* **1995**, *253*, 576-589.
20. Gryk, M. R.; Jardetzky, O. *J. Mol. Biol.* **1996**, *255*, 204-214.
21. Bass, S.; Sorrells, V.; Youderian, P. *Science* **1988**, *242*, 240-245.
22. Arrowsmith, C.; Pachter, R.; Altman, R.; Jardetzky, O. *Eur. J. Biochem.* **1991**, *202*, 53-66.
23. Wishart, D.S.; Sykes, B. D.; Richard, F. M. *Biochemistry* **1992**, *31*, 1647-1651.

24. Humphrey, W.; Dalke, A.; Schulten, K. *J. Molec. Graphics* **1996**, *14.1*, 33-38.
25. Altman, R.; Arrowsmith, C.; Pachter, R.; Jardetzky, O. In *Computational Aspects of the Study of Biological Macromolecules by NMR Spectroscopy*; Hoch, J. C.; Poulsen, F. M.; Redfield, C., Eds.; Plenum Publ. Corp.: New York, NY, 1991; pp 363-374.
26. Arrowsmith, C. H.; Czaplicki, J.; Iyer, S. B.; Jardetzky, O. *J. Am. Chem. Soc.* **1991**, *113*, 4020-4022.
27. Koehl, P.; Lefèvre, J.-F.; Jardetzky, O. *J. Mol. Biol.* **1992**, *223*, 299-315.
28. Steitz, T. A. *Q. Rev. Biophysics* **1990**, *23*, 205-280.
29. Jones, S.; van Heyningen, P.; Berman, H. M. *J. Mol. Biol.* **1999**, *287*, 877-896.
30. Günes, C.; Müller-Hill, B. *Mol. Gen. Genet.* **1996**, *251*, 338-346.
31. Günes, C.; Staacke, D.; von Wilchen-Bergmann, B.; Müller-Hill, B. *Mol. Gen. Genet.* **1995**, *246*, 180-195.
32. Jin, L.; Fukayama, J. W.; Pelczer, I.; Carey, J. *J. Mol. Biol.* **1999**, *285*, 361-378.
33. Drew, H. R.; Travers, A. A. *Cell* **1984**, *37*, 491-502.
34. Burge, C.; Campbell, A. M.; Karlin, S. *Proc. Natl. Acad. Sci.* **1992**, *89*, 1358-1362.
35. Paluh, J. L.; Yanofsky, C. *Nucl. Acids Res.* **1986**, *14*, 7851-7861.
36. Arrowsmith, C. H.; Pachter, R.; Altman, R. B.; Iyer, S.; Jardetzky, O. *Biochemistry* **1990**, *29*, 6332-6341.
37. Zhao, D.; Jardetzky, O. *J. Phys. Chem.* **1993**, *97*, 3007-3012.
38. Brooks, B. R.; Bruccoleri, R. W.; Olasfon, B. D.; States, D. J.; Swiminathan, S.; Karplus, M. *J. Comput. Chem.* **1983**, *4*, 187-217.
39. Brünger, A. T. *X-PLOR Version 3.1. A system for X-ray crystallography and NMR*; Yale University: New Haven, CT, 1993.

Chapter 22

The 3.6 Å Structure of the Reovirus Core Particle

Karin M. Reinisch[1] and Stephen C. Harrison[2]

[1]Department of Cell Biology, Yale School of Medicine, 333 Cedar Street, New Haven, CT 06520
[2]Harvard University and Howard Hughes Medical Institute, Fairchild Building, 7 Divinity Avenue, Cambridge, MA 02138

When reoviruses enter a host cell, the virion sheds its outer coat proteins to reveal a "core". An icosaheral protein shell contains the viral dsRNA genome and does not release it even when the core reaches its destination, the cytoplasm. Instead, the core contains the necessary enzymatic machinery to transcribe the dsRNA genome into mRNA and to cap the mRNA at the 5' end. The genome consists of 10 unique segments of dsRNA, each of which encodes at least one gene, and a core transcribes several genes simultaneously and repeatedly. Thus the core, in addition merely to containing the viral genome, is a remarkably complex enzymatic assembly. (See references [1] & [2] for a review.) Here we discuss the atomic-resolution structure of the reovirus core particle in terms of its overall architecture and the organization of its enzymatic activities.

The structure of the reovirus core particle at 3.6 Å resolution[3] (Fig. 1A) shows the icosahedral protein shell, which includes the capping complexes. The particle is ~750Å in diameter, and including the genome, it has a mass of 52 million daltons. A 20Å thin shell composed of 120 copies of a protein λ1 (142kDa[4]) directly surrounds the genome. On top of this shell, and stabilizing it, are 150 nodules of a protein σ2 (47kDa). Finally, at each icosahedral 5-fold axis, there are pentameric complexes of a third protein, λ2 (144kDa). The pentamers form hollow turrets that perform three of four capping reactions. λ2 has three

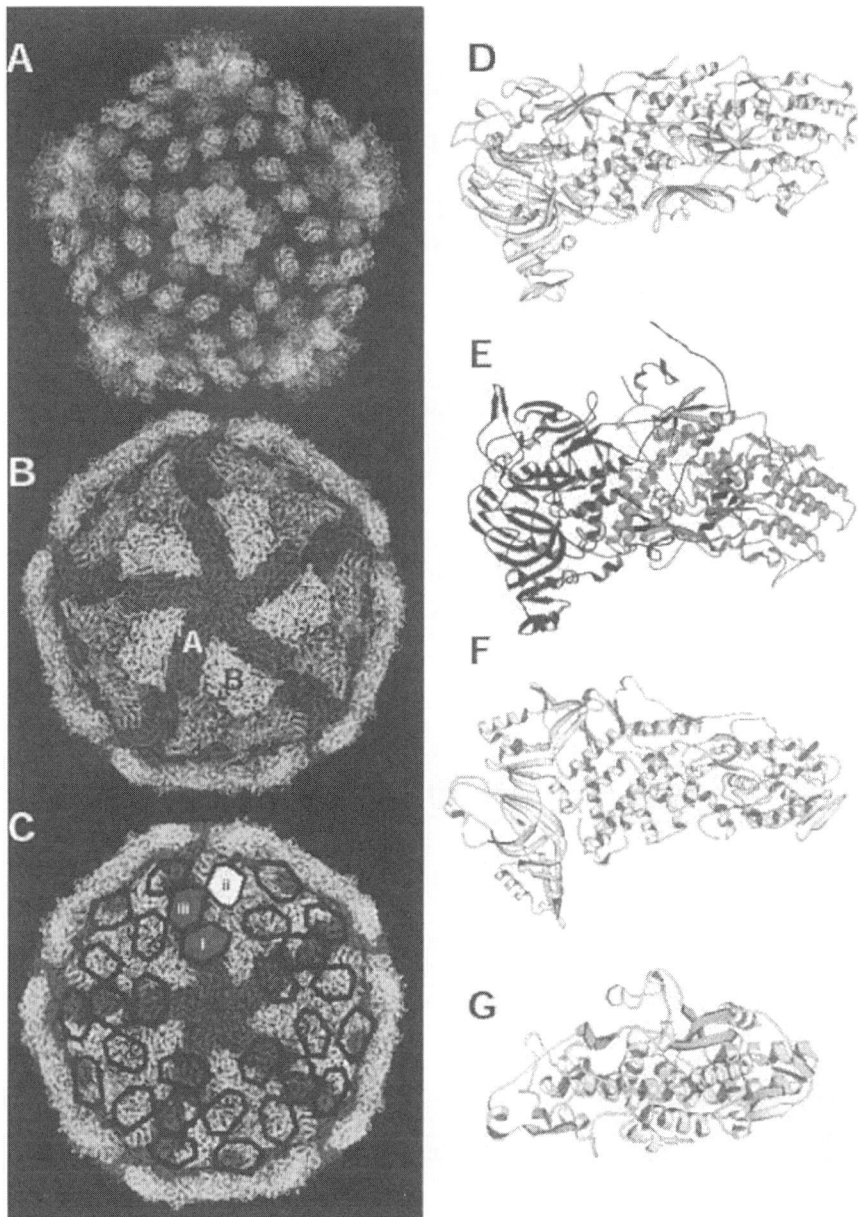

Figure 1.

Figure 1. A The Cα-trace of the reovirus core particle viewed down a 5-fold axis. One of twelve λ2 turrets is at the center of the image, and five more are visible at the periphery. Multiple copies of σ2 are shown as small nodules decorating the λ1 surface. **B. The λ1 surface.** Copies of λ1-A are shown in dark grey, and copies of λ1-B are shown in lighter shades. Five copies of λ1-A surround the 5 fold axis, and five copies of λ1-B are interdigitated between them to form a decameric unit. Twelve such units comprise the λ1 shell. Three copies of λ1-B surround the 3-fold axes. **C. The λ1 shell with footprints outlining the σ2 binding sites.** There are 150 σ2 binding sites; three of these are unique and unrelated by icosahedral symmetry. **D. A ribbon diagram of λ1-A.** The 5-fold axis is at the right of the molecule. **E. A ribbon diagram of λ1-B.** λ1-B is rendered in two colors to indicate its two subdomains and the location of the pivot point between them. The relative orientation of the subdomains about the pivot point differs in λ1-A and B. Residues 40-167 and 181-206 are also indicated in dark colors. The Zn^{++} is not shown. **E. VP3-A from the BTV core structure.** Its fold is similar to that of λ1. **F. σ2-I rendered so that the λ1 surface is toward the bottom of the page.** σ2-ii and –iii are very similar with small changes at the interface with the λ1 surface. This figure has been adapted with permission from Nature[3] (404: 960-967), MacMillan Magazines, Ltd. 2000.

enzymatic domains: a guanylyltransferase domain, a 7N methyltransferase, domain and a 2'O methyltransferase domain. While the interior of the core (the dsRNA and ~12 transcription complexes) is not ordered icosahedrally and is therefore not visible in the crystal structure, low angle x-ray scattering experiments[5] and low resolution electron density maps from crystallography suggest that the dsRNA is tightly packed into concentric layers that extend inward from the λ1 shell at 26Å intervals. Locally, the dsRNA could be packed hexagonally with a 30Å distance between adjacent RNA helices. Studies using electron cryomicroscopy[6] further show that a transcription complex is tethered under each turret, so that the transcriptase may extrude the 5' end of the mRNA out of the core and into the capping complex, thus ensuring that the 5' end is not lost in the tightly packed interior of the core. Such a packing is consistent with a model previously proposed for other members of the Reoviridae family[7,8], in which each segment of RNA is coiled into large spools around a polymerase complex, thus keeping the genes from tangling during their concurrent and repeated transcription.

λ1, the λ1 shell, and non-equivalence

The λ1 shell around the dsRNA genome is relatively smooth. Five monomers of λ1 (set A) radiate from the five fold axis, and a second set of five (set B) interdigitates with the first. Twelve such decamers together form the complete protein shell (Fig. 1B).

Thus far, this decameric arrangement of capsid proteins is unique to the Reoviridae[9,10] (although it has been proposed for other dsRNA viruses with 120 protein subunits in the capsid[9,11,12]). The arrangement is the same for the innermost capsid proteins in the bluetongue virus core[9] (BTV; an orbivirus) and in cytoplasmic polyhedrosis virus (CPV; a cypovirus) [10]. Furthermore, the fold of λ1, while without any recognizable sequence similarity, resembles that of the VP3 proteins in BTV[9] and CPV[10]. Figure 1D and E illustrate the folds of λ1 and of BTV VP3[9], respectively. λ1, like the equivalent proteins in the other viruses, is an elongated flat plate. Despite its large size, the plate consists of one domain rather than several concatenated domains and is thus well suited to its role as armor for the viral genome. In all three proteins, there is an alpha helical bundle nearest the 5-fold axis, a region of criss-crossing alpha helices at the middle of the molecule, and a beta-sheeted region furthest from the 5-fold axis. λ1 assumes two conformations (Fig1 B,D&E), where monomers in set A are related to those in set B by a reorientation of two subdomains (I: residues 253-470 and 923-1260 and II: residues 482-922) about a pivot point. The number and location of the pivot points is not conserved across the Reoviridae. Sets A and B further differ in the conformations of loops around their periphery. Most notably, residues 560-568 and 774-794 in λ1-A, which define the aperture at the 5-fold axis through which the mRNA is extruded into the λ2 turrets, have a different conformation in λ1-B or are partially disordered. Since the aperture at the 5-fold axis measures only 5.5Å in diameter, the residues bordering it probably undergo a conformational change in the course of mRNA extrusion.

What is most striking in the comparison of λ1-A and λ1-B, however, is that the contacts they make with neighboring molecules differ entirely in all but one small interface across a local 2-fold dyad (see Fig 1B). The portions of λ1-B contacted by A differ from those in A contacted by B. Thus, the λ1 shell, and the inner protein shells of the other studied Reoviridae, differ from all other viral capsids for which structures are known in that chemically equivalent species make almost entirely non-equivalent contacts. In such capsids, specifically those that have more than 60 protein subunits, even those subunits that are not related by icosahedral symmetry have mainly similar contacts with neighboring molecules.

The first 240 residues of λ1-A and the first 180 residues of λ1-B are partially disordered inside the core. For λ1-B, residues 181-206 form a classic zinc finger with two histidines and two cysteines to coordinate the metal. The zinc finger is tucked between copies of λ1-B on the interior of the core near the 3-fold axis. It cannot bind RNA in this orientation since the RNA would collide with the λ1 shell, but both this zinc finger and the corresponding one in λ1-A could bind to RNA in the course of viral assembly. Further, the λ1-A zinc finger dangles into the core, where it could bind dsRNA even in the assembled core. A set of λ1 residues 40-167 also nest under each λ1-B. They do not belong to this particular λ1-B since residue 167 is ~90Å removed from residue 181 but belong to some neighboring λ1, A or B. Thus, residues 40-167 form a network of arms on the interior of the λ1 shell that may contribute to capsid stability.

σ2 clamp and more non-equivalence

Decorating the exterior of the λ1 shell, there are 150 nodules made by σ2, a globular and predominantly helical protein. Each σ2 makes extensive contacts only with the λ1 shell. There are three distinct positions for σ2 within an icosahedral asymmetric unit.

These three unique binding sites represent the second example of non-equivalence in the reovirus core structure. While site iii is partially similar to site ii, sites i and ii are entirely different both in terms of the secondary structure and the pattern of charged/hydrophobic/polar residues that the λ1 surface presents to σ2. σ2-i lies over the middle of a λ1-A molecule, and σ2-ii bridges from the middle of a λ1-B across to the carboxy-terminal part of a λ1-B from another decamer. σ2-iii lies on the λ1 shell directly on an icosahedral 2-fold axis in one of two equally-likely, two-fold related orientations. Consequently, σ2-iii has not been built into the 3.6Å electron density maps, and instead a σ2-ii model has been docked onto that site. It is clear, however, that the various versions of σ2 differ only at the interface with the λ1 surface. The differences between σ2-i and ii are subtle, and the most drastic change is an unravelled helix (residues 39-46) in σ2-ii with respect to σ2-i.

Chemically identical protein subunits using the same surface to make completely different intersubunit contacts have not been widely observed in crystal structures. The list of proteins involved in such non-equivalent contacts includes capsid proteins in the Reoviridae family[9,10], the scaffolding protein in the X174 prohead[13], and HIV reverse transcriptase[14], wherein two monomers have non-equivalent dimer interfaces. All these examples are viral proteins and constitute a testament to the strong evolutionary pressure imposed by the compactness of the viral genome.

Two of the three sites explain why σ2 is essential in order for λ1 to form icosahedral particles, for recombinant λ1 will form icosahedral particles in mouse L fibroblasts[15] or insect cells (Max Nibert, unpublished[3]) only if co-expressed with σ2: in two of its binding sites (ii,iii in Fig1C) σ2 behaves as a clamp that holds together decamers of λ1. While σ2 does not have a clamp analog in the bluetongue virus core, some of the other turretted Reoviridae such as CPV[16,17] and aquareovirus[18] have proteins of a similar size positioned like σ2-i and –ii. There the subunit in binding position ii probably also functions to stabilize the innermost protein shell.

λ2 Capping Complex & the Active Sites

The transcription complex in the core interior passes the newly synthesized mRNA directly into the pentameric λ2 turrets (Fig 2A&B). The turrets catalyse three capping reactions: the addition of a guanosyl moiety to the 5' end of the mRNA and then the transfer of a methyl group from S-adenosyl-L-methionine (SAM) to both the N7 position of the newly added guanosine and the 2'O of the first template encoded nucleotide, a guanosine in each of the ten reovirus gene segments (reactions 2-4 below). [19]

(1) 5'-pppG-mRNA → 5'-ppG-mRNA + pi
(2) 5'-ppG-mRNA + GTP → GpppG-mRNA
(3) SAM + GpppG-mRNA → ^{me7}GpppG-mRNA + SAH
(4) SAM + ^{me7}G-pppG-mRNA → ^{me7}GpppG$^{me2'}$-mRNA + SAH

where SAH is S-adenosyl-L-homocysteine.

The turrets are about 120Å in diameter and 80Å in height and have a hollow interior (volume=2×10^5 Å^3). All the enzymatic active sites face the interior cavity. The top of the turret is partially closed by five flaps that, according to cryo-em reconstructions, have variable conformations[20]. In the intact virus, the flaps grip the viral cell attachment protein σ1, whereas in the core their conformation controls the size of the aperture at the top of the turret.

λ2 consists of 7 concatenated domains (Fig 2C). The most N-terminal domain (residues 1 to 385, Fig2E) is the guanylyltransferase[21] positioned directly on the λ1 shell. The guanylyltransferase is cup shaped, and five such domains form a ring about the icosahedral 5-fold axis. The interior of the cup contains the active site and lysines 190 and 171, which are necessary for guanylyl transfer[21]. The reovirus guanylyltransferase domain consitutes a novel fold and bears no structural resemblance to the PBCV-1 guanylyltransferase, for which a structure has been determined recently[22].

Residues 386-433 and 690-802 form a small, probably structural domain. The next two domains, residues 434-691 and 804-1022, are methyltransferase domains, and their folds are two different variations on the "universal" methyltransferase fold (Fig. 2D), a beta sheet with defined strand order and directionality sandwiched between alpha helices[23,24]. Soaking experiments with SAH confirm that for both domains the SAM binding site coincides with that in other methyltransferases. For the more N-terminal methyltransferase domain, SAH binding is accompanied by a conformational change, in which residues 519-524 and 579-587 rearrange in order to participate in binding. While each methyltransferase domain binds SAM independently, experiments indicate that 7N methyltransfer occurs only in λ2 pentamers and not in monomers. (Equivalent experiments for 2'O methyltransfer have not been reported.)

The carboxyterminus forms the flap, a series of three concatenated Ig-like domains. The first two domains resemble the V and C regions of an antibody light chain; the most carboxyterminal such domain is a truncated V domain. The

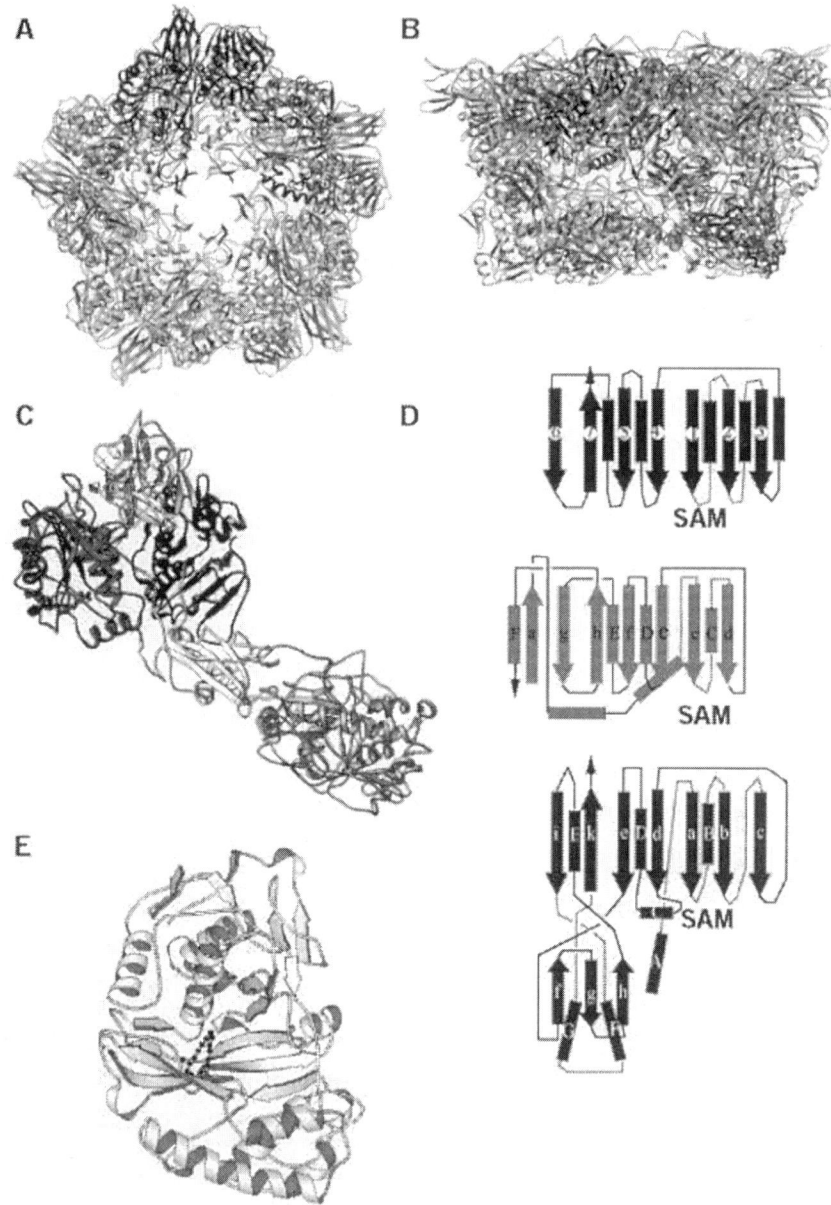

Figure 2.

Figure2. A. The pentameric λ2 turret as viewed looking down the 5-fold icosahedral axis onto the core. One of the λ2 monomers is rendered in black. *B. A view of the λ2 turret from the side*, 90° from the view in Fig1A, so that the λ1 surface is toward the bottom of the page. The same monomer is rendered in black. *C. The "black" λ2 monomer* in the same orientation as in (B). The monomer is shaded by domain. The guanylyltransferase domain is grey at the lower right. The two domains near the top left and rendered dark and darker grey are methyltransferase domains I and II ,respectively. The light grey domain at the top left corresponds to the flap. SAH binding sites are indicated but difficult to see. *D. Diagrams of the SAM-binding domains.* The universal methyltransferase domain is shown at the top. Methyltransferase I (light grey) and II (dark grey) are arrayed below it so that conserved structural elements are aligned vertically. The SAM binding site is indicated. *E. Ribbon diagram of the guanylyltransferase domain* looking into the "cup", so that the λ1 surface is behind it. Lysines 171 and 190 mark the location of the active site. This figure has been adapted with permission from Nature[3] (404:960-967), MacMillan Magazines, Ltd. 2000.

conformational flexibility of the flap probably derives from a hinge region near residue 1023.

Organization of the Active Sites.

The various active sites are ordered vertically in the core. The transcription complex is tethered inside the λ1 shell and below the λ2 turret[6]; the guanylyltransferase active sites are at the base of the turret; and one set of methyltransferase active sites is half way up, and another set is at the top of the turret. There is no biochemical evidence to establish which methyltransferase domain is the 7N and which is the 2'O methylase. One possibility is that the vertical ordering of the active sites reflects the temporal ordering of the reactions, so that since 7N methyltransfer always precedes 2'O methyltransfer, the methyltransferase near the middle of the turret methylates 7N and the methyltransferase near the top of the turret methylates the 2'O. Probably such a vertical ordering alone would not impose an order on the methylation reactions, and it seems likely that the reovirus 2'O methyltransferase, like VP39 from vaccinia[25], only methylates a cap structure that has been methylated previously at the 5' guanosine moiety.

What ensures the efficiency of the capping reaction? While the different active sites are close spatially, they are not contiguous. Nor is there any evident groove that links active sites. The capping efficiency probably results from the container like nature of the turret. Although the λ2 flap domains form a lid for the turret, efficient capping occurs even in the absence of the flap domains[26]: the polymerase pins the 3' end of the mRNA in place at the bottom of the turret, and the cylindrical turret walls restrict the 5' end in two dimensions. Thus reovirus traps the 5' substrate in a cavity densely occupied by fifteen active sites. Indeed, the major theme underlying the organization of active sites, including the location of the polymerase, seems to be that of not losing the substrate.

Conclusion

The reovirus core structure is among the largest structures determined to date by x-ray crystallography. We have focussed here on a description of the structure in terms of the non-equivalent interactions made by both lamba1 and σ2. The capsids of the reovirus family do not conform to the theory of quasi-equivalence proposed for icosahedral viruses by Caspar and Klug in 1962[27], and stable, non-equivalent interactions appear to evolve more readily than previously imagined. We have found that the arrangement and the fold of λ1 is universal for members of the reovirus family and that the arrangement and the structures of the other capsid proteins vary and reflect different replication and entry strategies. And finally, we have begun to understand how in this instance organizing multiple enzymatic functions into an efficient machine depends on not completely releasing the substrate until its processing is complete.

References

1. Fields, B. N. in *Fields Virology* (eds. Fields, B. N., Knipe, D. M. & Howley, P. M.) 1553-1555 (Lippincott-Raven, 1996).
2. Nibert, M. L., Schiff, L. A. & Fields, B. N. in *Fields Virology* (eds. Fields, B. N., Knipe, D. M. & Howley, P. M.) 1557-1596 (Lippincott-Raven, 1996).
3. Reinisch, K. M., Nibert, M. L. & Harrison, S. C. Structure of the reovirus core at 3.6A resolution. *Nature* **404**, 960-967 (2000).
4. Harrison, S. J. et al. Mammalian reovirus L3 gene sequences and evidence for a distict amino-terminal region of the λ1 protein. *Virology* **258**, 54-64 (1999).
5. Harvey, J. D., Bellamy, A. R., Earnshaw, W. C. & Schutt, C. Biophysical studies of reovirus type 3: iv low-angle x-ray diffraction studies. *Virology* **112**, 240-249 (1981).
6. Dryden, K. A. et al. Internal/Structures containing transcriptase-related proteins in top component particles of mammalian orthoreovirus. *Virology* **245**, 33-46 (1998).
7. Prasad, B. V. et al. Visualization of ordered genomic RNA and localisation of transcriptional complxes in rotavirus. *Nature* **382**, 471-3 (1996).
8. Gouet, P. et al. The highly ordered double-stranded RNA genome of bluetongue virus revealed by crystallography. *Cell* **97**, 481-490 (1999).
9. Grimes, J. M. et al. The atomic structure of the bluetongue virus core. *Nature* **395**, 470-478 (1998).
10. Zhou, H. Z. in *FASEB Summer Research Conference on Virus Assembly* (Vermont Academy, Saxton Rivers, Vermont, 2000).
11. Butcher, S. J., Dokland, T., Ojala, P. M., Bamford, D. H. & Fuller, S. D. Intermediates in the assembly pathway of the double-stranded RNA virus φ6. *The EMBO Journal* **16**, 4477-4487 (1997).
12. Cheng, R. H. et al. Fungal virus capsids, cytoplasmic compartments for the replication of double-stranded RNA, formed as icosahedral shells of assymetric Gag dimers. *Journal of Molecular Biology* **244**, 255-258 (1994).
13. Ilag, L. L. et al. DNA packaging intermediates of bacteriophage phi X174. *Structure* **3**, 353-363 (1995).

14. Kohlstaedt, L. A., Wang, J., Friedman, J. M., Rice, P. A. & Steitz, T. A. Crystal structure of 3.5A resolution HIV-1 reverse transcriptase complexed with an inhibitor. *Science* **256**, 1783-1790 (1992).
15. Xu, P., Miller, S. & Joklik, W. K. Generation of reovirus core-like particles in cells infected with hybrid vaccinia viruses that express genome segments L1, L2, L3, and S2. *Virology* **197**, 726-731 (1993).
16. Hill, C. L. et al. The structure of cypovirus and the functional organization of dsRNA viruses. *Nature Structural Biology* **6**, 565-568 (1999).
17. Zhang, H. et al. Visualization of protein-RNA interactions in cytoplasmic polyhedrosis virus. *Journal of Virology* **73**, 1624-1629 (1999).
18. Shaw, A. L., Samal, S. K., Subramanian, K. & Prasad, B. V. V. The structure of aquareovirus shows how the different geometries of the two layers of capsid are reconciled to provide symmetrical interactions and stabilization. *Structure* **4**, 957-967 (1996).
19. Furuichi, Y., Muthukrishnan, S., Tomasz, J. & Shatkin, A. Mechanism and formation of reovirus mRNA 5'-terminal blocked and methylated sequence $^{m7}GpppG^mpC$. *The Journal of Biological Chemistry* **251**, 5043-5053 (1976).
20. Dryden, K. A. et al. Early steps in reovirus infection are asssociated with dramatic changes in supramolecular structure and protein conformation: analysis of virions and subviral particles by cryoelectron microscopy and image reconstruction. *The Journal of Cell Biology* **122**, 1023-1041 (1993).
21. Luongo, C. L., Reinisch, K. M., Harrison, S. C. & Nibert, M. L. Identification of the Guanylyltransferase Region and Active Site in Reovirus mRNA Capping Protein l2. *Journal of Biological Chemistry* **275**, 2804-2810 (2000).
22. Håkansson, K., Doherty, A. J., Shuman, S. & Wigley, D. B. X-Ray crystallography reveals a large conformational change during guanyl transfer by mRNA capping enzymes. *Cell* **89**, 545-553 (1997).
23. Schluckebier, G., O'Gara, M., Saenger, W. & Cheng, X. Universal catalytic domain structure of AdoMet-dependent methyltransferases. *Journal of Molecular Biology* **247**, 16-20 (1995).
24. Hodel, A. E., Gershon, P. D., Shi, X. & Quiocho, F. A. The 1.85Å structure of the vaccinia protein VP39: a bifunctional enzyme that participates in the modification of both mRNA ends., (1996).
25. Hu, G., Gershon, P. D., Hodel, A. E. & Quiocho, F. A. mRNA cap recognition: dominant role of enhanced stacking interactions between methylated bases and protein aromatic side chains. *Proceedings of the National Academy of Science USA* **96**, 7149-7154 (1999).
26. Luongo, C. L. et al. Localization of a C-terminal region of λ2 protein in reovirus cores. *Journal of Virology* **71**, 8035-40 (1997).
27. Caspar, D. L. D. & Klug, A. in *Cold Spring Harbor Symp. Quant. Biol.* 1-24 (1962).

Author Index

Altman, Russ B., 309
Baron, Margaret H., 294
Bartlett, Rodney J., 150
Brocchieri, Luciano, 340
Carrillo, Michelle Whirl, 309
Chang, Jeffrey, 309
Dadashev, Vladimir, 79
Del Bene, Janet E., 150
Eaton, Gareth R., 2, 321
Eaton, Sandra S., 321
Epstein, Irving R., 103
Evans, David R., 249
Gindulyte, Asta, 79
Grimes, Russell N., 20
Guy, Hedeel I., 249
Harrison, Stephen C., 367
Hegstrom, Roger A., 177
Hosmane, Narayan S., 46
Jardetzky, Oleg, 340
Kaesz, Herbert D., 67
Krause, Kurt L., 270
Lipscomb, William N., 79

Ludwig, Martha L., 186
Maguire, John A., 46
Massa, Lou, 79
Matthews, Rowena G., 186
McKee, Michael L., 135
Miller, Mitchell D., 270
Nguyen, Kimloan T., 67
Ortiz, J. V., 118
Perera, S. Ajith, 150
Quiocho, Florante A., 216
Rees, Douglas C., 202
Reinisch, Karin M., 367
Seabra, Gustavo, 118
Squire, Richard, 79
Steitz, Thomas A., 231
Tossell, John A., 165
Waugh, Allison, 309
Wei, Liping, 309
Zakrzewski, V.G., 118
Zhou, Guo-Ping, 340
Zinn, Alfred A., 67

Subject Index

A

Ab initio electron propagator theory
 ionization energies and electron affinities, 119
 See also Electron propagator theory
Activation. *See* Methionine synthase (MetH)
Adenosine deaminase
 activated water or hydroxide coordination to Zn^{2+}, 221
 mode of binding of pentostatin, 221–222
 role and location, 219, 221
 structure determination, 221
 Zn^{2+} cofactor, 221
Administrator, Lipscomb as, 13
Aldose reductase, noncompetitive inhibitor, 222
Allostery
 Lipscomb's mechanism, 10–11
 principles for allosteric behavior, 11
 See also Tryptophan (trp) repressor–mtr operator DNA (ODNA) complex
Aluminosilicates
 ab initio calculations of properties and experimental determination, 165–166
 applicability and reliability of quantum mechanical calculations, 173
 calculated geometry of $Al_6O_6(CH_3)$, 172f
 calculated structures for $SiAlO_7H_6NaH_3O_2$ at different levels, 171f
 calculated structures of two isomers of $Si_2Al_2O_4H_8$, 168f
 computational methods, 166
 difference in energy between alternating and paired isomers of $Si_2Al_2O_4R_8$, 169t
 "drum", 173
 electrostatic valence sum rule, 167
 evaluating NMR shieldings, electric field gradients at O and vibrational spectra, 169
 isomerization reaction of neutral molecule $Si_2Al_2O_4H_8^{-2}$, 167
 limitations with calculations, 169
 Loewenstein's rule of cage or double ring structures, 170, 173
 model calculating energetics and properties for Al–O–Al bridging species, 167
 model for calculating Na NMR, 170
 occurrence, 166
 Pauling's rules, 167
 relative stabilities of Si–O–Si, Si–O–Al, and Al–O–Al bonds, 167
 species with -OH's terminating Si's, 170
 structural chemistry on computer and relative stability of isomers, 169–170
 structures of two different species with three-coordinate O as models for tricluster O's, 174f
 study of Al–O–Al linkages, 170
 tricluster species, 173
 water interacting with aluminosilicate melt, 170
Angioblasts, mesoderm cells forming, 299
Anterior visceral endoderm (AVE), mammalian embryos, 297–298

380

Antibiotics, ribosome as major targets of, 247
Anticancer agents, polyhedral boron clusters, 38–39
Anti-crowns, ortho-carborane, 31–32
Aromaticity/antiaromaticity, magnetic criteria judging, 140

B

Bailar twist, trischelate complexes, 8
Belousov–Zhabotinsky (BZ) reaction
 chemical oscillator, 104
 computer simulations, 114, 115*f*
 oscillatory cluster patterns in, with global feedback, 108, 110
 spiral waves in, 107*f*
 See also Nonlinear chemical dynamics
Berry pseudorotation mechanism, cyclopentadienyl group, 7–8
Biochemistry, problems of structural, 11–12
Biological function, proteins at molecular level, 321–323
Biology. *See* Molecular biology
Bonding, molecules, 6–7
Bond polarity. *See* Spin-spin coupling constants
Bone morphogenetic protein-4 (BMP-4), blood island formation, 297
Borane chemistry
 1930s and 1940s, 22–25
 cluster geometry and skeletal electrons for binding, 26–27
 correlation of structure with electron county in clusters, 26
 evolution of modern, 35
 examples of borane–carborane–carbocation continuum, 27
 incorrect notions about structures and bonding, 24–25
 nido/closo/arachno classification, 26–27

 Pitzer's protonated double bond structures, 24
 proposed structures by Alfred Stock, 23
 ripple effect of original Lipscombian structural ideas, 28
 synthesis, 28
 topological theory, 25–29
 Wade's rules, 26, 27
Boranes, medicine, 37–39
Boron cluster science
 benzene-anchored polynuclear metallacarboranes, 33
 borane–hydrocarbon border, 34–35
 boron neutron capture therapy (BNCT), 37–38
 C_4B_8 carboranes, 35
 carboranes as organic building block units, 32
 current applications of polyhedral boranes and carboranes, 35–39
 efficacy of compounds, 39
 examples of metallacarborane anticancer agents, 38
 extraction of metal ions from radioactive waste, 37
 hydride $B_{20}H_{16}$, 29
 hydrocarbons as platforms for polycluster systems, 33
 icosahedral barrier, 30–31
 macropolyhedral condensed boranes and supra-icosahedral clusters, 29–31
 medicine, 37–39
 mercuracarborane anti-crown ligand and Cl⁻ complex, 31
 new specialty materials, 36
 organocarboranes with icosahedral connecting units, 32
 polyhedral borane frameworks in organic chemistry, 31–33
 predictable pathways to macropolyhedral target structures, 29–30

question of stability of closed polyhedral boron cages, 30–31
rigid linear oligomers or carborods, 32
Boron fullerenes
applying *ab initio* method to proposed, 88
constraining geometries, 88
Descartes–Euler formula suggesting possible geometries, 89
dual of carbon, 100
geometrical correspondence between multicage carbon and boron analogs, 89, 90t
interest, 81, 82f, 99
multicage carbon compounds and boron analogs, 90t
optimization of geometries, 88, 91
quantum chemical calculations, 88
single cage carbon compounds and boron analogs, 86, 87t
stability factors, 89t
stability factors for multicage, 91t
structure, 82f
See also Boron nanotubes
Boron hydrides
$B_{20}H_{16}$ structure, 29
mystery surrounding, 23–24
patterns by Lipscomb, 4–5
Pauling lecture, 4
predicting possible ions, 5
structure-bonding theory and NMR spectroscopy, 28–29
three-center bond concepts, 6–7
X-ray crystallography, 5
Boron nanotori
carbon and boron dual, 98t
critical points and topology, 96–97
dual of carbon, 100
duals of carbon nanotori, 97–98
Euler–Poincare formula in terms of Betti numbers, 94
factors of Euler–Poincare formula, 98t
geometrical duals C_{960} and B_{480}, 98f

Morse theory, 96
topology, 97
Boron nanotubes
armchair geometry of carbon and, 93f
chiral geometry of carbon and, 93f
dual of carbon, 100
examples of carbon nanotubes and boron duals, 92f, 93f
factors of Euler–Poincare formula, 94t
relation of nanotube and lattice for carbon and boron, 94, 95f
zigzag geometry of carbon and boron nanotubes, 92f
See also Boron fullerenes
Boron neutron capture therapy (BNCT), polyhedral boron clusters, 37–38
Bose–Einstein (BE) condensate
analogous Bragg law in BE case, 181
analogy with Bragg diffraction, 179–183
exact wavefunction for pair of ideal gas, 178–179
formal resemblance to Bragg diffraction, 180–181
form of condensate orbitals, 179
Pauli principle for bosons, 182, 183
phase difference of two interfering macroscopic wave functions, 182
physical interpretation of phase, 182–183
production, 177–178
release of atoms from trap, 182
total wavefunction in momentum space, 180
usual theoretical description, 178
Bragg diffraction
analogy to Bose–Einstein condensates, 179–183
See also Bose–Einstein (BE) condensate

C

Ca aluminosilicate glasses, tricluster species, 173
CAD. *See* Multifunctional protein CAD
Calmodulin, binding target enzymes/proteins, 217–218
Carbamoyl phosphate (CPS) subdomains
  activity of isolated, 260t
  cloning and expression, 259
  dissection, 263–264
  glutaminase (GLN) domain determining function, 262
  hypothesis, 260–261
  pressure induced dissociation of hybrid molecules, 261–262
  rearranged CPSases, 262–263
  synthesis, 251f, 252
  synthesis by glutaminase (GLN)–CPS.A, 260f
  See also Multifunctional protein CAD
Carbohydrates
  polar vs. nonpolar interactions, 218–219
  stereo view of H-bonding and stacking interactions between D-galactose-binding protein and D-glucose, 220f
Carbon fullerenes
  interest, 81, 82f
  See also Boron fullerenes
Carbon nanotori
  critical points and topology, 96–97
  existence, 85
  imagining formation, 97–98
  production, 85, 86f
  quantum crystallography (QC), 100
  See also Boron nanotori
Carbon nanotubes
  description, 92
  geometries, 92f, 93f
  geometry of open single walled nanotubes (SWNTs), 84–85
  specifying single wall, 93–94
  SWNTs, 81, 83f, 84f
  See also Boron nanotubes
Carboranes
  benzene-anchored polynuclear metallacarboranes, 33
  borane–hydrocarbon border, 34–35
  C_4B_8 organic-inorganic hybrid, 35
  conversion into electron-precise heterocyclic ring, 53, 59
  current applications, 35–39
  medicine, 37–39
  organic building block units, 32
  organocarboranes with icosahedral connecting units, 32
  reactivities of carbons apart C_4B_8-, 51, 53
  reactivity with group 13 elements, 59–60, 64
  syntheses of C_2B_4- and C_4B_8-carboranes, 47, 51
  uses, 35–36
  See also Metallacarboranes
Carborods
  new specialty materials, 36
  organocarboranes, 32
Carboxypeptidase (CPA)
  structure determination, 10
  X-ray crystallography, 9–10
Carroll, Lewis
  imagination, 21–22
  quotations, 21, 29
Catalysis. *See* Methionine synthase (MetH)
Cation-double π interaction, cap messenger RNA (mRNA), 222–223, 224f
Cell fate
  anterior-posterior axis determination and specification, 305
  determination during development, 295

respecification of anterior ectoderm to hematopoietic and endothelial, 301–303
See also Embryo, mouse
Cesium
  carbons apart C_4B_8-carboranes, 51, 53
  crystal structure of dicesiacarborane complex and cesiacarborane unit in polymeric chain, 56f
Charged ligands
  interactions between phosphate-binding protein and phosphate, 225f
  ligand specificity and electrostatic balance, 224–227
  schematic of H bonds from sulfate-binding protein to sulfate, 226f
  stabilization of buried charges, 227–228
Chemical dynamics. See Nonlinear chemical dynamics
Chemical shielding, isotopic, nuclear magnetic resonance (NMR), 138
Chemical shielding tensor, nuclear magnetic resonance (NMR), 136
Chemical shift calculations
  density functional theory (DFT), 139
  rotational and vibrational average, 141
  See also Nuclear magnetic resonance (NMR)
Clusters
  computer simulations, 114, 115f
  experimental arrangement for global feedback, 109f
  family of cluster patterns vs. strength of feedback, 111f
  irregular, 110, 113f
  localized, 113, 114f
  oscillatory patterns with global feedback, 108, 110
  standing, 110, 112f
  See also Boron cluster science; Nonlinear chemical dynamics

Cobalamin cofactor. See Methionine synthase (MetH)
Cobalt
  composite size distribution of metal nanoparticles, 71f
  size distribution of powder obtained at 162°C, 75f
  transmission electron microscopy (TEM) images of metallic nanostructures, 70f
  XRD pattern of powder obtained at 162°C, 75f
  XRD pattern of powder obtained at 171°C, 76f
  See also Nanostructured phases
Colonel. See Lipscomb, William N.
Computer simulations, pattern formation, 114, 115f
Conformational rearrangements, methionine synthase, 198–199
Continuous wave (CW)
  electron paramagnetic resonance (EPR), 326–327
  impact on amplitude of CW spectrum, 329–330
Coupling constants. See Nuclear magnetic resonance (NMR); Spin-spin coupling constants
Covalency of bonds. See Spin-spin coupling constants
Crawford, Bryce, borane structural theory, 25–26
Crick, Francis, central dogma of molecular biology, 231–232, 233f
Critical points, Morse theory, 96
Crystal structures, molecules, 4
Cyclooctatetraene (COT) complexes, controversy over interpretation, 8
Cyclopentadienyl complexes
  Bailar twist, 8
  Berry pseudorotation mechanism, 7–8
  Ray–Dutt twist, 8
  structure speculation, 7

D

Dendrimers, new specialty materials, 36
Density functional theory (DFT)
  chemical shift calculations, 139
  description, 119
Deoxyribonucleic acid (DNA)
  differences between I-*Ppo*I with and without bound DNA, 284–285
  replication, 232, 234–235
  ribbon diagram showing I-*Ppo*I dimer structure complexed with DNA, 283*f*
  role for shifting magnesium in binding site, 288–289
  superposition of I-*Ppo*I–Mg^{2+} structure without DNA, 286*f*
  transcription, 235, 239–241
  See also Molecular biology; Nucleic acid cleavage; Ribonucleic acid (RNA); Tryptophan (trp) repressor–mtr operator DNA (ODNA) complex
Descartes–Euler formula
  defining boron fullerenes, 99
  generalized to Euler–Poincare, 94
  polyhedra, 80
Diamagnetic spin orbit (DSO), component of coupling constant, 152
Diborane
  Dyson orbitals, 126*f*
  electron propagator calculations, 123–125
  ionization energies, 125*t*
  structures, 124*t*
Dickerson, R.E., borane structural theory, 25–26
Dictyostelium discoideum, spiral waves, 107
Digallane
  Dyson orbitals, 129*f*
  electron propagator calculations, 125, 127–128
  ionization energies, 127*t*
  structures, 127*t*
  synthesis, 125
Distances in proteins. *See* Electron paramagnetic resonance (EPR)
Diversity of topics, Lipscomb, 14
Double stacking, cap messenger RNA (mRNA), 222–223, 224*f*
Drum aluminosilicates, 173
Duals, definition, 81
Dynamics, protein, 323–324
Dyson orbitals
  diborane, 126*f*
  digallane, 129*f*
  electron propagator calculations, 130, 132
  electron propagator theory, 120–121
  gallaborane, 131*f*

E

Eberhardt, W.H., borane structural theory, 25–26
Ectoderm
  patterning of anterior, 297–298
  patterning of posterior, 298
  See also Embryo, mouse
Ectodermal cell fate. *See* Embryo, mouse
Electron binding energies, measurements, 119
Electron diffraction, boron hydrides, 23
Electron microscopy, distances in proteins, 322
Electron nuclear dipolar interaction (ENDOR), electron-nuclear distances, 325
Electron paramagnetic resonance (EPR)
  accessibility of slowly relaxing spin to paramagnetic species in solution, 332

analysis of dipolar splitting significant relative to linewidths, 326–327
application of pulsed techniques, 328
distance between rapidly relaxing spin and slow-relaxing spin in immobilized sample, 329–331
distance between spins in slowly tumbling macromolecule, 331–332
distance between two slowly relaxing spins in immobilized sample, 325–329
electron nuclear dipolar interaction (ENDOR), 325
electron spin echo envelope modulation (ESEEM), 325
features, 324
future of methods for determining distances between spins, 332–333
half-field transitions, 326
impact of rapidly relaxing spin on spin-lattice relaxation for more slowly relaxing spin, 331
impact on amplitude of continuous wave (CW) spectrum, 329–330
impact on spin echo dephasing of slowly relaxing spin, 330
method, 324–325
pulsed techniques, 327–329
temperature dependence of effects of rapidly relaxing spin on slowly relaxing spin, 329t
See also Proteins
Electron-precise heterocyclic ring, conversion of carborane, 53, 59
Electron propagator theory
canonical Hartree–Fock orbital energies, 131
diborane, 123–125
diborane ionization energies, 125t
diborane structures, 124t
digallane, 125, 127–128
digallane ionization energies, 127t
digallane structures, 127t
Dyson orbitals, 120–121, 130, 132
Dyson orbitals for digallane, 129f
Dyson orbitals for gallaborane, 131f
Dyson orbitals of diborane, 126f
electron binding energies and corresponding orbitals, 120–123
gallaborane, 128, 130
gallaborane ionization energies, 130t
gallaborane structures, 128t
methods, 123
partial third-order (P3) method, 122–123
pole strength, 121, 122
Electron spin echo envelope modulation (ESEEM), electron-nuclear distances, 325
Embryo, mouse
activation of hematopoiesis by primitive endoderm, 303–304
anterior-posterior axis determination and specification of cell fate, 305
bone morphogenetic protein-4 (BMP-4), 297
common origin of blood and endothelial cells, 303
conservation of primitive endoderm signaling in hematopoietic induction, 304
determination of cell fate during development, 295
embryonic hematopoiesis not autonomous to mesoderm, 299, 301
embryonic induction, 295
embryonic stem cells, 298–299
epithelial-mesenchymal interactions in early post-implantation development, 297–298
epithelial-mesenchymal interactions in hematopoiesis and vasculogenesis, 298–299
explant culture assays for gastrulation stage, 300f
explant culture system, 304

gastrulation and induction of mesoderm, 295–296
homeobox gene *Otx2*, 302–303
induction assay, 300*f*
induction of hematopoietic mesoderm, 296–297
lineage tracing experiments, 301–302
patterning of anterior ectoderm, 297–298
patterning of posterior ectoderm, 298
respecification of anterior ectoderm to hematopoietic and endothelial cell fates by visceral endoderm, 301–303
reverse transcription–polymerase chain reaction (RT–PCR), 299, 302
transgenic embryon explant culture assay for induction of hematopoiesis by non-mesodermal signals, 299–304
unresolved questions, 306
visceral endoderm providing signal for embryonic hematopoiesis, 301
yolk sac, 295
Embryonic induction, 295
Endonucleases. *See* Nucleic acid cleavage
Energy transduction mechanisms
 mechanosensitive channel (MscL), 210–213
 membrane proteins, 210–213
 metalloproteins, 204–210
 nitrogenase, 204–210
 processes, 203–204
 See also Membrane proteins; Metalloproteins
Enzymes
 adenosine deaminase, 219, 221–222
 aldose reductase, 222
 myosin light chain kinase, 217–218
 X-ray crystallography, 9–10

See also Glutaminase domain (GLN); Methionine synthase (MetH); Multifunctional protein CAD; Protein-ligand interactions
Epithelian-mesenchymal interactions
 early post-implantation development of mouse, 297–298
 hematopoiesis and vasculogenesis, 298–299
 See also Embryo, mouse
Escherichia coli. See Multifunctional protein CAD; Tryptophan (trp) repressor–mtr operator DNA (ODNA) complex
Euler–Poincare formula
 boron duals of carbon nanotori, 99–100
 molecular nanotori, 96
 terms of Betti numbers, 94
Extraction, metal ions from radioactive waste, 37

F

Failure, occurrence, 7
FEATURE program
 computer program for proteins, 311
 method, 312
 See also Protein sequence-structure
Fe (iron)
 composite size distribution of metal nanoparticles, 71*f*
 size distribution of powder obtained at 250°C, 73*f*
 transmission electron microscopy (TEM) images of metallic nanostructures, 70*f*
 X-ray diffraction (XRD) pattern of powder obtained at 250°C, 72*f*
 XRD pattern of powder obtained at 264°C, 74*f*
 See also Nanostructured phases
Fermi-contact, component of coupling constant, 152

FeS clusters, redox proteins, 6
Fibroblast growth factor (FGF), mesoderm inducers, 296
Fish, tropical, Turing patterns, 107, 108f
Fluorescence energy transfer, distances in proteins, 323
Fractional bonds, molecules, 6–7
Freeze-out technique. *See* Nanostructured phases
Frog Xenopus, induction of mesoderm, 296
Fullerenes. *See* Boron fullerenes

G

Gallaborane
  Dyson orbitals, 131f
  electron propagator calculations, 128, 130
  ionization energies, 130t
  structures, 128t
Gallium
  crystal structure of carbons apart chlorogallacarborane and carbons adjacent hydridogallacarborane, 63f
  crystal structure of Ga(II)–Ga(II)-linked digallacarborane, 60, 64f
  crystal structures of full-sandwich compounds, 62f
  reactivity of C_2B_4-carboranes with Group 13 elements, 59–60, 64
  syntheses of half- and full-sandwich gallacarboranes of C_2B_4-cage systems, 61
  See also Metallacarboranes
Gastrulation
  homeobox gene *Otx2*, 302–303
  mesoderm, 295–296
Gauge-including atomic orbitals (GIAO)
  comparing calculated chemical shifts, 140

nuclear magnetic resonance (NMR), 137, 138
Gauge problem, nuclear magnetic resonance (NMR), 136–137
Genealogy, Lipscomb, 2
Global feedback
  experimental arrangement, 109f
  oscillatory cluster patterns, 108, 110
Glutaminase domain (GLN)
  attenuation of GLN activity, 258
  cloning GLN subdomains, 256–257
  construction of mammalian *E. coli* hybrid, 256
  coupling of glutamine hydrolysis and bicarbonate activation, 255
  kinetic parameters for glutamine hydrolysis, 255t
  mechanism of glutamine hydrolysis, 254
  organization, 254
  See also Multifunctional protein CAD
Group 13 elements, reactivity of C_2B_4-carboranes, 59–60, 64
Group 1 and 2 metals
  carbons apart C_4B_8-carboranes, 51, 53
  syntheses of tetracarbon-carborane compounds, 54, 55

H

Half-field transitions, electron paramagnetic resonance (EPR), 326
Haloarcula marismortui, ribosomal subunit structure, 244
Hamiltonian
  nuclear magnetic resonance (NMR), 136
  spin-spin coupling constants, 152
Hartree–Fock theory, description, 119
Hawthorne, M.F.
  carboranes as organic building block units, 32

synthesis of polyhedral borane anions, 28
Hematopoiesis
  activation by primitive endoderm, 303–304
  conservation of primitive endoderm signaling in induction, 304
  embryonic, not autonomous to mesoderm, 299, 301
  epithelial-mesenthymal interactions, 298–299
  transgenic embryo explant culture assay, 299–304
  visceral endoderm providing signal for embryonic, 301
  yolk sac, 295
  See also Embryo, mouse
Hoffmann, Roald, borane structural theory, 25–26
Homocysteine. See Methionine synthase (MetH)
Hückel molecular orbital theory, description, 119
Humor
  Lipscomb, 202–203
  science, 12
Hybrids, C_4B_8 carboranes as organic-inorganic, 35
Hydrocarbons
  borane–hydrocarbon border, 34–35
  platforms for polycluster systems, 33
Hydrogen bonds
  charged neutral, 227
  coupling constants as fingerprints of type, 162
  interactions between phosphate-binding protein and phosphate, 225f
  low barrier, 159
  polarity, 158–161
  relationship between covalency and coupling constant, 158–159
  schematic of, from sulfate-binding protein to sulfate, 226f
  stabilization of buried charges, 227–228
  See also Protein-ligand interactions; Spin-spin coupling constants
Hydrogen fluoride
  dipole moment, Fermi-contact term, and total spin-spin coupling constant as function of field strength, 156f
  electron populations on F and H, dipole moment, and Fermi-contact term and total coupling constant as function of field strength, 154t
  electron populations on F and H, dipole moment, and Fermi-contact term and total spin-spin coupling constant as function of field strength for HF, 154t
  relationship between bond polarity and spin-spin coupling constants, 153–155
  See also Spin-spin coupling constants

I

Icosahedral barrier, boron clusters, 30–31
Icosahedron B_{12}
  intrinsic stability, 29
  See also Boron cluster science
Individual gauge for localized orbitals (IGLO)
  comparing calculated chemical shifts, 140
  nuclear magnetic resonance (NMR), 137, 138
Induction
  hematopoietic mesoderm, 296–297
  mesoderm, 295–296
Induction assay, hematopoiesis by non-mesodermal signals, 299, 300f
Information transfer, mechanism, 11

Inorganic compounds. *See*
  Aluminosilicates
Interdomain signaling, CAD domains
  and subdomains, 253
Intramolecular dynamics
  power of nuclear magnetic
    resonance (NMR), 7
  rearrangements of atoms, 8–9
I-*Ppo*I endonuclease
  active site changes, 285–286
  background, 278–279
  binding to asparagine protein ligand,
    287
  cognate cleavage site for, 278*f*
  comparison to *Serratia*
    endonuclease, 286, 287*f*
  crystallization and X-ray data
    collection, 279–280
  differences between I-*Ppo*I with and
    without bound DNA, 284–285
  differences between I-*Ppo*I with and
    without bound magnesium, 284
  differences between two monomers,
    284
  final refinement statistics, 282*t*
  homing endonuclease, 278
  molecular replacement and
    refinement, 280–282
  properties of cleavage engine, 288
  ribbon diagram showing I-*Ppo*I
    dimer with DNA, 283*f*
  role for shifting magnesium, 288–
    289
  structural analysis, 283–286
  structural studies on magnesium
    complex of, 279–282
  structure convergence, 281–282
  superposition of I-*Ppo*I–Mg^{2+}
    structure without DNA with I-
    *Ppo*I–product complex, 286*f*
  superposition of structures without
    and with DNA, 285*f*
  See also Nucleic acid cleavage
Iron (Fe)
  composite size distribution of metal
    nanoparticles, 71*f*
  size distribution of powder obtained
    at 250°C, 73*f*
  transmission electron microscopy
    (TEM) images of metallic
    nanostructures, 70*f*
  X-ray diffraction (XRD) pattern of
    powder obtained at 250°C, 72*f*
  XRD pattern of powder obtained at
    264°C, 74*f*
  See also Nanostructured phases
Irregular clusters
  computer simulations, 114, 115*f*
  oscillation, 110, 113*f*

J

Jefferson, Thomas
  imagination, 21
  quotations, 25, 34

K

Kinetics, methionine synthase, 198–
  199
Kohn–Shame equations, derivatives of
  total energy, 119

L

Leadership, scientist contributions,
  13–14
Ligands. *See* Protein-ligand
  interactions
Lipscomb, William N.
  bonding in molecules, 6–7
  class on symmetry and MOs, 9
  Colonel, 3
  contribution to nuclear magnetic
    resonance (NMR), 135–136

controversy of interpretation of cyclooctatetraene (COT) complexes, 8–9
crystal structures of molecules, 4
cyclopentadienyl group, 7–8
diversity of topics, 14
dynamic intramolecular rearrangements, 8–9
failures, 7
FeS clusters, 6
genealogy, 2
humor, 12, 202–203
imagination in scientific discovery, 21–22
importance of nuclear magnetic resonance (NMR), 12
intramolecular dynamics, 7
leadership contributions of scientists, 13–14
Lewis Carroll and Thomas Jefferson, 21–22
mechanism of allostery, 10–11
mechanism of information transmission, 11
molecular cluster science, 39
Nobel Prize 1976, 3
patterns of boron hydrides, 4–5
Pauling lecture on boron hydrides, 4
photograph, 14
polyhedral boranes, metal clusters and cagelike molecules, 21
predicting possible boron hydride ions, 5
predicting rearrangement pathways for polyhedral molecules, 6
principles of allosteric behavior, 11
protein structure determinations, 10
publishing, 3
relationship between structures and properties, 3
retrospectroscopy, 5
role as administrator, 13
"Spectacular Experiments and Inspired Quotes", 5
structural biochemistry, 11–12
structural measurements and models of molecular electronic structure, 6
structure determination of carboxypeptidase, 10
teaching, 3–4, 9
teaching and research, 12–13
three-center bond concepts, 6
value of quotations, 5
vitality of work, 13
X-ray crystallography, 5
X-ray crystallography of enzymes, 9–10
Lipscomb and collaborators, borane structural theory, 25–26
Localized clusters
  computer simulations, 114, 115f
  oscillation, 113, 114f
Localized orbitals, local origin (LORG), nuclear magnetic resonance (NMR), 138
Loewenstein's rule
  absence of Al–O–Al linkages, 167
  double ring structures, 170, 173
  See also Aluminosilicates

M

Macropolyhedral condensed boranes, boron cluster science, 29–31
Magnesium
  change between I-*Ppo*I structures in Mg site, 285–286
  comparison of I-*Ppo*I nuclease to *Serratia* endonuclease, 286, 287f
  differences between I-*Ppo*I with and without bound Mg, 284
  final refinement statistics for I-*Ppo*I endonuclease structures with and without, 282t
  role for shifting, in binding sites, 288–289
  structural studies on I-*Ppo*I endonuclease–Mg complex, 279–282

syntheses of magnesacarboranes, 53, 57f
See also Nucleic acid cleavage
Magnetic properties
  programs calculating, 142
  See also Nuclear magnetic resonance (NMR)
Materials, proposed new. *See* Boron fullerenes; Boron nanotori; Boron nanotubes
Mechanism, information transmission, 11
Mechanosensitive channel (MscL)
  coupling mechanism, 213
  function, 210, 213
  helical wheel for MscL inner helix, 212f
  nonselective ion channel, 210–211
  organization as pentamer, 211
  packing, 213
  structural analysis for understanding of channel gating, 212–213
  See also Energy transduction mechanisms
Medicine
  anticancer agents, 38–39
  boranes and carboranes, 37–39
  boron neutron capture therapy (BNCT), 37–38
Membrane proteins
  function of mechanosensitive channel (MscL), 213
  helical wheel for MscL inner helix, 212f
  MscL, 210
  nonselective ion channel MscL, 210–211
  organization of MscL as pentamer, 211
  structural analysis of MscL for understanding of channel gating, 212–213
  view of Tb–MscL pentamer perpendicular and parallel to normal of membrane plane, 211f

See also Energy transduction mechanisms
Mercuracarborane, anti-crown ligands, 31–32
Mesoderm
  embryonic hematopoiesis not autonomous to, 299, 301
  gastrulation and induction, 295–296
  induction of hematopoietic, 296–297
  See also Embryo, mouse
Messenger ribonucleic acid (mRNA), recognition, 222–223, 224f
Metal carbonyls
  thermal decomposition, 68
  See also Nanostructured phases
Metal ions, extraction from radioactive waste, 37
Metallacarboranes
  anticancer agents, 38–39
  benzene-anchored polynuclear, 33
  cage closure reactions, 47
  cage geometries of carbons adjacent isomers, 51, 52f
  conversion of carborane into electron-precise heterocyclic ring, 53, 59
  crystal structure of carbons apart chlorogallacarborane and carbons adjacent hydridogallacarborane, 63f
  crystal structure of dicesiacarborane complex and cesiacarborane unit in polymeric chain, 51, 56f
  crystal structure of Ga(II)–Ga(II)-linked digallacarborane, 60, 64f
  crystal structures of full-sandwich compounds, 62f
  crystal structures of magnesacarboranes of C_4B_8-cage systems, 57f
  cuboctahedral structure, 47, 50f
  electron-precise heterocyclic ring, 59f
  magnesacarborane production, 53, 57f

procedures for syntheses of known carbons apart compounds, 54, 55
reactivities of carbons apart C_4B_8-carboranes, 51, 53
reactivity of carbons adjacent C_4B_8-carboranes, 53
syntheses of C_2B_4- and C_4B_8-carboranes, 47, 51
syntheses of *closo*-C_2B_4 and *nido*-C_4B_8-carboranes, 48, 49
syntheses of half- and full-sandwich gallacarboranes of C_2B_4-cage systems, 61
ten-vertex *arachno*-$(SiMe_3)_2C_2B_8H_8$ cage, 53, 58*f*
traditional *nido*-cage structure, 50*f*, 51

Metallacarborane sandwich polymers, new specialty materials, 36

Metalloproteins
coupling of ATP hydrolysis to electron transfer in other systems, 209–210
energy transduction in nitrogenase, 209
mechanistic considerations, 207–208
outlines of electron flux through nitrogenase system, 205
relative positions of metalloclusters in complex structures, 207–208
role of ATP in nitrogenase reaction, 207
structural models for nitrogenase metalloclusters and coordinating ligands, 206*f*
structures of component proteins, 205–206

Metals (group 1 and 2)
carbons apart C_4B_8-carboranes, 51, 53
syntheses of tetracarbon-carborane compounds, 54, 55

Methionine synthase (MetH)
activation domain and reactivation of cob(II)alamin form, 194–197
catalysis and reactivation, 188*f*
cobalamin-binding fragment and nucleotide-off conformation of cobalamin, 192–193
cobalamin cofactor in alkyl Co(III) form, 187*f*
conformation rearrangements and kinetics, 198–199
conversion of cob(II)alamin to cob(I)alamin species, 193*f*
converting homocysteine (Hcy) and methyltetrahydrofolate (CH_3H_4folate) to methionine and tetrahydrofolate, 187, 188*f*
corrin movement in switch to activation conformation, 196–197
distribution of conformations, 197–199
domain movements, 190
folate-binding module and related methyltransferases, 191–192
methyl transfer reactions, 186–187
model of cap-on conformation of C-terminal fragment, 195, 196*f*
modules and functions, 190–199
N-terminal fragment: zinc-dependent activation of homocysteine, 191
organization, 188, 189*f*
reactivation conformation, 194–195
reactivation reaction, 194
stereoview showing cobalamin displacement in activation complex, 197*f*
structure of cob(II)alamin form, 195, 196*f*
three modules comprising, 189*f*

Methyl chloride
dipole moment, Fermi-contact term, and total C–Cl spin-spin coupling constant as function of field strength, 157*f*
polarity of C–Cl bond, 155
See also Spin-spin coupling constants

Methyl fluoride

dipole moment, Fermi-contact term, and total C–F spin-spin coupling constant as function of field strength, 156f
polarity of C–Cl bond, 155
See also Spin-spin coupling constants
Methyl lithium
dipole moment, Fermi-contact term, and total C–Li spin-spin coupling constant as function of field strength, 157f
polarity and coupling constants, 155, 158
See also Spin-spin coupling constants
Methyltetrahydrofolate. *See* Methionine synthase (MetH)
Methyl transfer reactions. *See* Methionine synthase (MetH)
Minerals. *See* Aluminosilicates
Modeling. *See* Tryptophan (trp) repressor–mtr operator DNA (ODNA) complex
Molecular biology
aminoacyl-tRNA synthetases, 241
binding substrates, intermediates, and product analogues to ribosomal subunit, 246–247
catabolite gene activator protein, 235
central dogma, 231–232, 233f
complex of *E. coli* glutamine tRNA with *E. coli* glutaminyl-tRNA synthetase, 243f
Crick's central dogma, 233f
DNA replication, 232, 234–235
Haloarcula marismortui ribosomal subunit in rotated crown view, 245f
HIV reverse transcriptase, 239, 241
orthogonal views of T7 RNA polymerase complexes, 240f
proteins stabilizing overall rigidity of ribosome structure, 246
protein synthesis, 241, 244–247
ribosome, 244–247
RNA–RNA, RNA–protein, and RNA–metal ion interactions, 244, 246
schematic of polypeptide backbone of reverse transcriptase (RT) heterodimer, 242f
shuttle mechanism of editing in DNA polymerase, 236f, 237f
structure of CAP–DNA showing angles of DNA bending, 238f
T7 RNA polymerase, 235, 239
transcription, 235, 239–241
view of docked editing and sliding clamp complexes, 238f
Molecular dynamics (MD), protein structure, 341
Molecular electronic structure, structural measurements, 6
Molecular mechanics and dynamics, protein, 323–324
Molecular modeling. *See* Tryptophan (trp) repressor–mtr operator DNA (ODNA) complex
Molecular nanotori
critical points and topology of torus, 96–97
critical points of torus, 96f
Euler–Poincare formula and, 96
See also Boron nanotori
Molecular recognition
polar and nonpolar interactions, 219
See also Protein-ligand interactions
Molecular wires, new specialty materials, 36
Monosaccharide binding, crystallographic studies, 218, 220f
Morse theory
dimensional holes, 97
relationship between critical points and topology, 96
Mouse. *See* Embryo, mouse
Mouse system, naturally occurring mutants, 296

Multifunctional protein CAD
  active site of CAD glutaminase
    domain (GLN) domain, 258f
  activity of isolated CPS subdomains,
    260t
  attenuation of GLN activity, 258
  CAD structure and function, 250–253
  carbamoyl phosphate synthesis,
    251f, 252
  carbamoyl phosphate synthesis by
    GLN–CPS.A, 260f
  catalysis complex, 249–250
  cloning and expression of CPS.A
    and CPS.B, 259
  cloning of CPS.A and CPS.B
    domains, 259f
  cloning of GLN subdomains, 256–257
  construction of mammalian *E. coli*
    chimera, 264–265
  construction of mammalian *E. coli*
    hybrid, 256
  control mechanism of regulatory
    chimera R2, 265f
  coupling of glutamine hydrolysis
    and bicarbonate activation, 255
  CPS subdomains, 259–264
  determination of CAD domain
    structure, 251
  dissection of CPS subdomains, 263–264
  domain structure of CAD, 250f
  GLN, 254–258
  GLN domain determining function
    of CPS domains, 262–263
  hypothesis for CPS subdomains,
    260–261
  interdomain signaling, 253
  kinetic parameters for glutamine
    hydrolysis, 255t
  kinetics of glutamine hydrolysis of
    mammalian GLN–*E. coli* CPS
    hybrid, 256t
  mammalian *E. coli* regulatory
    chimera, 264f
  mammalian GLN–*E.coli* CPS
    hybrids, 257f
  mechanism of glutamine hydrolysis,
    254
  organization of GLN domain, 254
  pressure induced dissociation of
    GLN–CPS.B dimer, 261f
  pressure induced dissociation of
    hybrid molecules, 261–262
  reactions catalyzed by CAD,
    250f
  rearranged CPSases, 262f
  rearranged molecules, 262f, 263
  regulation of CAD activity, 253
  regulation of CPS.A, 265–266
  regulatory subdomain, 264–266
  structural studies, 252–253
Myosin light chain kinase (MLCK),
  calmodulin-binding, 217–218

N

Nanostructured phases
  analysis methods, 68–69
  composite size distribution of metal
    nanoparticles, 71f
  decomposition temperature and
    particle size, 69
  information storage, 67
  isolating metal particles, 69
  protecting metal powders from
    oxidation, 67
  TEM images (transmission electron
    microscopy) of metallic
    nanoclusters, 70f
  thermal decomposition of metal
    carbonyls, 68
  typical experiment, 68
  X-ray diffraction (XRD) pattern and
    size distribution of Fe powder
    obtained at 250°C, 72f, 73f

XRD of Fe powder obtained at 264°C and size distribution at 250°C, 74f
XRD pattern and size distribution of Co powder obtained at 162°C, 75f
XRD pattern and size distribution of Ni powder obtained at 170°C, 77f
XRD pattern and size distribution of Ni powder obtained at 182°C, 78f
XRD pattern of Co powder obtained at 171°C, 76f
Nanotori, carbon and boron dual, 98t
Nanotubes. *See* Boron nanotubes
Nickel
  composite size distribution of metal nanoparticles, 71f
  size distribution of powder obtained at 170°C, 77f
  size distribution of powder obtained at 182°C, 78f
  transmission electron microscopy (TEM) images of metallic nanostructures, 70f
  X-ray diffraction (XRD) pattern of powder obtained at 170°C, 77f
  XRD pattern of powder obtained at 182°C, 78f
  See also Nanostructured phases
Nitrogenase
  biological nitrogen fixation, 204
  coupling of ATP hydrolysis to electron transfer in other systems, 209–210
  energy transduction, 209
  Fe-protein, 206
  mechanistic considerations, 207–208
  relative positions of metalloclusters, 207–208
  role of ATP, 207
  structures of component proteins, 205, 206f
  See also Energy transduction mechanisms
Nobel Prize
  Lipscomb, 3

various, 2
Noncompetitive inhibitor, aldose reductase, 222
Nonlinear chemical dynamics
  Belousov–Zhabotinsky (BZ) reaction, 104
  clusters, 110, 113–114
  computer simulations, 114, 115f
  experimental arrangement for global feedback, 109f
  family of cluster patterns vs. strength of feedback, 111f
  irregular clusters, 110, 113f
  localized clusters, 113, 114f
  mechanism of bromate–chlorite–iodide oscillating reaction, 106t
  oscillatory chemical reactions, 104
  oscillatory cluster patterns in BZ reaction with global feedback, 108, 110
  patterns and waves, 105, 107
  patterns of tropical fish, 108f
  spiral waves in BZ reaction and in aggregating slime mold, 107f
  standing clusters, 110, 112f
  taxonomy of chemical oscillators, 105f
  Turing patterns, 107
  Turing patterns in chlorine dioxide–iodine–malonic acid reaction in unstirred gel reactor, 108f
Nonlinear optical (NLO), new specialty materials, 36
Nuclear magnetic resonance (NMR)
  application, 139–142
  averaging chemical shifts over rotational and vibrational motion, 141
  calculation of nuclear spin-spin coupling constants, 140–141
  chemical shielding tensor, 136
  comparison of calculated chemical shifts using gauge-including atomic orbitals (GIAO) or

individual gauge for localized orbitals (IGLO) method, 140
computing chemical shifts using single origin, 138
contributions of Lipscomb, 135–136
converting calculated absolute shielding to chemical shift, 139
density functional theory (DFT) approach, 139
distances in proteins, 322
effective spin Hamiltonian, 136
gauge problem, 136–137
GIAO method, 137, 138
IGLO method, 137, 138
importance, 12
interaction of magnetic field with atoms and molecules, 135
localized orbitals, local origin (LORG) method, 138
magnetic criteria judging aromaticity/antiaromaticity, 140
methods, 137–139
model for calculating Na NMR, 170
power for elucidating structure, 7
programs calculating magnetic properties, 142
progress in protein structures, 341
protein dynamics, 341–342
QM:MM/QM:QM method, 138
relativistic spin-free "scalar" effects on light nuclei, 141
second order LORG (SOLO) method, 138
sensitivity of chemical shifts, 139
solvation effect, 141–142
spin-spin coupling constants, 141
structure and dynamics of simplest allosteric system, 342
structure-bonding theory and NMR spectroscopy of boron hydrides, 28–29
sum-over-states (SOS–DFPT) method, 139
theory, 136–137
transition metal systems, 141

See also Spin-spin coupling constants; Tryptophan (trp) repressor–mtr operator DNA (ODNA) complex

Nucleic acid cleavage
amino acid alignment of nucleases homologous to *Serratia* endonuclease, 273*t*
analysis of I-*Ppo*I endonuclease structures, 283–286
background on I-*Ppo*I endonuclease, 278–279
background on nucleases, 271–272
background on *Serratia* family of endonucleases, 272–273
cleavage sites in common nuclease reactions, 271*f*
cognate cleavage site for I-*Ppo*I endonuclease, 278*f*
comparison of I-*Ppo*I nuclease to *Serratia* endonuclease, 286, 287*f*
crystallization and X-ray data collection for I-*Ppo*I endonuclease, 279–280
crystallographic studies of *Serratia* endonuclease, 273–278
electrostatic potential surface on van der Waals surface of *Serratia* nuclease, 276*f*
enzymes binding catalytic metal ion with single asparagine protein ligand, 287
final refinement statistics for I-*Ppo*I endonuclease structures, 282*t*
mechanism of *Serratia* nuclease, 277–278
mechanisms of enzyme catalyzed phosphodiester cleavage, 271–272
molecular replacement and refinement of I-*Ppo*I, 280–282
oxygen-phosphate cleavage reaction, 271
properties of cleavage engine, 288
ribbon diagram of nuclease dimer, 275*f*

ribbon diagram showing I-*Ppo*I dimer structure with DNA, 283*f*
role for shifting magnesium, 288–289
Serratia nuclease at 2.1Å, 274–276
structural studies on I-*Ppo*I endonuclease–magnesium complex, 279–282
structure of catalytic site, 276, 277*f*

O

Oligosaccharide binding, crystallographic studies, 218, 220*f*
Organic chemistry, polyhedral borane frameworks, 31–33
Oscillatory chemical reactions
Belousov–Zhabotinsky (BZ), 104
mechanism of bromate–chlorite–iodide oscillating reaction, 106*t*
taxonomy of chemical oscillators, 105*f*

P

Paramagnetic species. *See* Electron paramagnetic resonance (EPR)
Paramagnetic spin orbit (PSO), component of coupling constant, 152
Partial third-order (P3) method, electron propagator theory, 122–123
Patterns
Belousov–Zhabotinsky (BZ) reaction, 105, 107
Turing, 107, 108*f*
Pauling, Linus C.
boron hydride lecture, 4
electrostatic valence sum rule, 167
Nobel prizes, 2
Pauli principle for bosons, phase, 182, 183

Pentostatin, mode of binding, 221–222
Photograph, Lipscomb, 14
Physarum polycephalum. *See* I-*Ppo*I endonuclease
Pole strength, electron propagator theory, 121, 122
Polyhedral boranes
boron neutron capture therapy (BNCT), 37–38
current applications, 35–39
frameworks in organic chemistry, 31–33
mercuracarborane anti-crown ligand and Cl complex, 31
new specialty materials, 36
uses, 35–36
Polyhedral molecules
predicting rearrangement, 6
question of existence of boron cages, 30–31
Polyhedron, regular, 80*f*
Protein–ligand interactions
calmodulin (CaM) binding to targets, 217–218
cap mRNA recognition, 222–223
carbohydrates, polar vs. nonpolar, 218–219
charged ligands, ligand specificity and electrostatic balance, 224–227
charged ligands, stabilization of buried charges, 227–228
charged neutral H bond, 227
double stacking interactions and van der Waals contact, 224*f*
H-bonding between phosphate-binding protein and phosphate, 225*f*
noncompetitive inhibitor of aldose reductase, 222
schematic of H bonds between sulfate-binding protein and sulfate, 226*f*
single-stranded RNA, 223–224
stereo view of H-bonding and stacking interactions between D-

galactose-binding protein and D-glucose, 220f
zinc discovery in adenosine deaminase, 219, 221–222
Protein sequence-structure
 database growth, 310
 environments of 1COL-A (colicin) and 1CPC-A (phycocyanins), 317
 expense of threading technique, 314, 317
 FEATURE computer program, 311, 312
 generating and scoring actual and sample structures, 313
 generating sample alignments, 312–313
 growing database, 310
 methods, 312–313
 periodicity in globin alignment scores, 318
 scores for perfect alignments for target proteins, 313–314
 scores for sampled alignments of 2GDM and 1BAB-A, 316f
 scores for samples alignments of 1HLB, 315f
 scores for shifted alignments of 1BAB-A, 2GDM, and 1HLB, 315f
 SCWRL (side chain placement with a rotamer library) program, 313, 317
 sensitivity of scoring functions to alignment errors, 314
 sensitivity of scoring method to distance of template backbone, 317–318
 success of alignment methods, 310
 threading methods, 310–311
Protein synthesis
 aminoacyl-tRNA synthetases, 241, 243f
 messenger RNA, 241
 ribosomes, 244–247
 See also Molecular biology
Proteins

deciphering biological function and structure at molecular level, 321–323
dynamics, 323–324
electron microscopy, 322
electron paramagnetic resonance (EPR) methods, 324–332
estimating distances between parts and assemblies of, 321–322
fluorescence energy transfer, 323
insights about function, 323
molecular mechanics and dynamics calculations, 323–324
multifunctional, 249
nuclear magnetic resonance (NMR) techniques, 322
progress in structures, 341
structure determination, 10
understanding catalytic function, 341–342
X-ray diffraction, 322
See also Electron paramagnetic resonance (EPR); Multifunctional protein CAD; Tryptophan (trp) repressor–mtr operator DNA (ODNA) complex
Pulsed techniques, electron paramagnetic resonance (EPR), 327–329

Q

Quantum crystallography (QC), carbon nanotori calculations, 100
Quantum mechanical calculations. See Aluminosilicates

R

Radioactive waste, extraction of metal ions, 37
Ray–Dutt twist, trischelate complexes, 8

Rearrangement, predicting pathways for polyhedral molecules, 6
Regulatory subdomain
   construction of mammalian E. coli chimera, 264–265
   regulation of subdomain CPS.A, 265–266
   See also Multifunctional protein CAD
Reoviruses
   black λ2 monomer, 374f
   Cα-trace of core particle, 368f
   comparison of λ1-A and λ1-B, 371
   diagrams of SAM-binding domains, 374f
   function, 367
   λ1, λ1 shell, and non-equivalence, 370–371
   λ1 shell with footprints outlining σ2 binding sites, 368f
   λ1 surface, 368f
   λ2 capping complex and active sites, 372–376
   organization of active sites, 376
   pentameric λ2 turret, 374f
   ribbon diagram of guanylyltransferase domain, 374f
   ribbon diagram of λ1-A, 368f
   ribbon diagram of λ1-B, 368f
   σ2 clamp and more non-equivalence, 371–372
   structure of core particle at 3.6Å resolution, 367, 368f, 370
Replication
   deoxyribonucleic acid (DNA), 232, 234–235
   shuttle mechanism of editing in DNA polymerase, 236f, 237f
   structural studies and sequence comparisons among polymerases, 234–235
   structures of binary and ternary complexes, 232, 234
   structures of DNA polymerase and complexes, 234

view of docked editing and sliding clamp complexes, 238f
See also Molecular biology
Research of Lipscomb
   structure-property relationships, 3
   teaching and, 12–13
Retrospectroscopy, Lipscomb, 5
Reverse transcription–polymerase chain reaction (RT-PCR)
   expression of endoderm- and ectoderm-specific genes, 299
   expression of hematopoietic markers, 302
Ribonucleic acid (RNA)
   cap messenger RNA (mRNA), 222–223
   double stacking interactions and van der Waals contact, 224f
   non-specific recognition of single-stranded, 223–224
   See also Nucleic acid cleavage; Reoviruses
Ribosomes
   binding substrates, intermediates, and product analogues, 246–247
   Haloarcula marismortui subunit in rotated crown view, 245f
   major targets of antibiotics, 247
   proteins stabilizing overall rigidity, 246
   RNA–RNA, RNA–protein, and RNA–metal ion interactions, 244, 246
   structure studies, 244
   See also Molecular biology
Roussin's black salt, Fe_4S_3 clusters, 6

S

Science, humor in, 12
Scientific discovery, imagination, 21–22
Scientist, leadership contributions, 13–14
Serratia endonucleases

amino acid alignment of nucleases homologous to, 273*t*
background, 272–273
binding to asparagine protein ligand, 287
comparison to I-*Ppo*I nuclease, 286, 287*f*
crystallographic studies, 273–278
electrostatic potential surface superimposed on van der Waals surface, 276*f*
general properties, 273
homologs in eukaryotes and prokaryotes, 272–273
mechanism, 277–278
nuclease at 2.1 Å, 274–276
precession photograph of OKL zone from crystals of, 274*f*
properties of cleavage engine, 288
ribbon diagram of nuclease dimer, 275*f*
role for shifting magnesium, 288–289
structure of catalytic site, 276, 277*f*
See also Nucleic acid cleavage
Side chain placement with rotamer library (SCWRL), 313, 317
Single-stranded ribonucleic acid (RNA), non-specific recognition, 223–224
Smad2 gene, specification of anterior aspect of embryo, 297
Solvation effect, chemical shifts, 141–142
"Spectacular Experiments and Inspired Quotes," Lipscomb review, 5
Spin dipole, component of coupling constant, 152
Spin-spin coupling constants
 calculation of nuclear, 140–141
 ClH:NH$_3$ complexes, 159, 161
 components, 152
 dipole moment, Fermi contact term, and total, for HF as function of field strength, 156*f*
 dipole moment, Fermi contact term, and total C–Cl, for CH$_3$Cl as function of field strength, 157*f*
 dipole moment, Fermi contact term, and total C–F, for CH$_3$F as function of field strength, 156*f*
 dipole moment, Fermi contact term, and total C–Li, for CH$_3$Li as function of field strength, 157*f*
 electron populations on F and H, dipole moment, and Fermi-contact term and total, as function of field strength for HF, 154*t*
 equilibrium Cl–N and Cl–H distances, Cl–N spin-spin coupling constants, and proton chemical shifts for ClH:NH$_3$ as function of field strength, 160*t*
 external electric fields changing structure of H-bonded complexes, 159
 fingerprints of hydrogen bond type, 162
 first-order perturbed wavefunction, 152
 Hamiltonian, 152
 hydrogen fluoride (HF) molecule, 153–155
 indicator of covalent bond character, 151
 low-barrier hydrogen bonds (LBHBs), 159
 method of calculation, 151–153
 methyl fluoride (CH$_3$F) and methyl chloride (CH$_3$Cl), 155
 methyl lithium (CH$_3$Li), 155, 158
 N–H–N hydrogen bonds, 151
 O–O coupling constant for O$_2$H$_5^+$ structure, 161
 optimization of structures of molecules, 153
 polarity in hydrogen bonds, 158–161
 relationship between covalency and, 158–159

structures of small neutral complexes (H$_2$O)$_2$ and (HF)$_2$, 161
two-bond, across X–H–Y hydrogen bonds, 158
See also Nuclear magnetic resonance (NMR)
Standing clusters
  computer simulations, 114, 115*f*
  oscillation, 110, 112*f*
Stem cells, embryonic, 298–299
Stock, Alfred
  lower boron hydrides, 22
  proposed borane structures, 23
  speculating about structure, 22
Structure-bonding theory, nuclear magnetic resonance of boron hydrides accompanying, 28–29
Structure determination
  carboxypeptidase, 10
  proteins, 10
  proteins at molecular level, 321–323
  See also Protein sequence-structure; Reoviruses; Tryptophan (trp) repressor–mtr operator DNA (ODNA) complex
Structure-property relationships, Lipscomb, 3
Sugars, polar and nonpolar groups, 218
Supra-icosahedral clusters, boron cluster science, 29–31

T

Teaching
  Lipscomb, 3–4, 9
  research and, 12–13
Threading
  expense of technique, 314, 317
  methods, 310–311
  See also Protein sequence-structure
Three-center bond, concepts in boron hydrides, 6–7
Topological theory
  boranes, 25–29
  correlation of structure with electron count in clusters, 26
  examples of borane–carborane–carbocation continuum, 27
  structure-bonding theory and NMR spectroscopy of boron hydrides, 28–29
  Wade's rules, 26, 27
Topology, Morse theory, 96
Transcription
  catabolite gene activator protein, 235
  HIV reverse transcriptase (RT), 239, 241
  orthogonal views of T7 RNA polymerase complexes, 240*f*
  schematic of polypeptide backbone of RT heterodimer, 242*f*
  structure of CAP–DNA complex showing angles of DNA bending, 238*f*
  T7 RNA polymerase, 235, 239
  See also Molecular biology
Transforming growth factor-β (TGF-β), mesoderm inducers, 296
Transition metal systems, chemical shifts and spin-spin coupling constants, 141
Trischelate complexes
  Lipscomb teaching, 9
  pseudorotation, 8
Tropical fish, Turing patterns, 107, 108*f*
Tryptophan (trp) repressor–mtr operator DNA (ODNA) complex
  amide proton chemical shift changes upon complex formation, 347*f*
  asymmetry in binding affinity, 359–360
  chemical shifts of protons of mtr operator DNA, 344*t*
  chemical shifts of protons of mtr operator DNA in complex with trp repressor, 345*t*

computational methods, 361, 364
contact nuclear Overhauser effects (NOEs) between protein and mtr ODNA, 348, 349*t*
distance and dihedral angle constraints, 348, 351
instability of DNA binding domain, 354, 358
long range NOEs, 346, 348
low definition for residue Met66 and Leu75, 359
mechanisms of allostery and induced fit, 360–361
models, 356*f*, 357*f*
mtr operator proton assignments, 343, 346
NOEs between repressor and co-repressor tryptophan, 346, 348
NOEs within protein structure, 346
nuclear magnetic resonance (NMR) spectroscopy method, 361
potential direct H-bonds between trp repressor and mtr DNA per dimer, 355*t*
preparation of trp repressor and operator DNA, 361
proton assignments in complex, 343–346
refinement output statistics, 352*t*
representation of relative positions of methyl group of T14 to Ala77 and Ala80, 362*f*, 363*f*
root mean square deviation (RMSD) by residue from average of 30 structures of, 353*f*
sequential NOEs in complex, 350*f*
strongly and weakly binding halves, 358–359
structure calculation, 351, 354
structure of complex, 348, 351–354
summary of NOE and other constraints, 352*t*

system, 342–343
topology, 258
trp repressor proton assignments, 346
Tumor therapy, boron neutron capture therapy (BNCT), 37–38
Turing patterns, chlorine dioxide–iodine–malonic acid reaction, 107, 108*f*

V

Vasculogenesis
epithelial-mesenthymal interactions, 298–299
yolk sac, 295
See also Embryo, mouse
Visceral endoderm
respecification of anterior ectoderm to hematopoietic and endothelial cell fates, 301–303
signals for embryonic hematopoiesis, 301

W

Water, interaction with aluminosilicate melt, 170
Wavefunction, pair of ideal gas Bose–Einstein condensates, 178–179
Waves, Belousov–Zhabotinsky (BZ) reaction, 105, 107

X

X-ray crystallography
boron hydrides, 5
carboxypeptidase (CPA), 9–10
enzymes, 9–10
progress in protein structures, 341

X-ray diffraction
  boron hydrides, 23
  distances in proteins, 322
  polyboranes, 25

Z

Zinc, discovery in adenosine deaminase, 219, 221–222
Zopolrestat, binding, 222